BECOMING A GEOGRAPHER

Space, Place, and Society

John Rennie Short, Series Editor

Other titles in the series include:

At Home: An Anthropology of Domestic Space,
Irene Cieraad, ed.

The Global Crisis in Foreign Aid,
Richard Grant and Jan Nijman, eds.

New Worlds, New Geographies,
John Rennie Short

Putting Health into Place: Landscape, Identity, and Well-Being,
Robin A. Kearns and Wilbert M. Gesler, eds.

BECOMING A GEOGRAPHER

PETER GOULD

Syracuse University Press

First Edition 1999
99 00 01 02 03 04 05 7 6 5 4 3 2 1

The paper used in this publication meets the minimum requirements
of American National Standard for Information Sciences—Permanence
of Paper for Printed Library Materials, ANSI Z39.48-1984 ∞™

Library of Congress Cataloging-in-Publication Data

Gould, Peter, 1932–
Becoming a geographer / Peter Gould.
p. cm. — (Space, place, and society)
Includes index.
ISBN 0-8156-0566-8 (cloth : alk. paper)
1. Gould, Peter, 1932– 2. Geography—Social aspects. 3. Geography—Philosophy.
4. Geographers—United States—Biography. I. Title. II. Series.
G69..G68 A3 1999
910'.92—dc21 98-48672

Manufactured in the United States of America

Contents

Figures

Table

Maps

Color Plates

Preface

WHEN JOHN SHORT initially approached me about "a book," I was still in the throes of indexing one, editing and writing another, and preparing yet a third for a French-speaking audience. I think the look of horror on my face must have come through somehow in the letter of my reply. This didn't faze John one bit: Scots geographical editors are not only tough, but crafty and wily, ready to offer soothing blandishments that in retrospect look more and more like forbidden fruit. Which is my way of saying two things: that this book is all John's fault, but that I am grateful for the temptation.

Choosing essays in retrospect has a funny feel to it: as you walk through a wood of your own planting over the years, blazing a tree here and there for more careful consideration, you meet yourself coming the other way, often with a slightly abashed grin on your face. Some geographers can trace a strong and coherent theme through a lifetime of their research and writing. Not me. There have been too many interesting ideas, topics, perspectives and methodologies to explore over the past forty years, and I have been congenitally afflicted with a low boredom threshold. I offer no apologies: one should be allowed to be what one is. As a peripatetic geographer, I take comfort that the Greek-rooted adjective was originally coined to describe another teacher exploring other ideas as he walked around the Lyceum.

Two people, from a cast of many, must be singled out for special thanks, or there is no justice in this world. Rose*marie* Hibler, graced with the patience of the damned, did all the word processing of the text (for a constantly wavering and indecisive author) with never a word of complaint. Kevin Reese, an undergraduate at Penn State in the Deasy GeoGraphics Lab, did all the maps, figures and plates with a skill that indicates that cartography in the twenty-first century is in safe hands.

To thank everyone else—colleagues, opponents, students, and other commentators from all walks of life—is impossible. Yet thinking of them, and how they have opened up new possibilities for me, *is* to thank them. All of them have enriched a privileged life, and, after all, you only get one. No rehearsal time: as Shakespeare's "poor players" we are always on stage, and it's always first night. One does what one can in one's own way.

BECOMING A GEOGRAPHER

Introduction

> The art of Biography
> Is different from Geography.
> Geography is about maps,
> But Biography is about chaps.
>
> —Edmund Bentley
> *Biography for Beginners*

Autobiographical introductions should probably start at the beginning and take the reader through to the present end, the always penultimate end before the final one we never can write. This one starts about fifty years later, before settling down to the more conventional chronology with a few flashbacks. In the mid-1980s to early 1990s, I had some of the most rewarding teaching experiences of my life with Deryck Holdsworth, a colleague at Penn State. New graduate students took a required seminar their first semester—a sort of introduction to graduate work, the department, and the faculty. At the end of one particular year, volunteers to organize things for the next were a bit sparse, so Deryck and I took one step forward, having listened to some dissatisfactions voiced by previous students, while sharing similar ideas about what a real introduction to *graduate* study should be like.

For a number of years the seminar was an exhausting success, built around a succession of weekly readings that ranged over contemporary issues both in and far beyond geography. Just choosing and structuring the weekly readings made enormous demands upon us, for we had to read avariciously and far and wide to come up with a judicious mixture of tried-and-true articles and highly contemporary issues at the start of each academic year. At first the students did not quite know what was going on, hammered one week by David Harvey's "Manifesto"[1] (a standard question was "Can you find one sentence you don't agree with and explain why?"), and the next week by John Imbrie's Fourier analyses of 400,000 years of seabed cores.[2] But this was precisely one of the points we wanted to get across, that there was a genuine quantum leap between the rather nanny-supervised school and undergraduate work, and the opening of intellectual horizons, and the linking up of apparent disparate topics, that mature and professional graduate work should be. Even concerned faculty regularly betray their real attitudes to young men and women, often referring to their undergraduate students as "the *kids* in my class," as though they were children to be chivvied constantly with pop quizzes. In marked contrast, graduate students are professional colleagues, young colleagues to be sure, but men and women who have made, at least for the moment, a commitment to the fascinations of geographic inquiry, inquiry that is often embedded in the issues of the larger society and world. We learnt, by feedback typical of any department, that a few of our colleagues were appalled ("Some of the readings aren't even *geographic*!"), but it was the students who asked that the weekly readings be extended into the spring semester.

One of the tried-and-true readings given quite early that first semester was Torsten Hägerstrand's delightful autobiographical essay that not only looked back to reflect on the sources of his own extraordinary geographic thinking, but structured, over a vertical axis of time, the events, books, places, and people that had informed his professional life.[3] The reading became a favorite, not the least because Deryck and I asked the students to think through what their own "graphic autobiography" would look like. The next week we collected them, including our own, and made up a packet for each person in the seminar as a basis for a "fun" discussion. So my own diagrammatic biography from the seminar (fig. 1) is the framework for this introduction. I cannot touch on everything—this is an introduction to a book, not a book in itself (and anyway, some things should be left unsaid, or held for another time). But there is a lot of reflection and casting back in that simple figure, with many of the connecting arrows taken out to prevent cluttering things up. Just looking at it brings back a joyful and thankful remembrance of things past. Now back to conventional chronology.

Personal Frictions of Distance

As far as I can tell, the Goulds have been good horse thieves in the West Country (Somerset and Devon) since the eleventh century. Fortunately, not too many of us got caught to prevent the name from appearing with some regularity in churchyards and on church walls around Chard, Somerset. Various family legends tell of Goulds going off to the sinful city of London to seek their fortunes, and at least the journey of my great-great-grandfather John must have

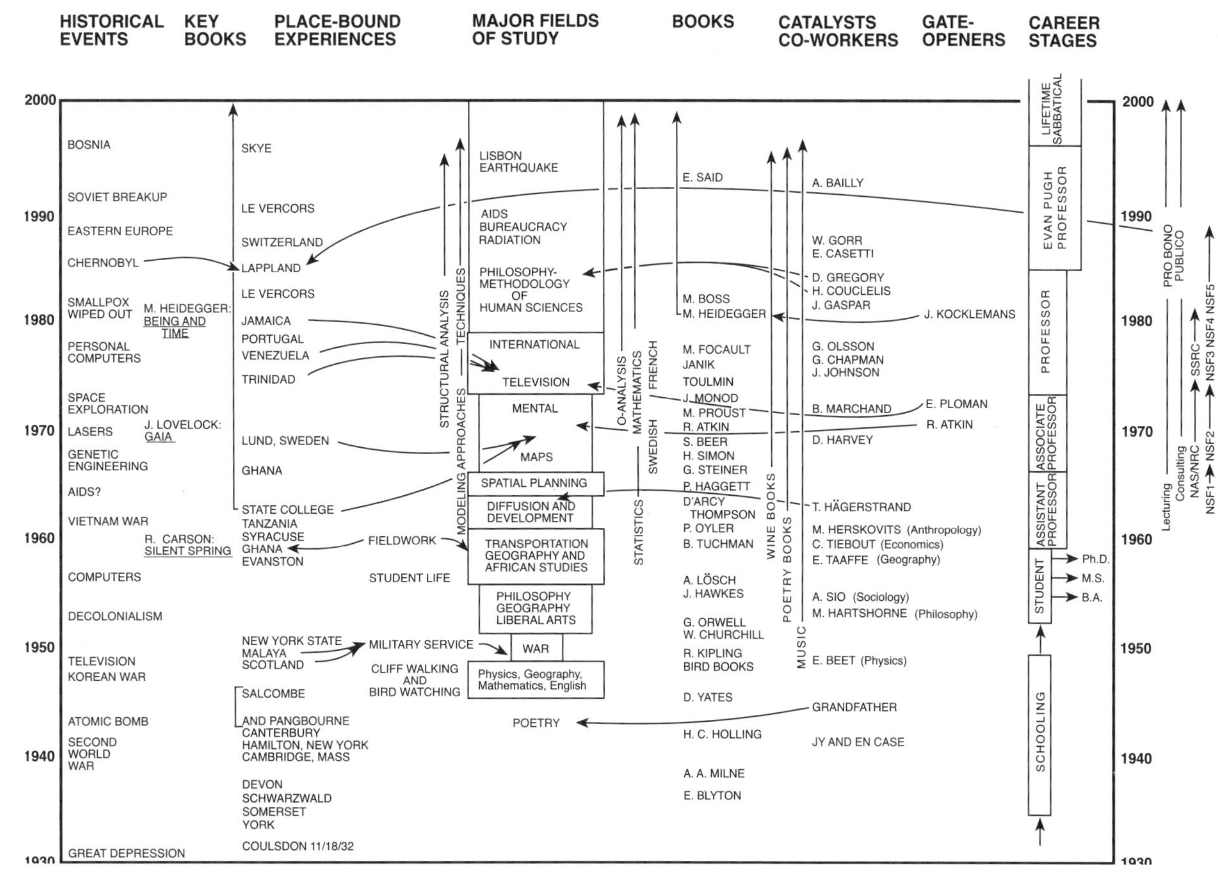

Fig. 1. The Hägerstrandian graphic biography of the author, with many of the connecting lines removed for the sake of clarity.

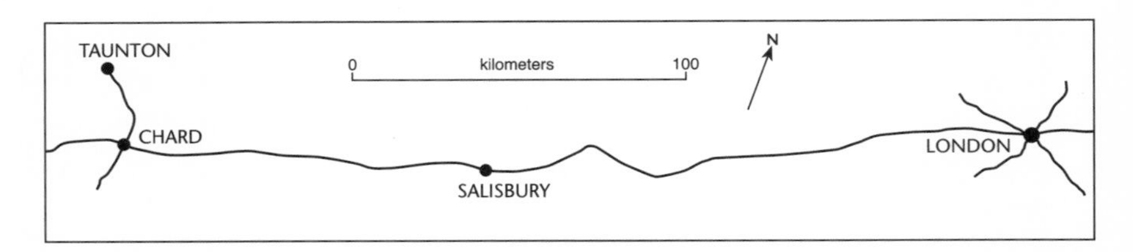

Map 1. The major travels of John Gould in the early nineteenth century.

succeeded (map 1), for he returned to Chard quite prosperous to father three boys with Mary Hutchings, including my great-grandfather Abraham in the summer of 1838. Unfortunately, the family has been going down financially ever since, not the least because Abraham ventured not only to London, but crossed the Channel to the even more sinful city of Paris (map 2). There, as a widower, he was caught, according to my great-aunt Florence, by a "French adventuress who ruined him." I have an old photo of them outside his London house, sitting in a very smart carriage pulled by two beautiful bay horses, so the

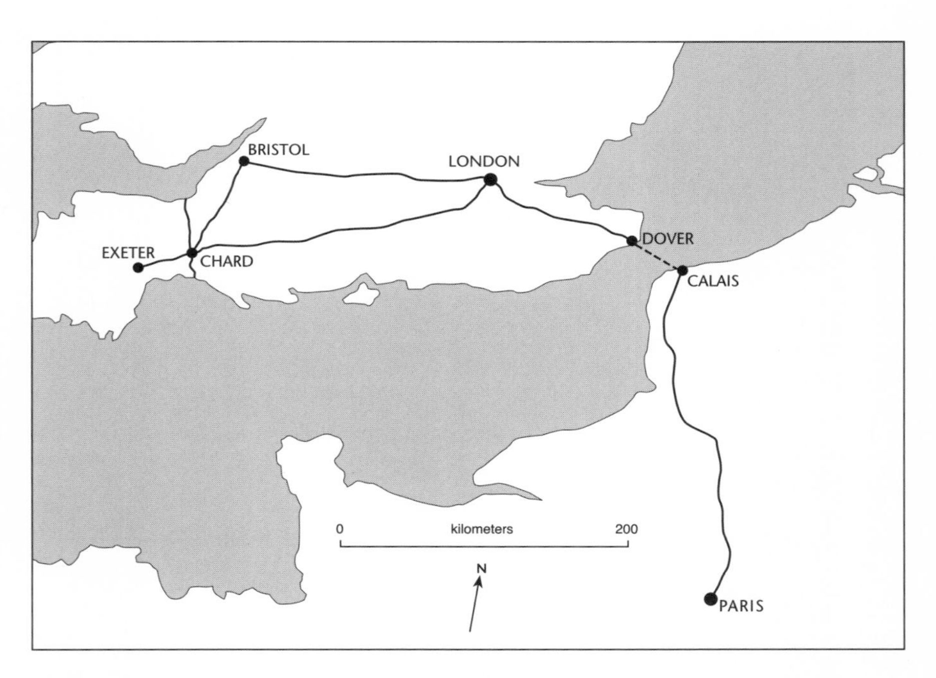

Map 2. The major travels of Abraham Gould in the mid- and late nineteenth century.

process of ruination must have taken some time. Certainly his four children, including my grandfather Charles, were able to lead comfortable lives, especially before the First World War. As a young man, Charles was apparently considered lazy, idle, and shiftless, and in the late 1880s was sent away to the Continent with a trunk and five gold sovereigns to seek his fortune. He went to Italy (map 3), spent the five sovereigns, and supported himself for a year or two picking grapes, while learning Italian love songs and the mandolin. He had a beautiful voice, and while the words of some of the love songs picked up from his fellow grape pickers were hardly suitable for after dinner in the drawing room, they were in Italian, so no one understood them—or pretended not to.

Charles must have mended his ways and been taken back into the bosom of his family, because a few years later, while on a fishing trip to the Schwarzwald, he and his brother Frank met two German sisters walking along the riverbank and married them shortly thereafter. As a result, first cousins tried to blow the hell out of each other during the First World War, and second cousins tried to do the same during the Second World War. "I often wondered," said my father, Ralph, "whether the Focke-Wulf Kondor we shot down on convoy off Cape St. Vincent was Helmut or Wolfgang." Few can teach the tribes of Eu-

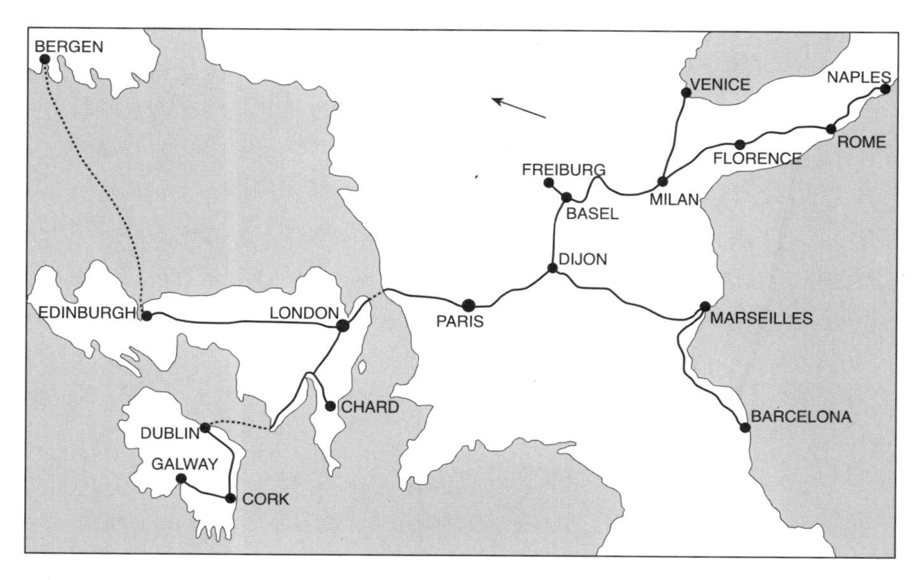

Map 3. The major travels of Charles Gould in the late nineteenth and early twentieth century.

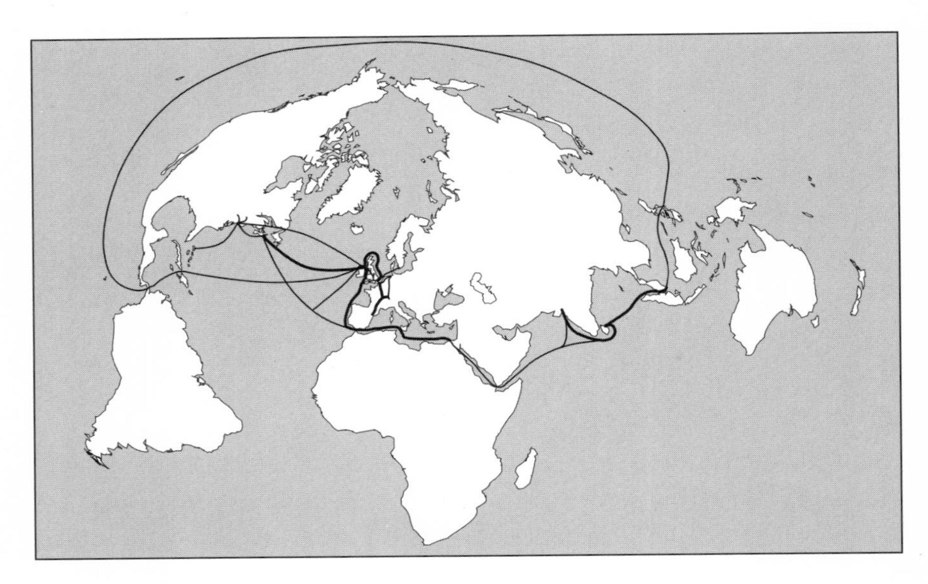

Map 4. The major travels of Ralph Gould in the twentieth century.

rope anything about internecine strife. Partly as a result of the Second World War, my father traveled widely (map 4). He joined the navy in 1939 and served as a stoker second class, shoveling coal into the boilers of an old tub armed with a pop gun off Scotland. This backbreaking service allows me to claim proletarian origins, which are so much the rage these days. He ended as Lieutenant-Commander RNVR, a decorated corvette captain, and Flag-Lieutenant (ADC) to the British admiral in Germany after the European and Far Eastern wars were over.

I appeared in November 1932, at the depths of the Great Depression, in Coulsdon, Surrey, a shameful location for a Westcountryman that I can only ascribe to the poor geographic taste of my parents. As Peter Haggett, Dick Chorley, and John Western will tell you, Westcountry origins are a necessary, though not sufficient, condition for becoming a geographer. I have been trying to overcome this handicap ever since, traveling over this wonderful planet of ours always with delight, and sometimes for professional reasons of one sort or another (map 5).[4] There are still lots of gaps, especially in the southern hemisphere, but I hope to have time to fill some of them. In five generations, the whole meaning of distance and space has changed for the Goulds. Making rough, but I think quite ready, estimates of major travels by each of us over the past 150 years (fig. 2), I can see that John, Abraham, and Charles seemed to

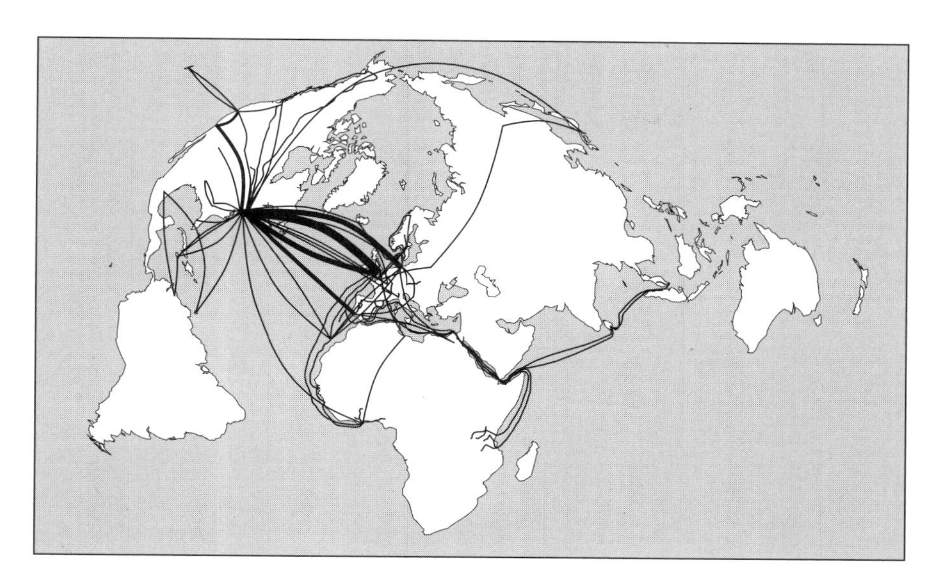

Map 5. The major travels of the author in the twentieth century.

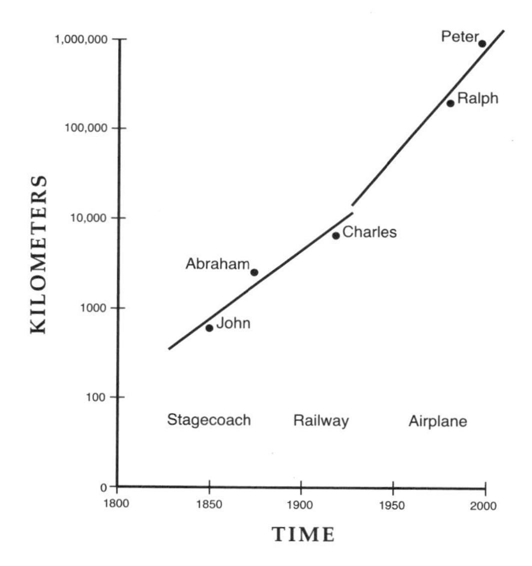

Fig. 2. The changing "elasticity" of travel space over five generations of the Gould family.

be in a space expanding quite rapidly as the stagecoach gives way to the railway, but my father and I are obviously children of the air age, as the line tilts to accelerate the change.[5] Of course, it cannot go on, at least not on a finite earth, and quite soon that line has got to tilt back and curve towards some asymptotic upper limit. I wonder if those lines are trying to become the familiar S-shaped logistic curve?

A *Moveable Feast?*

I know now that times were very hard in the 1930s for my mother and father, and we moved frequently from one place to another as my father sought various jobs to keep us going. Which is perhaps why the idea of spaces, rather than places, has always appealed to me more. Spaces are things you share with others, where new and exciting things come into view. Places are private and personal and nobody else's business. We always seemed to be moving, and most of the places are only faintly remembered blurs. But not all of them. When I was about four, things must have become literally unsupportable in a family sense, and I went with my mother to live in the Schwarzwald, the place where Charles had met my German grandmother Mathilda Kern walking along a riverbank. Her brother, Great-Onkel August, was a wonderful man, very rich before the First World War, having been responsible, as one of those new electrical engineers, for much of the electrification of southwestern Germany. He was totally wiped out financially by the catastrophic inflation of the Weimar Republic and gave his lovely house, the Bernshof, to the city of Freiburg on the condition that he and Tante Lisbeth could live in it until they died. They rented a suite to the famous mathematician Ernst Zermelo, of the "axiom of choice" fame, who was always referred to as the "Herr Professor." When I was a child at the Bernshof, I used to tap on his door after my own breakfast, and we gave the birds on the veranda theirs, after which we did "numbers" accompanied by a small cup of black coffee and caviar, a most nourishing breakfast supplement for a small child. "Numbers" meant learning the integers, adding them together on fingers, and making the figure 8 without taking the tip of the pencil from the sheet of paper.

In those days young children were to be seen rather than heard, especially at table, where my manners were hardly up to court *(Hof)* standard. One day at lunch Onkel August, sitting at the head of the table and seeing a much-scolded small boy peering over the table top, gave me a big wink, picked up his empty but gravy-smeared plate, and in a large rotary motion cleaned it with his tongue. Tante Lisbeth at the other end was absolutely horrified: "August, how could you! And in front of Das Kind!" Das Kind thought it was hilarious and tried to follow this splendid, literally in-your-face example and was promptly sent out of the room for his efforts. This convinced him that there was no justice in this world. The next day, Onkel August and I built a waterwheel that really worked, convincing me that the world had certain compensations.

Most of the eighteen months in the Schwarzwald my mother and I spent in the mountains around St. Märgen and St. Peter, where some of the larger farmhouses were providing long-term room and board, I suppose at a comparatively low price when you were paying in English shillings. I remember picking blueberries with wooden combs, and falling in love with Katrina, the farmer's daughter, with whom I was desperately sick after we ate unripe cherries together when we had been told not to. *Genesis* (3:6) obviously got it right, and I am certain the unspecified "fruit" must have been cherries.

Being violently sick with your loved one forms an indissoluble bond, and at the age of five I was sent with Katrina to one of the village schools. We sat at small desks and had nice books to read, with colored pictures on the right and text on the left. These were not primers: I have the impression that all the five- and six-year-olds already knew how to read, which probably says something about parents and older brothers and sisters reading to children during long winter nights in front of big farmhouse fires before bedtime. The problem was not reading, for I was already reading Enid Blyton's *Sunny Stories*, a weekly storybook for children that seemed to appear in the post from England, and I have no recollection of ever learning how to read. You just did.[6] The problem was that the text of the school book was written in strange and outlandish letters (Gothic) that no young Englishman could possibly condone. The pictures on the opposite page were fine: lots of little Hitler Jugend drummer boys marching along smartly waving those red-and-white flags with that funny black thing on it, a nice smiling Führer holding a pretty blonde Hitler Mädchen in his arms, and so on. Which probably says something about the speed with which the Ministry of Education in Berlin took over the entire German school system. After all, this was a tiny village school high in the mountains. Still, the letters of the text were not real letters at all, and I walked out of the schoolroom vowing never to return to an institution using such strange characters. And I never did.

It must have been at this point, about 1938, that my father got a job with Shell Mex and BP, going from £3-10-0 a week to £315 per year, and we came back to England to live in a small house outside Barnstaple in North Devon.

Apart from the pain of school—where adding pounds, shillings, and pence was never explained to me, so I always got the wretched things wrong and could never figure out why[7]—I recall that year as a very carefree one of walking up lanes to fetch milk in small metal cans, learning to ride a bicycle, and generally wandering around farms where sometimes the farmer would let you ride the huge horses pulling the hay carts from the fields. At one point, we had a small Jewish boy to stay with us for a few weeks, the son of a couple who had lived up the road from the Bernshof, and who had somehow got out to make a new life in a land that must have seemed very alien to them. I do not think their little son was very happy either, and I remember the love in his mother's eyes when she came down from London to fetch him.

These sorts of things pointed to the fact that the Second World War was not far off, and in his book, *One of Many,* my father describes coming back from work that third of September, 1939, to be greeted by my mother, waiting for him by the gate, with a laconic "So it's come." And for a second time in a generation, what would become of Onkel August, Tante Lisbeth, and the Bernshof, a place where lame birds of my family with broken wings went to mend themselves? The next months are faint memories of leaving Devon, Bristol, air raid sirens, small black dots of German bombers over the docks, my grandfather's house at Hampton Court (later bombed and gutted), and then a visit to a place in London where people spoke English, but in a funny, twangy sort of way.

By the Rude Bridge[8]

The strange accents came from people called Americans, people I had never met before, who lived in a large house in Grosvenor Square. Today an even more splendid embassy is there. I remember large rooms, the walls lined with tables, which you passed by in a certain order while people gave my mother various papers and stamped things with inky thumps. And then, one evening in the third week of August 1940, at the height of the Battle of Britain, when sometimes every Spitfire and Hurricane in the south of England was committed, we went to a huge hall to leave for America. It was at that point that it turned out that only *I* was leaving for America; I next saw my mother four years later. She joined the Royal Navy immediately upon my departure.

The children being evacuated were assigned to groups of about eight, each in the charge of a young woman, one of many volunteers. We were put on a train lit only by dim blue lights and taken to Liverpool, where we boarded in a dank and gray dawn the Canadian Pacific liner the *Duchess of Atholl*, sunk later in the war, I believe, carrying troops the other way.[9] The boat was packed with hundreds of children and their escorts, and in a numb sort of way we did what we were told. Children do. I spent much of the time in a terrified state, not because of the general circumstances, which were really quite exciting, but because some older boys had told me to go up one deck and press the elevator button. So I did, only to be seized from behind by the liftboy, who had been hiding to catch the others running up and down the stairs pressing buttons and making his life miserable. "I'm going to tell the captain," he said, "an' he'll getcha!" For the rest of the voyage I awaited the terrible summons, which never came. I suppose the captain had other things to think about, such as zigzagging at the highest speed to avoid the U-boats. Shortly afterwards, one of the other liners carrying child evacuees was torpedoed, and after that attempts to transport children stopped.

The young woman in charge of our group on the boat gave each of us a small book called *The Token of Freedom*. I have it to this day, with Britannia on the cover, armed with shield and trident, saying goodbye to a young girl and boy, and a solicitous America leaning down to welcome them. On the first page it says "This token of freedom was given to me . . . ," and there, in my childish handwriting I had written in "By some one i didnot no," followed by the printed words, "when I was 7 years old by someone who loved these words and knew what they meant and knew why I must cherish them and hold them sacred so long as I live." The sixty-three pages that follow contain poetry and prose from Pericles to Shakespeare, Milton, Bunyan, Lincoln, and Walt Whitman. They meant nothing to a seven-year-old boy, but I shall hold the words sacred so long as I live, and even now they can bring tears to my eyes. The small book was subscribed by "American friends of freedom," and it recalls a time of such innocence, high ideals, and selfless generosity that I do not think can ever come again.

After a short stop at Quebec, the boat docked at Montreal, and many of us got on a special train to New York City. It stopped just for me at Troy, in the Hudson Valley, on a late and very hot August afternoon: I was the only child getting off and the platform was bare except for one figure a long way off. The lady was to become my American mother thereafter, and her family mine. Her home in the small village of Van Hornesville, just up from Fort Plain on the Erie Canal in the Mohawk Valley, was another world, a world of peace and quiet—a grist mill, a country store, apples ripening on trees, and cold water spouting constantly into a large wooden tub. It still is. At that moment it served as the family vacation home: for the first two years, we lived in Cambridge, Massachusetts, where I walked to Shady Hill School every day and sometimes visited Lexington and Concord on school trips. I quickly learnt that the bad guys were the ones who wore the red coats. In the third grade, we did Vikings all year long, learning about Thor, Loki, and Odin, making round Viking shields in shop, and papier-mâché drinking horns, fitted for paper cup inserts, from which we swigged milk like true Vikings. In the fourth grade, where every young man fell in love with the beautiful teacher Anne Carter, we did the Greeks, learning the myths, and all about Zeus, Apollo, gray-eyed Athena, and wine-dark seas. In shop we made Greek shields (after all, the Trojans had started the whole thing), and then papier-mâché goblets, fitted for the same paper cups, from which we quaffed wine as smooth as our daily milk. I still think it is a marvelous way to teach children, and many of the Greek gods and goddesses are friends of mine to this day, helped by further introductions from Robert Graves.[10]

These were the days before television destroyed family life, and reading bedtime stories was always something you looked forward to as a child. Some of the most fascinating were by Holling C. Holling, whose text was invariably on the left-hand page, decorated with lots of small and fascinating detailed drawings, with an intriguing colored picture on the right illustrating something at just that point in the story. *Paddle-to-the-Sea* was the first I remember, about an Indian boy who carves a figure in a small canoe at his home on the shores of Lake of the Woods in Minnesota. The figure, Paddle-to-the-Sea, makes his way over many years through the entire Great Lakes system, and then along the St. Lawrence, where an Indian hunting guide finds him and murmurs, "You Little Traveler! You made the Journey, the Long Journey." In between is one of the finest geography lessons a child will ever have, matched only by Holling's other books, like *Tree-in-the-Trail* (the Santa Fe trail), *Minn-of-the-Mississippi*, and *Seabird.*[11]

In 1942 we moved to Hamilton, New York, only an hour's drive from Van Hornesville. By this time the attack on Pearl Harbor had brought America fully into the war, and grown-ups had grim and resolute faces when they talked about it. Little banners with blue stars, one for each son or daughter serving in the armed forces, were hung in windows, and sometimes the blue stars turned to gold. But for children, life went on as usual. School, in a small New York town with a small university, was not too bad; but whether it was winter or summer, being out of school was much more fun. We all read a lot, especially when the weather was bad, and we were always read to at bedtime. With

Norman Stauffer, I cultivated my love for magic tricks, a delight still with me, and a temptation I give in to from time to time when Mr. Tannen on Broadway supplies me with an effect I cannot resist.

In 1944 my mother was posted to New York as the WRNS (Women's Royal Naval Service) officer in charge of the Fleet Mail Office. The Wrens on her staff were responsible for reading and censoring every letter sent by British naval personnel, and I found myself spending time with her, visiting her office, and taking tea frequently with a group of elderly but lively ladies at the Barbizon Plaza Hotel. All through the war years, they arranged daily teatime sessions for any tannin-starved Commonwealth person serving in the armed forces. One day I visited HMS *Ajax*, lying along a wharf in the Hudson River, now all refitted and as good as new after the damage she had suffered, along with the *Exeter* and *Achilles*, in the engagement with the *Graf Spee*. It was a cold and overcast day, and all I can really remember is the feeling of cold gray metal everywhere. Life on a warship in North Atlantic service was not a glamorous life. I suppose, looking back, this was an arranged transition stage for me, although I had no inkling at the time of what transition was ahead. I spent a memorable fall term at Hackley School at Tarrytown, just north of New York on the Hudson, and the Christmas of 1944 with a wonderfully kind family in Greenwich, Connecticut. Almost every weekend they opened their large and lovely home to British naval personnel of all ranks, and my mother and I became very close to them.

And then, in the last days of December 1944, it was announced that I was returning not to Hackley but to England. It was all arranged, cut and dried, taken for granted, and when you are twelve years old there is not much you can do about it. My mother and I boarded a small aircraft carrier, HMS *Thane*, one of the Royal Navy's "flat tops" constructed to ferry planes over to Britain, and there I met eight other English boys who had arrived from the West Indies. I felt totally out of it: they all spoke with an English accent, the first people of my own age I had heard with those funny voices in four years, and they talked all the time about cricket and had never heard of Bob Feller or Joe DiMaggio.

We were all placed in camp beds in the bottom of a huge hold below the water line that you got to by climbing down a long vertical ladder welded to the bulkhead. Obviously the ship had done what it could for a group of young boys, but I recall vividly the pale watery-blue reconstituted powdered milk we put on our cornflakes. The contrast with a warm and lovely Greenwich house at Christmastime was very stark, and I remember vividly when the whole thing hit me. The ship pulled out of the dock with a number of others, ready to steam to some secret rendezvous off the coast of Nova Scotia to form a huge convoy across the Atlantic. As we moved slowly down the Hudson, the Statue of Liberty came into sight, and I could bear it no longer. I crept away from the others and found a hidden corner by a Bofors gun, where I wept my heart out and swore silently to myself that one day I would be back.

A Miserable Year

A few miles off the Clyde (for we were to land at Glasgow), a torpedo hit the stern of the ship, crippling her completely. Fortunately, the shattered mess deck had just been cleared after lunch, and quick action with the watertight doors saved us from sinking. We were all down in our hold and felt the explosion as a terrible, shuddering blow that left us all on the floor. Grabbing our awkward, hard cork life vests, we raced up the ladder where a sailor guided us to the open flight deck. All the guns were manned with helmeted sailors, and a terrible sight greeted us. The U-boat had picked out two tankers first, and they were going up like roaring torches. Corvettes were racing about, and planes from the coast arrived quickly to look for any signs of the submarine. But they never got it. They must have been very brave and cool men down there to take out three ships like that in the midst of a dozen or more escorting vessels, with antisubmarine planes so quickly available. A pity they were fighting for a cause such as theirs.

My mother must have had a week or two of leave, for we stayed in a small thatched cottage in Ugley, just north of Bishop's Stortford, with my grandparents, who had been bombed out of their previous home. The cottage had gaslight, and each evening we lit the glowing mantles in the kitchen and living room, the only warm rooms in the house.

At the end of my mother's leave I was sent to King's School, Canterbury, at that time evacuated to the St. Austell Bay Hotel near St. Austell, Cornwall, to start what was the most miserable year of my life. English boarding schools are brutalizing experiences, and I recall reading recently the comments of an English businessman who had been released after five years in an Iranian prison to the effect that if you can survive a British public (i.e., private) school education, an Iranian prison does not seem so bad.

The monitors, boys of seventeen or eighteen, were the lords of creation, and all the younger boys tried to imitate their drawly voices, while practicing an affected gesture that flipped a carefully cultivated strand of lank hair from the forehead. As an alien from outer space, with a Central New York accent, I quickly copied these ways of speaking simply as protective coloration. (It has since been almost impossible to shake them off). But the effort to make my speech inconspicuous did not always work. In one divinity lesson I was asked to read a passage describing Jesus riding into Jerusalem on an ass, pronouncing it, I thought correctly, as riding on his "arse." A titter of laughter was stopped cold by a glance from the master, but looking up I saw the corners of his own mouth twitching, and the horror of what I had said suddenly struck me.

Once you could get away from the formal part of the school, things were better. We spent weekend afternoons on the cliffs looking for gulls' eggs, or walked the beaches hoping to find washed up K-rations. Nobody really cared what we did as long as we were back for the evening meal. One afternoon I took the short bus ride into St. Austell and saw a movie, the first I had ever seen entirely on my own. It was an American musical, in color, with beautiful June Haver and some tap-dancing male star, a typical Hollywood production. But at the end I found tears rolling down my cheeks, not because the story was sad (because it obviously had a happy ending), but because of the color, warmth, and, above all, those lovely but now-to-be-despised Yankee voices. I quickly pulled myself together, because boys don't cry, and took the bus back to prison.

In the autumn of 1945, the school moved back to Canterbury, right next to the cathedral, where the school held its services in one of the chapels. The smell of orange pippins, apples sold by greengrocers along the street outside the main gate, always brings back that autumn in Canterbury. The teaching was appalling: physics consisted of being given a nice new notebook, into which you copied . . . notes (!), all the while sitting in a steeply banked, dark-paneled lecture room, which, we were informed, was a tremendous honor and privilege because William Harvey, who discovered the circulation of the blood in 1628, had also warmed the same benches. Meanwhile, the lectures, given slowly so we could copy them down, remained absolutely meaningless. Mathematics was the same, since I had been thrown into a form already doing simultaneous equations, without any first steps in algebra at all. These strange equations were things you did things to in a totally mechanical—and, you hoped to God, correct—way. No one ever explained that you could plot the lines on a piece of graph paper, and the solution was where they crossed!

Younger boys were considered the scum of the earth. They were often dirty because the lords of creation, coming haw-hawing from the playing fields, would wallow in baths nearly overflowing with hot water, leaving none for the younger boys under those wartime conditions of fuel rationing. Bullying, which was rampant, became utterly cruel and sadistic in the hands of the house monitors. At the end of each term, it was considered great sport to beat a boy "for slackness." They invariably chose one poor boy who was rather quiet, and whose mother used to visit from time to time, looking rather pinched and forlorn, and dressed in carefully mended and rather dowdy clothes. He was informed on the last day of the term that he had been "slack" and was bent over a chair at the end of the house monitors' common room. Then each of the six louts, the cream of England's educational system, took a running swipe at him with a heavy walking stick. Getting undressed for bed in a large dormitory that night, we saw the terrible and livid bruising damage the blows had done, and we heard him crying himself softly to sleep, trying not to make too much noise. Blubbering was considered frightfully *infra dig*, but I am sure most of us prayed that something or somebody would protect us from the same fate. It would have been unlikely. All of this was known and condoned by the housemaster, a smarmy Anglican "Reverend" with the most affected, drawly Cambridge accent I have ever heard.

The effects of the school, not the least the filthy state I was in, convinced my parents to take me away. At the time, almost everyone in my family seemed to be in the navy, including an older cousin at a naval college at Pangbourne. I passed interviews, Board of Trade eyesight exams, and was fitted with the uniform of a naval cadet at Gieves in Plymouth. By this time we were living in a small cottage close to the end of the Salcombe Estuary, South Devon, and I can remember the almost mortifying embarrassment of putting on a naval uniform the first time at the age of thirteen, taking the ferry to the town, and walking through it to the bus station thinking that everyone was looking at me. By the time I was on the train I was over it, and it never bothered me again.

Pangbourne, as a naval college, was disciplined, but in all proper senses of the word. Bullying, while it inevitably occurred from time to time, was never condoned; showers were available every day; there was a parade and uniform inspection before morning prayers and announcements; you folded all sporting gear in a special way to make a layered stack on the shelf of your locker; and you pulled and tucked the crested cover on your bunk bed so tight a half crown would bounce on it. Come rain or shine, everyone started the day with a run and assigned duties before breakfast, and every boy did some sort of sport every day, whether on a team or not. After classes and the evening meal there were two hours of "prep" (homework) in your home classroom. The subjects were all standard ones, but marine navigation substituted for Latin, and there were periods of Seamanship when you learned semaphore, the flags of the alphabet, Morse Code, and how to tie all the usual knots, as well as the complicated ones like the monkey's fist and Spanish bowline. The latter provided two loops for a man's legs if you had to lower him over the side. It is a pretty, symmetrical knot, and the Spaniards are reputed to halter two horses simultaneously with it.

Sunday, and sometimes Saturday, afternoons were blessed times of freedom, for spare time during the week had to be snatched from brief moments between classes, parades, meals, sports, chapel, and other compulsory events. At the beginning of each month you were given your sweet coupons from your ration book allowing you to buy six ounces of candy with an allowance that was carefully eked out of 20 shillings per term held by the housemaster, supplemented by "Bursar's bobs," a strange naval legacy that provided each boy with one shilling each week after lining up on a special parade. We walked to the village, where you could buy a small brown Hovis bread loaf for sixpence farthing (6¼), always receiving a small coin with a wren on it in change. Or we walked miles across the fields to a small sweetshop at Tidmarsh to purchase three two-ounce bars of Cadbury's chocolate. Other weekends I spent making model airplanes with small diesel engines running on a mixture of ether and oil, or birdwatching with a small group of boys similarly fascinated with these lovely creatures. They are still my favorites. Even as I write this, there are two nuthatches and red-bellied, hairy, and downy woodpeckers queuing up for their turn at the suet and seed mixture hanging from the old shagbark hickory outside my window.

But the best times of all were the holidays when I took the train from Pangbourne to Reading, where you changed to the Great Western Railway's express to Penzance, changing once again at Brent for Kingsbridge and the bus to Salcombe. I spent most of those days in a dinghy pram with a small Seagull outboard, walking the lovely wild cliffs of Bolt Head or exploring by bicycle as far as I could go and get back before nightfall. By the time I was sixteen, I had saved enough to buy a beautiful BSA target rifle with a heavy barrel and a Martini action. It was much too precious (and heavy!) to take into the fields, and it never left the rifle range at school, where I suppose my name is still engraved on a large silver cup three years in a row. It was a skill that was to save my life some years later. For the fields I had a simple, single-barreled shotgun, and I shot rabbits for the family pot as well as the two dogs. This duo consisted of Schnapps, a long-haired dachshund who could get through anything, and Whisky, a whippet mongrel (known by some of the locals as "The Lurcher"), who could outrun any rabbit put up by his small companion.

It was on those cliffs, watching for early migrants in the spring, that I decided to become a geographer. It was no sudden awakening on the road to Damascus, but I can remember the moment quite well. For some time I had been thinking about what I was going to do with my life, while I suppose everyone around me was assuming I would go into the Merchant Navy. But although I could have navigated a ship from London to Sydney without bumping into too much on the way (the school certificate navigation exams were set by the Elder Brothers of Trinity House—something like that), the thought of spending the rest of my life loading or unloading cargo at one port after another had no appeal at all. Of course, I had no idea of what becoming a geographer actually meant, but it was obvious you needed something beyond Higher School Certificate, and that meant going to a university. None of my family had done that before.

I was told that Cambridge had the best geography, but you needed Latin to get in, something I had never done at Pangbourne. So I started *amo-*, *amas-*, *amat-*ing under the tutelage of the Padre (the British naval term for "Chaplain"), a tall, gaunt man with an El Greco face whom we called "Creeping Jesus." It was a hopeless case all around. For him, teaching Latin to this career-changing boy, who was obviously no scholar, was a cross to be borne as a Christian duty. As for me, my heart sank every time I opened the hated books and looked at all the memorization that seemed so utterly useless. I went up to Cambridge twice, to take the Latin exams in a huge hall with a sea of individual desks at regulation distances apart and dons swanning around in flowing gowns, proctoring everyone. I failed twice too. On both occasions I stayed with the Chaplain of Sidney Sussex, who arranged on the second occasion for an interview with a geographer at St. Catherine's College. I know now that Gus Caesar was a geographer beloved and respected by a whole generation that passed through his tutorials, but my experience and interview with him was totally disheartening. I loved geography and knew by that time that all I wanted to be was a geographer, although still not really knowing what geographers did. The interview was short: Caesar fixed me with his eye and said of course I knew German. I confessed I did not, at which point he raised his eyebrows with a pitying look that implied there really was no hope for anyone who did not read German fluently. I mean, how could one possibly read geography! I left on that cold December day feeling totally inadequate and hopeless. I was put out of my misery by His Majesty's Government and the law that every man, upon reaching eighteen, would serve for a minimum of two years in the armed forces. It was called National Service.

From Too Cold to Too Hot

With my background of five years as a naval cadet, I could have gone into the Royal Navy as a midshipman, but with the navy you had to serve for three years instead of two, and when you are just eighteen three years seems (and actually is!) an enormous part of your life. Anyway, I was so discouraged at this point that I just let the winds of the time blow me where they would. So in early January of 1951, I reported to Plummer Barracks of the Wessex Brigade at Plymouth, Devon. For many of my fellow recruits army life was a terrible shock. For me it was a bit like going back to school—the same sort of discipline, folding clothes, making beds, and so on, and if you did these things correctly and with feigned enthusiasm, then nasty little corporals generally shouted at the others. The food was execrable and demonstrated the well-known English capacity for taking wholesome and nourishing ingredients and turning them into stomach-churning slops, which were not improved by being served out of ice-cold and grease-congealed pans onto ice-cold metal plates. For this, and other generally useless forms of abuse, we were paid the sum of 28 shillings each week, which was whittled down to 21 by income tax, national health contributions, and something called "regimental dues." Out of the remainder, you were expected to buy all your necessary toiletries, boot polish for the everlasting lacquering of your "dress boots," brass polish for brass fittings that should have disappeared after Waterloo, and Khaki Blanco for webbing suitable only for some sultan's guard whose soldiers were never required to fire a shot in anger.

After a few weeks I was pulled out with some others to what was known as the "Cadre Platoon," soldiers who might go forward to the War Office Selection Board (WOSB) for tests, interviews, and perhaps Officer Cadet School. Those selected were given no special treatment in basic training (if anything, it was made tougher), but we were resented by others, particularly the sergeant of the guard, who inspected you minutely on those rare occasions when you could leave the barracks on a Sunday late afternoon. We inspected each other before trying to get past the swine, rubbing wet soap on the inside of all creases before we ironed them to a razor's edge. Whenever I could save 18 shillings, I went to a small restaurant where you could get a *prix fixe* dinner for eight shillings and a half bottle of dubious wine for the same, leaving two shillings for the bus and a tip. It was my last hold on civilization, and I dined by myself, usually with a book to read. One evening the colonel of the barracks and

...lso came to dine, and were being shown a table near mine. I did not ...what to do—sit there? stand to attention? run like hell?—but he obvi-... saw my embarrassment as I started to rise to my feet, and he waved me ...k saying "Don't worry, enjoy it while you can."

The WOSB tests and interviews were quite fun in their own way, and certainly a change of pace from taking a brengun (a light machine gun) apart and putting it back together for the umpteenth time. At WOSB you did various paper and pencil tests, and commanded five or six others to "solve a problem," something like how to get a heavy baulk of wood and all of you across a ditch or over some obstacle. The interview was with a ruddy-faced Colonel Blimp, who asked me why I wanted to be commissioned into the infantry (I had been assigned to the Royal Army Education Corps by the guy handing out clothing and cap badges, and no one ever questioned this). Of course, I did not say that I thought army life as a private soldier was the absolute pits and the sooner I could wangle my way out of it the better. Instead, I made appropriate martial and patriotic noises, which were obviously appreciated for they triggered lots of "Splendid . . . jolly good chaps, haw-haw-haw . . . Queen of battles, of course," and other half-swallowed nonsense. Judging from some who failed, the interview demonstrated little except the degree to which class and accent ruled in Britain at the time. Anyway, I spent four months at Officer Cadet School, in beautiful Eaton Hall near Chester belonging to the Duke of Somethingorother, and was commissioned in the Gordon Highlanders, mainly because it pleased my grandfather who had lost his only son, Richard, seconded from the Gordons to the Glider Regiment, in the Sicily landings.

After kilting and sporraning up, I was posted to Fort George, the Highland Brigade Training Center at the end of a single road leading to a point on the Moray Firth near Inverness. There, at the tender age of nineteen, and under the benign eye of an experienced sergeant who had gone through El Alamein, I ostensibly trained raw recruits in the art of warfare, showing them how to shoot and how to put fuses in grenades, and also lecturing them on the physiological signs and general dangers of gonorrhea and syphilis, as well as the approved methods of prevention. All of this was done by a virgin officer with a straight face to recruits from Glasgow, some of whom sewed razor blades inside their caps, nonregulation attachments that apparently came in handy when a cap was turned inside out in a bar fight. If they were caught during an inspection "their feet didn't touch" as the sergeant marched them to the guardhouse.

Teaching raw recruits how to shoot in early winter, with a cold wind from the North Sea whistling up your kilt, is no fun. The tedium and pettiness of the officer's mess at Fort George was extreme, and life was enlivened only by blowing up grenades on the range that had failed to explode. As the officer in charge this was your job, while others crouched out of harm's way in the throwing pits. As a Gordon, I was called for a few days to the regimental depot at Brig-o-Don in Aberdeen. There I again made some appropriate martial noises of the sort that clearly reverberated in senior infantry officers, and begged to be sent to the first battalion on active duty in Malaya. I had another year of service left, and Malaya seemed much more exciting than freezing your nether parts off in the Highlands.

The martial noises seemed to work, and I soon boarded a troopship that stopped at Gibraltar, Malta, Port Said, Aden, and Colombo before reaching what was for me the final outpost of Empire at Singapore. The Gordons were "up country," with headquarters at Ipoh and all the companies scattered around half of Perak. It was the start of a year as platoon commander in A Company stationed then at a small village near Ayer Kuning. The whole story of jungle patrols, ambushes, air drops, leeches and mosquitos can only be told at another time, but two incidents stand out. One was an occasion when, quite by accident, we came across a base for about twenty or thirty guerrillas late in the afternoon. They must have pulled in their sentries, thinking it was too late for any patrols to be so deep in the jungle. All hell broke loose, the main danger being shooting at each other in thick cover with everyone wound up tighter than a violin string. I moved quickly through the base with Munsan, one of the two Dyak trackers in the platoon, and followed a track leading out the other side, when I saw two guerrillas turning and lifting their guns to shoot. Munsan dropped to one knee and killed one with one shot, while I forgot my American carbine was on automatic and got the other with a full magazine because I was so frightened I just hung on to the trigger. Munsan looked at me in a way that is difficult to describe, but later we exchanged a skean dubh, a useless, but highly ornamental dagger you stick in your sock in Highland evening dress, for a most practical parang, Munsan's second best, beaten out from the spring of a Model T Ford. Second best or not, it had two heads to its credit, both Japanese who had been foolish enough to patrol the jungle in Dyak territory on Borneo during the Second World War. The Dyaks ambushed patrols silently with blowguns, and then took the heads as souvenirs. I have the parang to this day, and keep it sharp for slashing weeds in the garden and lopping heads off intruders.

The second occasion was gentler, even prosaic. We were emerging one afternoon on patrol from deep cover to an open patch of lalang (a tall, broad-leaved grass), when just on the edge of the clearing a small, about chest-high rhinoceros looked at us and then turned and trotted off quickly into the thick jungle cover. The leading scout, I (in second place), and the man behind me all saw it, and when the battalion went down to Singapore for six weeks of rest and retraining, I went to the Singapore Museum and told the curator in charge of zoology. He expressed his disbelief in rather scornful terms, saying the rhino was extinct in the peninsula, and hinted that the 160 proof black naval rum carried on patrol (one ounce per man per day) had been imbibed inappropriately. But all three of us saw it, and my zoological gaze was not affected by black naval rum.

"Rest and retraining" in Singapore was euphemistic. You cannot retrain for jungle warfare on the open parade grounds of Changi Barracks, once the notorious Japanese prison camp, and few got much rest. As "bloody ensigns," who had no idea how to comport themselves at regimental dinners, all the junior officers were on the tennis courts at six each morning, practicing the steps of the Highland fling, foursome reel, and ghilly cullum (an individual dance doing rather careful fling steps around a sword crossed with its scabbard). All this went on under the eagle eyes of the colonel and the pipe major. Apart from making beehive charges out of plastic explosives, those dances were the only really enjoyable things I learned in two years.

The practices paid off at fairly frequent regimental dinners, when everyone was in tropical Highland evening dress and the long table groaned with silver captured from the French in the Peninsular War and at Waterloo. Pipers made intelligent conversation almost impossible, and everyone sat with appreciative looks on their faces. After the port had passed clockwise, and the madeira counterclockwise, the pipe major strode in in full regalia and saluted the colonel, who had ready a replica of the Waterloo glass, a tall, thin wineglass reputed to have been carried in the coattails of a Gordon officer through that particular bit of carnage. The colonel filled the glass to the brim with whisky, the pipe major drained it, and then the tables were pushed to the side and dancing to the pipes began. It was all rather fun.

At the end of the year, I returned on the same troopship, spending Christmas on board. Just before lunch, all the officers assembled in the ward room for a preprandial drink and to listen to the traditional Christmas Day broadcast of the Queen. When "God Save the Queen" was played, everyone stood rigidly to attention with stiff upper lips, save one Irish officer of the Gurkhas who had been drinking steadily since breakfast. In the middle of the anthem, he broke down and sobbed loudly, explaining to all that as he was Irish he didn't ha-ha-

have a qu-qu-queen! Everyone looked frightfully British and embarrassed, and the Queen's speech was listened to with frozen faces, eyes looking everywhere but into someone else's, while quiet snuffles and blown noses indicated that the merits of Republicanism were being reconsidered.

I returned to Aberdeen but was told that there was no point in staying the four or five days before my two years were officially up, so I was issued a railway warrant and took the train back to Salcombe. Things looked pretty bleak. A South Devon fishing village, sunny and alive with tourists in July, is virtually empty, gray, and cold in January. Cambridge held no prospect, for studying Latin to jump through a hoop after leading thirty men in jungle warfare for a year seemed even more idiotic and numbing than it did before. Bristol, it was said, also had a geography department, and as my father was living near Bath at the time, I applied and had an interview with someone who taught surveying and mapping. I told him that more than anything else in the world I wanted to study geography and be a geographer, still not really sure what geographers did, or why they got paid, except that there seemed to be some at universities managing to keep body and soul together. His response, which had a very subjunctive tone of doubt to it as he perused my meager credentials, was that *if* I got in I should take surveying "because they always need surveyor chaps in the colonies." Hardly a reassuring vote of confidence.

Fortunately, help—quite unbelievable and wonderful help—was on the way. I received a letter from my "American parents" saying I could enter Colgate University on a one-year foreign scholarship and live with them. I accepted at once, and remember spending hours with a wonderful catalog that arrived later in the mail. True to its liberal arts tradition, the undergraduate curriculum was built around a succession of "core courses" that exposed everyone to science, philosophy, literature, the arts, music, and so on, with the junior year focusing upon domestic issues and politics, and the senior year on something like America in the Modern World. This was no smörgåsbord of general education requirements that often are nothing more than departments trying to grab as many student credit hours as they can for the bean counters in the central administration. The core requirements were carefully constructed courses taught by faculty who believed in such endeavors. The core course load declined as you went through the four years and your major courses took up more time. There was a one-man geography department, consisting of Shannon McCune, who had been born of missionary parents in Korea. I was the second student of his ever to go on to the life of a professional geographer, and I think he was delighted with a student who knew so clearly that this was what he wanted to do. Later Ted Herman took over, another kind man who gave me great freedom to explore things on my own.

There are a few times in your life when you feel the blackboard has been wiped clean, that what you make of events is up to you, and that those around you and judging you have no preconceptions. Perhaps wisps of nostalgia cloud events when we reflect upon them years later, but Colgate for me was a truly golden time of so many things opening up that I never knew existed. I was given one year's credit for the work I had done in England, with the exception of the core course in philosophy and religion, for which I had an outstanding scholar and teacher, Holmes Hartshorne. I avoided the fraternities: after two years in the army the utter childishness of their silly little ceremonies seemed, literally, puerile. I still think so.

I worked very hard, and with utter delight, and managed to supplement the original scholarship with others. I worked in the summers in camps for teenagers, or as a research assistant to a sociologist. One summer I had to spend some weeks in Geneva, New York, recording sociological details about salesmen who worked for a rose company. The object of the study was something about rapid turnover, but when I returned to the campus I was shown how to do arithmetic on a mechanical calculator, and how to go through a whole series of steps, all carefully laid out on a preprinted sheet, that ended up with a decimal number that could vary between −1.0 and +1.0. If you got values toward the extremes it was considered a good thing, while numbers close to zero in the middle were disappointing. It was, of course, a correlation coefficient, and I can remember asking why they did not plot the *X* and *Y* on a piece of graph paper, something I had been doing in physics experiments since I was thirteen. But no, the coefficient was the thing. Still, the *idea* of measuring how closely two things were related came in handy later.

It was in my second year at Colgate that I came across John Wellington's *Southern Africa*,[12] a large, two-volume work that had just been published, and a very different geography book from many of those then available. I had already read all of John Gunther's "Inside" books, including the one on Africa,[13] which had also appeared during that second year at Colgate. As there was no official regional course, I asked if I could do an independent study, although it is difficult to explain to myself, let alone others, why, from a small New York town Africa appeared so intriguing. Before this there had been nothing in my background, no connection, to lead me to Africa. Perhaps it was because all the colonial systems were showing signs of stress after the Second World War, something that would be reflected a few years later in Harold Macmillan's

"Winds of Change" speech,[14] perhaps it was the sense of enormously rapid human geographic change against a backcloth of great physical geographic variety. I do not know—things happen.[15] But when it came time to apply to graduate school, and I received fellowship offers from Chicago, Wisconsin, and Northwestern, I chose Northwestern not simply for its geography, but because of its superb Africa Program. Shortly after choosing Northwestern, I received letters from both Chicago and Wisconsin asking *why* I had turned them down. The implication was, How could I be so misinformed and make such a dreadful mistake! I never answered their letters, and I still think it is an inappropriate question to pose to a young student.

Northwestern was somewhat of a mixed bag, and had it not been for Ned Taaffe I think I might have left (although for what?). I came to graduate school totally starry-eyed, only to go through a summer field camp of utter banality.[16] When we were not looking at Archaean boulders by the roadside, pretending to be awestruck by their ability to make geological hammers bounce off their surfaces, we were plunging soil augers into loess that must have been plowed for nearly a century. However, during the last two weeks I became intrigued with my own research project, which looked at the way the idea of the Soil Bank was spreading amongst the farmers of Grant County, Wisconsin. During the fall term, in the research seminar when we had to write up our results, I actually fitted a quadratic distance decay function by least squares, using the Doolittle Method on a mechanical desk calculator. This was in the fall of 1956, and I claim this was the first quadratic ever fitted by a geographer. I await, without too much angst, for all the refutations to pour in. It all seems a long time ago. Anyway, Ned Taaffe told me to grit my teeth for the remainder of the year, and then I could do more or less as I liked.

The Africa Program was superb, led by the anthropologist Melville Herskovits, who, with his wife Frances, had opened up African studies in America in the 1920s. He seemed to know everyone and everything, and had Ford Foundation support to invite distinguished scholars in the departments contributing to the program. Which is why I was able to take courses not only with Herskovits in anthropology, but with Kenneth Dike in history and Pius Okibo in economics. Pius stunned everyone that first day by arriving immaculately dressed, with a stiffly starched white collar, a splendid cravat (it was hardly a mere tie!), and lecturing without notes in a most fer-right-fully sper-lendid Oxford accent, acquired, I suppose, during his doctoral years. Everyone in the program took the general African seminar, a weekly evening affair held throughout the year, often with distinguished visitors like Louis Leakey of archaeological fame, or Tom Mboya, who had world statesman written all over him until he was brutally murdered in his own country of Kenya. I remember going with some other African Program students, after Mboya's address to the seminar, which was open to the public and held in a large auditorium, to a private home owned by a black couple, with lots of African American guests. Tom Mboya was close to the end of a grueling lecture tour and was almost gray with fatigue. He sat in a large armchair in a corner, with everyone crowded around listening to him, when one of the guests said, "Say, Man, where did you get your confidence?" Mboya looked him straight in the eye and said, "Where'd you lose it?"

Apart from the field camp, and the research seminar the first term, the only compulsory step in the geography graduate program was . . . the Pressure Problem! You picked up your research problem at eight o'clock one morning, and it had to be finished and all written up exactly two weeks later. You had not the faintest idea what it might be until you tore open the envelope, rather like John Wayne as a submarine commander opening his sealed orders to sink the Japanese fleet. I tore my envelope open and read "The Geography of West Wilmette." "You bastards!" I thought. "I know exactly what you want, and I'll be damned if I'll give it to you!" What was required by that august faculty was a traditional "geography" of the north Chicago suburb of West Wilmette, the classical checklist of geological structure, soils, precipitation, vegetation, settlement history, agriculture, . . . the whole miserable shitlist to rub my nose in the fact that I had not taken courses with some of them, and to teach me a lesson for all that locational analysis in economics, and the quantitative and theoretical transportation stuff with Ned.

This was the spring of 1958, and already ideas about central place structures, the restructuring of "shopping space" by the new malls, and spatial economic behavior were around. I spent the first day and night drawing a detailed base map, and had one hundred copies printed in light blue. I used these not only as work maps in the field but also to draw the final presentation copies for the paper (the light blue lines not inked over would not photograph for the final copies). I spent days interviewing poor housewives, asking about their shopping behavior, and plotted hundreds of lines marking out the nested umlands (the shopping catchment areas) of the traditional shopping centers, showing how these were changing under the impact of the then-new malls just beginning to spring up. With the exception of Ned, this was not what was expected, and for some it hardly constituted anything geographic at all. I learned later that all the "that'll-teach-'ims" wanted to fail me, but since I had already been awarded a

Ford Foundation Fellowship to do dissertation field work in Ghana, they did not quite have the guts to do it. So I got a C, a barely passing grade for a graduate student, and the only one in my entire university career.

The oral comprehensives went fine, except for one elderly, then well-known member of the committee who was determined to show me up because I had not stood in line for his courses in economic geography and Latin America. The former used his text, and I had no desire to take part in another dreary round of rote memorization to regurgitate on the exams. (The contrast with Charlie Tiebout, using August Lösch's *Economics of Location* over in the economics department, was like night and day.) The exam was held in his large office: "Well," he said, slapping a three-inch tome on his desk, a newly arrived and obviously unopened volume on the forestry resources of the Soviet Union, "what do you know about the movement of lumber in the Soviet Union, eh?" "You swine," I thought, knowing damn all about flows of lumber in the bloody Soviet Union. "But I know you haven't opened the book, and you don't know anything about them either!" So I dredged down deep to the residues of a rather boring Soviet course I had had as an undergraduate and visualized the map of the Soviet Union. Visions of old newsreels of Finnish ski troops gliding through thick pine forests to ambush Soviet columns came to me, so there must be some coniferous trees up there worth cutting, and surely the Urals also had some timber besides spindly birches? So with total confidence, and bluffing all the way, I talked learnedly about "major flows of pine lumber from the Finno-Karelian Shield to the Leningrad-Moscow-Kiev industrial triangle, augmented by substantial flows of hardwoods from the Urals," and so on, rather desperately hoping that no one, least of all the old codger, would know whether I was right or wrong. When I had exhausted the resources of my imagination, dropping names like Khabarovsk, which, according to me, sent massive flows of timber to Vladivostok (God only knows why), there was complete silence for a few seconds. Finally, the old boy grunted "Hmm, well . . . yes . . . not too bad . . . no more questions."

The Ford Foundation Fellowship was probably awarded out of pity, after I appeared at an interview in Chicago a day after the infectious period for chicken pox. My wife drove me down in a snowstorm, and I remember the two interviewers backing away to the other side of the hotel room when I appeared at the door pale and blotched. Pity or not, the fellowship enabled me to go to Ghana for the field and archival work on my dissertation.[17]

It was a wonderful year, not the least just getting there and back. You took one of the Queens from New York to Southampton, and then an Elder Dempster passenger ship from Liverpool to Takoradi, stopping on the way at Las Palmas in the Canaries, Bathurst in Gambia, and Freetown in Sierra Leone. There is a special smell of wood fires mixed with pepper, an odor that you can smell far offshore, that always brings Ghana and its lovely people back to me. They deserved so much better than the succession of military coups, *rondes de petite violence,* that jailed the opponents, only to let them out once the cash in the Swiss bank accounts had been satisfactorily transferred. But this was still 1958–59, only a year after Independence, with Kwame Nkrumah yet to elevate himself from mere President to Asayagefo ("Supreme Ruler"), and if you had had to bet on success, Ghana was the surest of all. The country could feed itself, with enough left over to export to neighboring countries; and it had large cash reserves in hard currencies, as well as exports of cocoa, manganese, palm oil, gold, and diamonds, with the promise of the Akosombo dam on the Volta turning Ghana into a major aluminum producer. I was to return seven years later as a transportation consultant to USAID to find the finest road system in Black Africa transformed into a deeply rutted, often impassable mess. The Public Works Department, and other crucial areas of government, had been totally depleted of maintenance funds, while three huge presidential palaces, triumphal arches, athletic stadiums for political rallies, and other accoutrements of wealth and power had consumed national funds belonging to the people, only to produce an economy in total disarray.[18]

When we returned to Northwestern, our daughter was born, and we ran out of money. We had had a very lean life in Ghana, on a very stringent budget, using up most of what we had previously saved out of my Northwestern fellowship and my wife's salary as a teacher. It never occurred to us to ask the Ford Foundation to augment the fellowship they had granted, although it turned out that they would have done so. Things looked pretty bleak, and we sat down one evening and calculated that we might just make it to the end of the fall term, when I was desperately trying to finish the dissertation, if we could find $600 from somewhere. I went to Mel Herskovits and asked if there were *any* funds available as a student loan. "How much do you need?" he asked, well aware that we were now a family of three. "I think we can do it on $600," I said. "Nonsense!" he replied, and my heart sank. "Don't be ridiculous, you'll need at least $1,800 to get to the end of the year!" He gave us that grant outright, bless him, and we made it to my first job in January 1960 broke, but just free of debt.

On the way out to Evanston, I had had an interview at Syracuse University. Preston James and others were looking for one of those newfangled "quantifiers," and someone with a mud-on-your-boots regional interest seemed desirable. I was offered an assistant professorship at $6,000 per year; provided I had defended the dissertation by Christmas. By the time I got to Evanston, the written offer was an instructorship at $5,500: George Cressy, I learned later, had thought my hair (very short by today's standards) was too long and that this indicated intellectual unreliability. (We did not have money for haircuts, and used rather inexpertly a razor-comb device.) Ned Taaffe and Ed Espenshade were not at all happy that one of their new Ph.D.'s was only worth an instructorship, even if he was going to a geography department of Olympian stature, and I gather that a phone call restored the assistant professorship, but not the promised salary. I am not sure to this day how the three of us managed it, but make it we did, and I started teaching at Syracuse that winter term.

Syracuse turned out not to be a happy place for us, but that is another story to be told at another time. In those years, the university's Maxwell School had a remarkable program supported by the Ford Foundation, under the direction of a fine colonial historian, whose family became close personal friends of ours. It placed highly selected young Americans for two-year internships in what were essentially positions of colonial service in countries around the British Commonwealth that were already preparing for Independence, or were clearly about to feel the "Winds of Change" very soon. Some of these new graduates took on enormous responsibilities in those years, far beyond anything they had imagined.

This particular international program brought up the idea of a formal East African Studies Program, and these were the days when the Ford Foundation seemed willing to bankroll almost any sort of regional studies venture, providing it had an interdisciplinary basis. The upshot was the start of such a program, and because of my West African experience in Ghana, I went to Tanzania, with a political scientist who had done his own dissertation work in Uganda. This time it was the *Nieuw Amsterdam* to Rotterdam, and then to London, where the four of us took the *Braemar Castle* of the Union Castle Line to Mombasa, stopping at Gibraltar, Genoa, Port Said, and Aden. I commend this civilized way of traveling at other people's expense, for it allows you to catch your breath after a grueling academic year, do lots of background reading on the boat, and arrive fresh, rested, and without debilitating jet lag after a long flight (and living in a space that would properly be outlawed if provided for animals).

We were able to choose almost anywhere we liked as a home base, and the political scientist and his family had chosen, very sensibly, Moshi, in the shadow of Kilimanjaro. The town was at a height above the malarial zone, and the mountain water was very pure. We had two small children by this time, and on weekends we would pile into our Volkswagen and drive as far as we could up the mountain, walk up to the first hut used by climbers, and then through the last remaining trees to the alpine meadows. The United Nations trusteeship had not yet been given up, and we lived in a house near the police compound that had belonged to a former Education Commissioner. Dagga, our Alsatian police dog, had flunked out of police school, refusing to tear the padded instructor to shreds, sashaying up to him instead wagging her tail. However, she had the look of a fierce guardian, and a bark that scared the living daylights out of you.

While I was in Tanzania, I tried to undertake three studies of diffusion: membership in a progressive farmer's club on Kilimanjaro at the person-to-person level; the diffusion of cotton cooperatives in Mwanza District south of Lake Victoria at the regional and institutional level; and the diffusion of "modernization" at the national level. Despite some frustrations at the lack of cooperation, it was a happy and memorable year, our two children for the most part healthy, and becoming quite blasé about elephants, giraffes, hippos, and other wildlife, which were simply part of the natural scene on weekend drives to Amboseli and other nearby parks.

Our thoughts about going back to Syracuse were not terribly enthusiastic, but as it turned out, we did not have to worry about it. One day, a letter arrived out of the blue from Alan Rodgers and Pierce Lewis. They had gone to a small "short course" on quantitative methods in geography, taught in the summer by Ned Taaffe, who by this time was the chair of the geography department at Ohio State. The department at Penn State was being reorganized and somewhat enlarged, and they wanted someone to teach these still rather new approaches. Was I interested? Truthfully, I had never heard of Penn State in those days: the "big boys" were places like Berkeley, Washington, Syracuse, Chicago, Michigan, and some other Big Ten and West Coast universities. When I wrote to Preston James about the offer, I received a lukewarm reply indicating that a senior member of his department might even be relieved to see me go. So I accepted the offer at Penn State sight unseen on both sides. It was the best decision I ever made in my life, and we have been content to stay in Happy Valley, bringing up three children, all Penn Staters, in a rural environment of ridges and valleys, forests and streams, a mix of small towns and Amish farms.

Happy Valley

The 1960s were a time of rapid expansion in many universities, and Penn State was no exception. Geography was in the College of Mineral Sciences (later the College of Earth and Mineral Sciences), one of the old colleges (along with Agriculture) in the first of the land grant universities set up by Abraham Lincoln under the Morrell Act in 1859. The origins are a tradition I cherish, a wonderful and farsighted nineteenth-century faith in the power of education in a democratic society, much the same sort of faith displayed by the thirteen men who founded my own alma mater of Colgate in 1819, on a bare hillside above the village of Hamilton, New York. Although geography courses appeared from time to time in the nineteenth-century catalogs, it was not until 1935 that a fully fledged department was founded by Raymond Murphy, before he went on to Clark University and the editorship of *Economic Geography*. By 1963 we were a small department of eight active members and eight majors; today that generation has retired, and the department has eighteen active members and about one hundred fifty majors. The graduate program also expanded somewhat, but has been pretty stable for the past couple of decades.

Looking back, I see ourselves blessed with two things. First, location in a fine college of high morale that by any measure you like to name is the best in the university. Geography started here, and wild horses would not drag us to any other, although with an original and strong emphasis on human geography, we were sometimes a bit of a puzzlement to mining engineers and geophysicists. Still, we went from twenty-ninth to fourth to second to first in the National Academy rankings, so some of our colleagues in other departments began to assume we were not quite so flaky as we appeared. Not that rankings within the top ten mean much, and the faculty in any department at that level always have self doubts about whether they are as good as they are perceived to be. Still, it certainly helps across the campus and with the bean counters.

The second blessing has been our students, most of them with an intellectual independence to match their intelligence, something less appreciated today, when genuine intellectual curiosity tends to be more and more warped by what is considered fundable by the various Establishments in Washington. Some would even like to inculcate students early in their careers with the "marionette" spirit, i.e. jerk to the money strings, and never mind about the ideas. But there are few pleasures in life as rich as sitting down with a young colleague and being used as a sounding board for a new idea, topic, or way of looking at the world. Most of the time, I find myself struggling to keep up, but they say getting out of breath for short, intense periods is good for the intellecto-cardio-vascular system. Zeus bless them every one!

Picking out incidents in a geographically rich life is difficult, and this introduction must not turn into a book of its own. On these occasions, when thirty-eight years of academic life melt into each other, the lines of Dylan Thomas's lovely narrative poem "A Child's Christmas in Wales" always come to me:[19]

> One Christmas was so much like another,
> . . . .
> that I can never remember whether it snowed
> for six days and six nights when I was twelve
> or whether it snowed for twelve days and
> twelve nights when I was six.
> . . . .
> . . . , and I plunge my hands in the
> snow and bring out whatever I can find.

Of course, the years have not really been "so much like another," not with new students, new colleagues, sabbaticals in Sweden and France, and many other exciting journeys to see new places and to talk to new people and old friends. So let five local incidents serve, simply because upon reflection they are what are granted to me at this moment.

One day a National Science Foundation proposal was rejected, a proposal that one of our new Ph.D.'s and I had submitted to be the first, and I think to this day *only*, attempt to apply the theoretical ideas of spatially dynamic models to an actual and concrete set of empirical data. The evaluating committee did not approve of this. The NSF, it was implied, was for *basic* scientific research, and actually testing what was an extraordinarily seductive model, generating spatial patterns that any geographer itched to tell a plausible story about, hardly came within the general guidelines. Well, OK, decisions are hard to make; but what caught my eye was one of the evaluations, which, under the strict rules of the Freedom of Information Act, must be forwarded to an applicant. Unbelievably, this one read "Professor Gould earns too much money." Now this may have been quite true: I am still amazed sometimes that someone, somewhere, is willing to pay me for what I love doing, something I would try to continue to do even if I had to dig ditches or flip hamburgers for a living.[20] But I find it difficult to respect an institution and its evaluating committees when the first thing they do is flip to the budget pages, compute your annual salary from the proportion of salary requested, and then decide on the scientific merits of your proposal according to whether their salary is as big as yours. Except for dissertation support for doctoral students, that was the last time I submitted a proposal to the NSF.

The second incident brings a much warmer glow. For many years a group of faculty across the university got together and held informal, and by the nature of our various backgrounds, interdisciplinary seminars, usually around a particular book or theme. This tradition still continues, and the Tuesday Night Seminar usually takes a formidable book, generating difficult questions, to read slowly and carefully over a semester. One year we took Alan Janik's and Stephen Toulmin's wonderful *Wittgenstein's Vienna*,[21] planning to use it to raise some of the many and varied themes that came into such extraordinary conjunction in that earlier *fin-de-siècle* world. Originally planned for a semester, the seminar lasted two years as we persuaded, begged, even shamed various deans into giving financial support for fares and modest honoraria. Alan Janik gave an early presentation, and Stephen Toulmin closed the seminar two years later. In between, we had a parade of extraordinarily stimulating visitors in art, music, literature, and science to augment our own formidable array of local talent, people who were willing to give a presentation and then became hooked on the seminar themselves.

The third incident is even more personal. From 1980 until just a few years ago, I attended almost every semester a formal or informal class or seminar on Martin Heidegger given by the philosopher Joseph Kockelmans. The thinking of Heidegger is not easy, and I am convinced you need a fine scholar and teacher to help you on your own way through the paths of the forest. Sometimes the text would be a single essay, over which we usually took many weeks; sometimes it was a volume of lectures. I sat through his course on *Being and Time* twice, and when I taught the eight o'clock class on spatial analysis three times a week I would be up at 6:10 to read other volumes at a quiet breakfast before taking off for the university at 7:20, gradually and, most important, slowly working my way through a volume, trying reflectively to follow Heidegger's own thoughts. Joseph would always work with the English, German, and sometimes the appropriate Greek texts in front of him, and he was always generous and helpful if you ran into difficulties and wanted to pose a question. These times, an hour or two in a late afternoon, were like little oases of thought in a too often frenetic, and sometimes idiotically stressful, academic life. It was an enormous privilege to learn from him, not the least because he had an absolute gift for helping you by giving you a concrete example of just the right difficulty that was meaningful to you.

The fourth incident is really telling tales out of school; but I cannot resist, and I shall withhold all names to protect the gullible. Interspersed throughout our beginning graduate seminar at Penn State are times when individual faculty members make presentations about their field, their research, their ideas, or simply themselves. Sometimes these do not turn out to be the most stimulating things at the end of a long day when people are feeling that they have had enough and are getting peckish. So one year I decided to do something different, and after shuffling a few things around at the head of the seminar table, I announced that I wanted to discuss "Science, Experimentation, and Hypothesis Generation." Polite but wan smiles, forlorn looks into which it was easy to read the "Oh, no, not another one!" "These are important ideas," I declared, "and I want to conduct an experiment with you." A few eyes flickered open. "I'm going to leave the room for one or two minutes—I'll be down the hall, call me when you're ready—and I want you to choose two of you that you all feel you can absolutely trust, people who could not possibly have colluded with me in any way before this seminar." Rustle of interest. Who is this guy? Has he gone round the bend—finally?

So out of the room I went, and waited about four or five minutes (it takes a while to determine who you really *do* trust) for them to sort out their two choices. Back I came, and the chosen two were a foreign student in physical geography (unlikely she would be in collusion with a human geographer?), and an army captain on his way to West Point (incorruptible, honor, duty, country, made of stern stuff, not a chance of collusion!). "Right," I said to our newly arrived foreign student, "I'd like you to choose one of these books," and here I took a couple from a pile I had in front of me, "and then go and sit on that chair in the corner with your back toward me. I'm going to go to this corner to get as far away from you as I can."

By this time, the students were definitely interested. "Now," I said, "examine the book carefully, leaf through it, and when you're ready tell us how many pages it has." After leafing through the book she announced "238." At which point I said, "OK, from now on don't say one word, or make any gesture with your head, until I tell you." Total silence, a rigid head looking at the blank corner. "Now," I said to America's first line of defense, "choose any number you like between 1 and 238." He thought a minute and then said "127." "All right," said I to my silent and rigid experimental partner in the corner, "turn to page 127, and concentrate all your attention on the first word of that page. Try to block out any other thought except that word, and then imagine you are throwing it across the room to me." Yeah, like telepathy's real, said some of the looks on the other faces!

And then, in a quiet running commentary to let the others know what was happening to me, slowly, faintly at first, but with concentration ever stronger, came a word. A bit difficult at first, because of the slight foreign accent. "Ah, that's better, buh . . . no beh . . . bee? . . . be-cow . . . becow? ah! 'because.' The word is 'because'! OK, you can turn around and speak now. What was it?" The word was "because," and I congratulated her on her telepathic abilities. Had she done it before? No? Remarkable! And to the rest, "All right, you guys, what is the explanation? What hypotheses can you generate to account for this clear demonstration, under conditions of impeccable experimental design, for the power, admittedly simple and limited in this case, of telepathy?"

They managed to generate four: it was real; there *had* been collusion; Gould had memorized the first word on every page of all those books ("Naw, the guy's quite bright, but not that smart"); and it was a trick ("OK, wiseguy, tell us how!"—this from one of the students, not me). Back and forth the discussion went, helped by a rhetorical gesture or two from me that implied "You mean you don't even believe what you have just witnessed with your own senses!" At the end we voted in public, a raising of hands. Over half of them voted for "It was real." After all, seeing's believing. A few, unwashed and unregenerate skeptics, were still asking me weeks later, "Come on, how did you do it?" But I did not tell them then that they could do the same if they were prepared to part with $29.95 to Mr. Tannen on Broadway.

Closure time: not long ago, an old friend, Peter Haggett, joined us for a departmental celebration, a lovely renewal for the department because many years before he and Brenda and their children had spent a fall semester with us. This was another lovely fall, the colors of the trees splashing the campus and brightening all the surrounding hills. We were walking slowly across the campus, talking of the changes we had seen in our thirty or so years as geographers and academics, not the least the terrible pressure now placed on young faculty, and increasingly on graduate students, to raise funds, no matter what the intellectual costs. I remember the name of Clarence Glacken came up, the author of one of the truly classic works in our field,[22] and how it would be impossible today for such a book to be written, with such enormous care and magnificent scholarship. After all, how could you get a grant to fund such an endeavor over such a period of time? (I mean the guy would never get tenure.) "You know, Peter," said the other Peter to me, "we were a blessed generation. We came in at the right time, and we're getting out at the right time."

And I could only agree. But in between, it's been fun!

African Beginnings

ADMINISTERING a country the size of Ghana, roughly the same area as the United Kingdom, is no easy task, especially when it is made up of many different people of various identities and languages. Administering Tanzania, a country four times as large, and roughly the size of Texas and New Mexico combined, is even more difficult. In colonial and UN trusteeship days (Ghana and Tanzania respectively), young men from Britain's major universities would serve in various branches of the Colonial Service, and after a period of apprenticeship those in Administration would be posted as district officers to areas often far from the major towns.[1] Much of their time was spent "on trek" as they made regular visits to all parts of their districts, looking after the people for whom they were responsible. Their responsibilities were many and often heavy, made up of a mixture of agricultural, financial, judicial, public health, and educational functions, most of which generated regular reports and records, the "bumf (paper-generating forms) makers," and the bane of many district officers, who usually felt they had more important things to do.

In the days after Independence, many administrative adjustments had to be made, and recruiting for national administrations often became difficult. Many university graduates tended to be drawn toward high-paying professions like the law; few relished the day-to-day independence and heavy responsibility that was often the lot of local administrators, particularly those posted to remote rural areas whose people had ways and languages different from their own. Having a common national language, with the direct and immediate communication it can provide, was often an important consideration, and Ghana and Tanzania chose to move in different directions. Immediately before Ghana's Independence in 1957, a British commission considered with great care the question of making English the first language of instruction in all of Ghana's schools. Highly sensitive to how such an imposition might be interpreted in those delicate pre-Independence days, they refused to make such a recommendation. Yet one of the first changes made by Kwame Nkrumah, Ghana's first president, was to make English the official language of the country. He was well aware of the unifying effect of a common tongue, as well as the crucial access a world language gave to contemporary education, commerce, science, and the arts.

Tanzania chose a different course, adopting Swahili, a partial *lingua franca* developing from strong Arab trading influences, but offering little in the way of access to a larger global community. As a result, children now start with their own language and then move to Swahili. At this point, access to higher levels of education is totally blocked unless a world language like English is acquired, generally at a much later age than children in Ghana, or children learning French in the huge Francophone areas of Africa. People make choices . . . or have choices made for them.

It was in these sorts of contexts—those of language, career choices, different cultures, and, not the least, preferences for places and people—that two small research projects on the space preferences of university students in Ghana and Tanzania were carried out.

As for wheat growing on Kilamanjaro, this research had a very different origin. In 1962, just before the end of the UN trusteeship, I went to Tanzania to carry out three studies of diffusion, ostensibly in cooperation with a political scientist. The local scale study of the diffusion of membership in a Progressive Farmers Club on Kilamanjaro was delayed and delayed, and to fill in the time I started to investigate the extensive growing of wheat in the drier, rain shadow slopes of the huge mountain. It seemed such an incongruous crop, growing almost on the equator and next door to thousands of small holdings producing bananas and coffee. I think both the wheat farmers and the agricultural officers at Moshi and Arusha were amused by my interest, but they made me welcome and gave me complete access to their files. And what a gold mine of "decision making" it turned out to be!

At the time, I carried out the research for my own amusement and to satisfy my own curiosity. I was also interested in game, decision, and learning theories, but even then I always liked to have something empirical and concrete to help me learn about such abstract approaches. This was in the days when doing such research for fun was still considered legitimate, and even quite honorable. I wrote up the results, perhaps for a paper somewhere, sometime, never thinking that anyone else would really be interested. I realize that such an apparently casual attitude verges upon the unbelievable today, but in the 1960s universities still resembled universities, rather than for-profit research institutes where a person's worth is measured in the dollars of research funds raised, rather than the teaching and scholarship achieved.

Admiring some of the people at Michigan doing research on game theory, I sent one of them a copy to see if he had any comments, criticisms, or suggestions. No reply. I was a bit disappointed, but obviously the poor guy was busy and probably had scores of papers from other neophytes across the social sciences. I forgot about the thing completely until one day, about a year later, a bound copy of *General Systems Yearbook* and a box of reprints arrived in the mail. The fellow had obviously assumed I was submitting the essay for publication in the annual review of which he was the editor.

Strange: despite the formality of the theoretical frameworks used to shape an empirical account, the paper still conjures up the smell of the fine red laterite powder that permeates every dusty file and report in so many offices across Africa.

Place Preferences in Ghana and Tanzania

> No duty the Executive had to perform
> was so trying as to put the right man
> in the right place.
>
> —Thomas Jefferson
> In J. MacMaster, *History of the People of the United States*

ONE OF THE most difficult tasks facing African governments in the post-colonialist years is how to restructure and replace the system of local administration with dedicated and well-educated civil service officers. University graduates tend to respond to equally pressing national needs in science, engineering, and medicine, or are attracted to financial and other rewards that they perceive in business, law, academic life, and other professions. A major problem is that the many possibilities for posting in the civil service are perceived in very different ways. Some districts are greatly desired, while others appear to be avoided at all costs. In brief, the different images of places emerge as important components of the choices and decisions people make, and the great differences between the images raise questions about the sorts of things that underlie and shape their formation.

Extracting Shared Preferences

By sampling university students who are thinking about career choices and possibilities, we can see whether these images of residential desirability on government service, or "posting preferences," are highly individualistic, or whether they are shared to some, perhaps considerable, degree. We usually start by asking the students to consider a map of their country, with the administrative districts and the towns serving as the principal centers clearly marked and labeled.

Revised and rewritten from my article "A Mental Map of Ghana," *African Urban Notes* 6 (1972): 4–5, and "The Structure of Space Preferences in Tanzania," *Area* 4 (1969): 29–35

They are then asked to think of themselves in the position of having an absolutely free choice about the place they would most prefer if they were posted on government service, and then their second choice, third choice, and so on, working directly on the maps in front of them. Most people find it quite easy to rank order their clear likes and dislikes, the subsets of the highly preferred and greatly disliked districts, but they become less certain about exact rankings in the more indifferent middle range.

Nevertheless, and despite some uncertainty, the rank order lists often display a high degree of agreement or correlation with each other. All these correlations between pairs of preference maps can be arrayed in a square matrix with the students as the rows and columns. It is a typical correlation matrix, with ones (1's) along the diagonal (at least all the students are perfectly correlated with themselves!), and symmetric correlations as the off-diagonal terms. With component and factor analysis, we can extract a scale of greatest agreement, and on such a scale each district has a weighted score that we can map and use as a spot height for drawing contours of residential preferences. Perhaps we should call them "isoprefs," for like the more familiar isolines on a topographic map, they indicate the collective hills of great desirability and the low troughs or valleys of shared dislike.

The Preference Surface of Ghana

In Ghana (map 6), the surface of posting desirability is generally high in the south of the country, and matches to a considerable degree many other maps of road density and other indicators of modernization.[1] All the major urban areas are peaks of spatial preference, generally connected by high ridges between them. Accra, the capital, and Takoradi, a major port, are the highest peaks along a ridge of high desirability running parallel to the coast, with other quite distinct ridges running north along the railway towns to Kumasi, the major town and capital of Ashanti, and to Ho in eastern Ghana. In marked contrast, the north is generally perceived as highly undesirable. In between the most and least desirable areas lies the Barren Middle Zone, an area of few people for both historical and environmental reasons. Historically, it was the area constantly raided by African slavers both from the south and the north. Even as late as 1896, Nkoransa was subject to the raids of the notorious Samory,[2] and slaving was one of the reasons Britain extended control over the area known as the Northern Territories. In the far north, population densities rise again as the Sahel is approached with its highly seasonal rainfall regimes, but the people

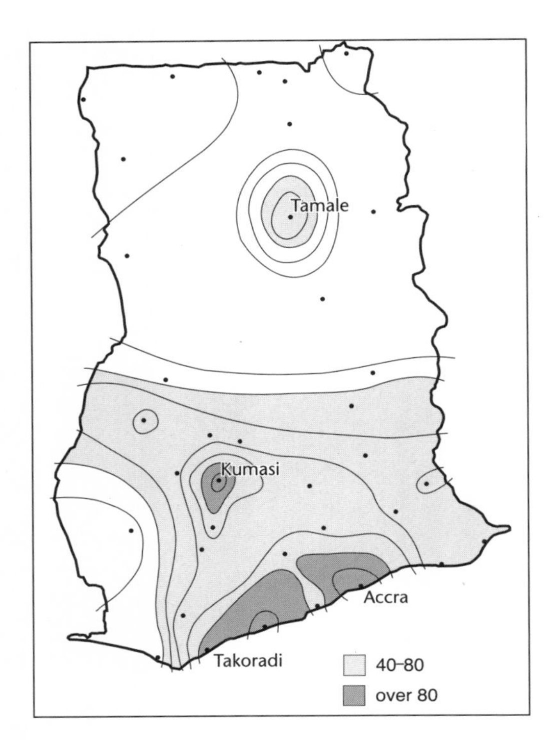

Map 6. The residential preference surface for Ghanaian university students contemplating posting on government service.

and their languages and customs are seen as very different by most of the university students who might enter government service. Only Tamale, the capital of the North, and Bolgatanga, a major market town, rise as small peaks of moderate desirability above the general trend.

Such marked regional disparities in the residential images raise the question of what shapes the mental topography of these highly educated, quite literally elite, young people of the country as they consider government service. Perhaps we can get some clue by examining a second geographic experiment, this time in Tanzania.

The Preference Surface of Tanzania

At the University of Tanzania, in the capital Dar es Salaam, students were presented with the same sort of question about posting on government service as their peers in Ghana, but they were also asked to augment the information about their locational preferences in two other ways. After first ranking their

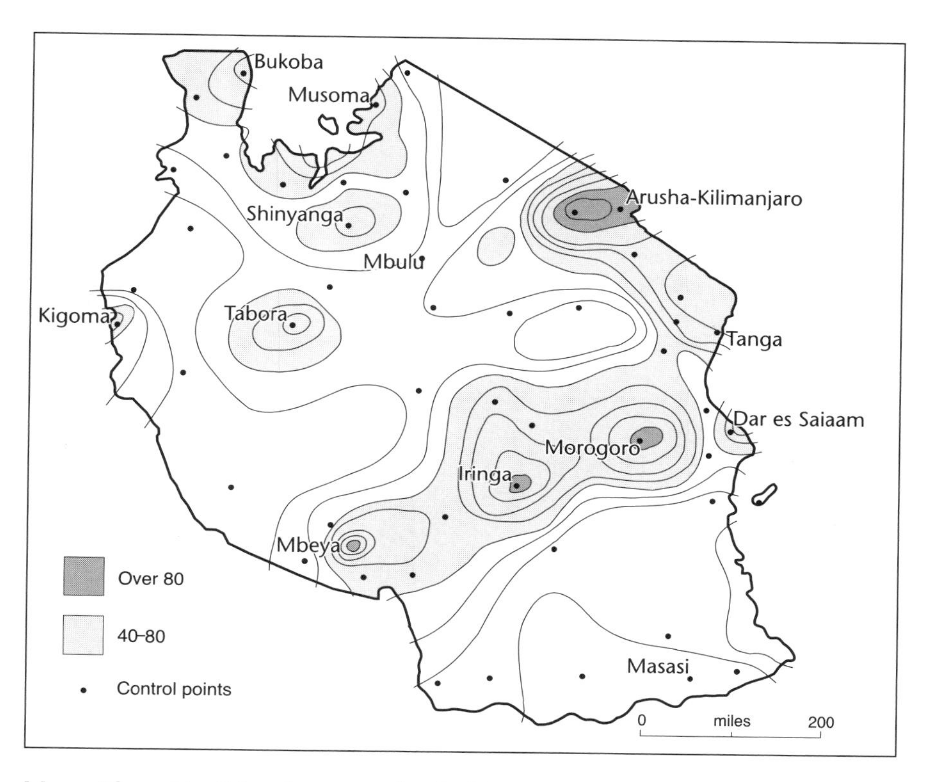

Map 7. The residential preference surface for Tanzanian university students contemplating posting on government service.

preferences 1 to 60 on the map, they were then asked to place in their district of first choice an annual salary with which they would be "reasonably satisfied." They then placed in each of the other districts salary figures they felt might just compensate them for not getting their most preferred assignment. Finally, they were asked to evaluate each of the districts on five simple scales of the type known as the "semantic differential." Although the whole exercise was somewhat arduous, we ended up with each of the districts evaluated by three, somewhat different measures: an ordinal scale (ranked preferences), an interval scale (salary assignments), and nominal scales (semantic differential).

As in Ghana, there was a high degree of agreement about the districts greatly desired or disparaged (map 7), and the space preference surface once again picks out the major urban areas, with ridges of high desirability following the rail and road alignments. It also matches, to a remarkably high degree, a modernization surface made up completely independently and based upon a composite of 26 indicators.[3] From most of the peaks on the preference surface, gradients are steep to the low valleys and plains that are regarded as very unpleasant places to find oneself for several years on government service. So far, the mental map for Tanzania's students seems to be shaped very much along the lines of the one generated by their peers in Ghana. But there is one major difference: while Ghana's capital, Accra, was the highest peak on the surface, Tanzania's capital, Dar es Salaam, only scored in the moderately liked range of 64 out of 100. It is also a clearly detached node on the coast, an area generally perceived as undesirable, with the exception of Tanga, a port and railway terminus at the end of the line from the rich coffee-growing area of Kilimanjaro. Dar es Salaam's moderate score seems to arise from its very equivocal perception by the students. It is a major urban area, to be desired for its cosmopolitan life, but at the same time it is also an extremely expensive place to live, and some of the students obviously did not feel at home there.

So the question arises again: just what is it that seems to shape these perception surfaces that are the outcome of generally very high levels of agreement between university students? Each person was asked to evaluate each district on the following five scales (fig. 3), designed to tease out how the student perceived and evaluated the economic costs of living, a district's general accessibility, the surrounding landscape, the facilities available, and what the local people were like. It turned out that the scales themselves were highly correlated, and a mental map constructed on their basis matched to a very high degree the one derived from the ranked preferences. But while four of the semantic differential scales pretty much collapsed to a single, overall scale, one stood out by the way it pointed to a second, separate dimension that influenced the students' choices. It was the "local people" scale, registering how well the students felt they could fit in and be comfortable with the local people. Over one-third of the districts actually had negative scores on this scale, most of them in the south and along the coast, with a tier of districts in the west equally disliked. A further exception was the northern area peopled mainly by the Masai, a people perceived as very different from the rest. In brief, when people think about government posting, such things as economic conditions, accessibility, surroundings, and facilities seem to go hand in hand, and so produce a mental map shaped very much like the preference surface based on rankings. But while the question of the local people is not totally independent of these other four effects, it definitely seems to give considerable pause when the question arises of how one will carry out the many responsibilities of government serv-

	Scale	
SALARY: Expensive place to live in, saving from salary hard.	3 2 1 0 -1 -2 -3	A very cheap place to live in, I can easily save from salary.
TRAVEL: Very easy to get there, movement in or out no problem.	3 2 1 0 -1 -2 -3	Very difficult to get to, and very hard to get in or out.
SURROUNDINGS: Very pleasant, a extremely nice place to live.	3 2 1 0 -1 -2 -3	Very unpleasant surroundings, not a nice place to live.
FACILITIES: Many things to do, spare time is no problem.	3 2 1 0 -1 -2 -3	Not much to do in spare time, far from the centre of things.
LOCAL PEOPLE: Friendly and very easy to get on with.	3 2 1 0 -1 -2 -3	Very difficult for me to fit in and get on with people there.

Fig. 3. The five scales used by Tanzanian university students to evaluate each district in terms of their own, quite personal perceptions.

ice. In their everyday duties, civil servants, like anyone else, need the goodwill and cooperation of the people they are dealing with. And if people, with their own languages and customs, are seen as very different from oneself, then there may be considerable hesitation when it comes to being assigned to their district.

Smoothing the Space Preferences

Assigning people to places is a problem of geographic allocation with many psychological as well as economic implications. So when we are faced with the task of assigning government personnel, is there any way the highly variable surfaces of residential desirability in Ghana and Tanzania might be smoothed out to perceptually flat and homogeneous areas? In other words, is there some way that people serving their country in undesirable places could receive compensation from those who choose to serve in highly desirable districts? Could this be done by relating salary scales inversely to the hills and valleys of the preference surfaces? Of course, we assume there is some simple trade-off between money and other properties of the districts. Could we think of holding a spatial auction, and so see the values at which the districts went under the hammer? Or do we simply assign people on an authoritarian "take it or leave it" basis, and not even think that their job performance might be related to how they feel about the quality of their lives?

ꔀ

Wheat on Kilimanjaro

The Perception of Choice within Game and Learning Model Frameworks

> It is always advisable to plant a range of varieties to spread the risk of total crop loss from rusts.
>
> —N. R. Fuggles-Couchman
> *Wheat Production in Tanganyika*

> For mixed strategies are appropriate even when we know that the other player is not in any way concerned with one's behavior.
>
> —R. Duncan Luce and H. Raiffa
> *Games and Decisions*

Introduction

IN THIS PAPER, game and learning models from a larger body of decision theory form conceptual frameworks within which to examine the choices of a group of farmers faced with an environment that affects their rewards from year to year. Its partial formality and apparent pretense to specific numerical accuracy should not mislead. It is written very much in a spirit of speculation, to provide a framework for examining some traditional but still intractable geographic materials, arranging these in such a way that questions about a spatial pattern of land use, considered as a cartographic expression of human decisions, may be seen more clearly. The "theoretical contents of facts"[1] is difficult to perceive unless the facts are examined and arranged within a formal structure. Thus, to reverse the usual metaphor,[2] this paper represents a pouring of old wine into new bottles. It does not have the temerarious claim that it solves any problems, representing as it does a hopeful groping toward the right

Somewhat revised from my article "Wheat on Kilimanjaro: The Perception of Choice within Game and Learning Model Frameworks," *General Systems Theory* 10 (1965): 157–166. The specialist may wish to consult the 54 footnotes of the original essay. Only direct quotations are referenced here.

questions. In the long run, it may be that these are more important than the answers.

Wheat on Kilimanjaro

Rising in a great cone from the grassland plains that stretch from the Rhodesias to the Ethiopian border, the huge volcanic mass of Kilimanjaro dominates the area around it (map 8). On the northwestern slopes of the mountain, much of the land has been divided up into large wheat estates, owned by European settlers who have been supplying an increasingly large proportion of Tan-

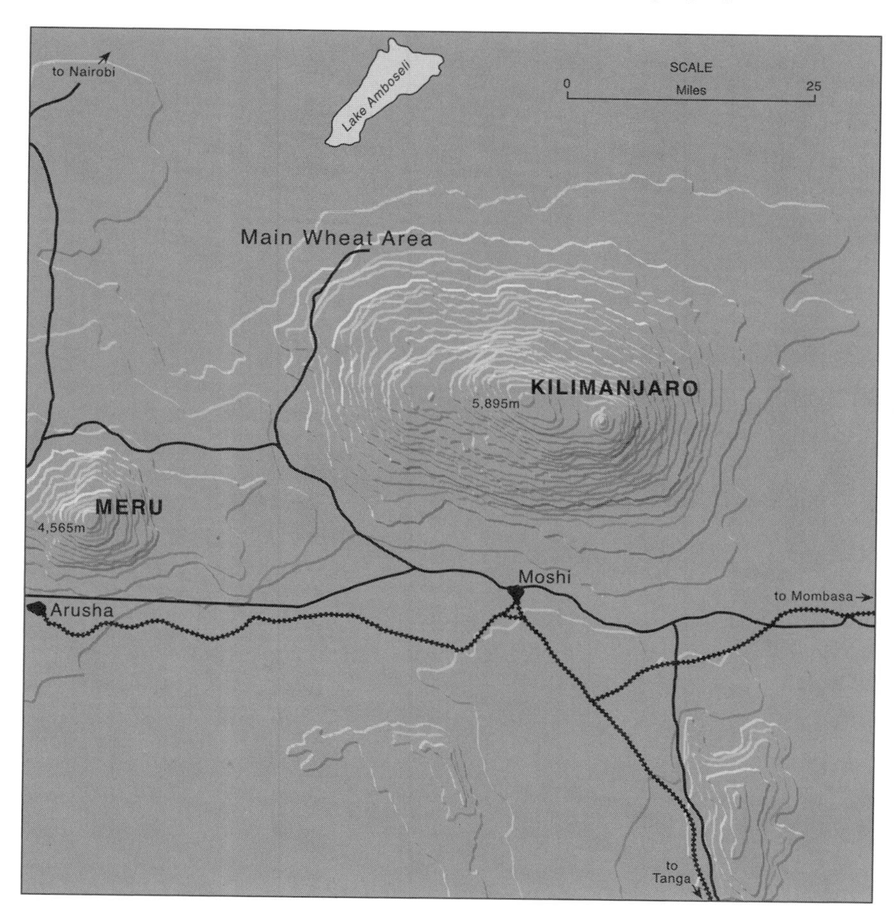

Map 8. Kilimanjaro rises to a height of 5,895 meters from a plain of roughly 800 meters (Moshi). The major wheat-growing areas lie on the northwestern slopes, served by partially paved and laterite roads to the major estates.

ganyika's needs for the past two decades. Unlike Kenya, where wheat has been grown since the 1920s, wheat growing in Tanganyika only started on a large-scale basis during World War II, when traditional sources of supply were cut off, and the country made a strong drive toward self-sufficiency. The Northern Wheat Scheme was started, and in addition to aiding farmers with equipment, seed, and advice, it guaranteed them a minimum price during the initial years.

Wheat Growing as a Game

Since patterns of land use and wheat choices are the result of decisions that the farmers have made either individually or collectively, it may be fruitful to cast a description of the patterns within a decision theory framework. Intuitively, the theory of games would seem to hold promise, and it is within this framework that wheat growing will be considered initially. In game theoretic terms, the wheat farmers on Kilimanjaro will be considered as a single player in a game against the environment, or Nature. The game model, in which two players face each other, with a number of strategies open to each, is a severe one, and more than one person has commented upon the "paucity of practical applications."[3] It is essentially a *normative* theory that points up specific courses of action under conditions of highly particular assumptions. "It states neither how people do behave nor how they should behave in an absolute sense, but how they should behave if they wish to achieve certain ends."[4] The assumption is made that the farmers wish to maximize their income by growing wheat in the face of uncertain and unpredictable conditions. Each year they have a number of wheat varieties from which to choose, and a number of locations on the side of the mountain where wheat may be planted. Various mixtures of these choices may be considered as strategies in a game, but the return the farmers receive at the end of each harvest depends not only upon the locational and varietal strategies they have chosen at the beginning of the season, but also upon the strategies the environment has thrown at them during the course of the year.

The Strategies of the Environment

Rainfall

The most obvious strategy open to the farmers' opponent in the wheat-growing game is the rainfall. Not only is the *average* rainfall on Kilimanjaro extremely variable from place to place, but it may also vary greatly from year to year, particularly in the lower, drier areas. For example, in the heart of the wheat-growing area some years may have up to three times the rainfall of the driest years, while down in the plain, only six miles away, the wettest year was recorded as 11.2 times the driest. One can, of course, start bringing in evapotranspiration rates, available moisture, and so on, but it seems best to simplify in a paper of this sort.

When varying amounts of rainfall are regarded as components of game strategies to be chosen by the farmers' opponent Nature, the way in which the farmers perceive the likelihood of sufficient moisture during the crop season is particularly pertinent. The conversations of the farmers are permeated with intuitive but nonformalized notions of probability, and range from the stoically humorous "Well, it just depends if Muntu [God] puts his finger in the ground" to the thoughtful but biased perception of regularity and order in rainfall cycles, which do not exist. Nearly all the farmers record rainfall daily, either on an official or unofficial basis, but the existing data have never been examined in a formal and systematic way.

Wheat Rusts

From the very beginning, wheat rusts have attacked the fields on the slopes of Kilimanjaro, and in certain years they have hit so severely that some farmers have had to write off the entire crop. Over the years, the farmers and agricultural personnel in government service have found that certain types of wheat are less susceptible to one of the many varieties of black, yellow, and brown stem and leaf rusts, but such knowledge is not sufficient to eliminate the danger. New mutations of rust appear continually and attack the new wheats developed especially for their resistance to previously known varieties. The appearances of new mutations are completely random and unpredictable, and over the years reports from agricultural officers and experimental staff at plant-breeding stations are replete with advice to spread the risk of attack by planting many varieties. Paradoxically, the most dangerous years may be those when no mutation appears. Then a particularly successful wheat of one year may be planted extensively the next, only to be severely damaged by an entirely new mutation of rust.

Dioch Swarms

Multimillion bird swarms of the Sudan Dioch *(Quelea Quelea Aethiopica)* are nothing new: ever since 1881, when famine conditions were induced in Ugogo District, scattered reports tell of the devastation the migrating birds leave in their wake. In some areas the birds have forced African farmers to abandon millet and sorghum as subsistence crops, and a rice-growing scheme during the war was brought to its knees by the avaricious clouds of finches. Today the bird swarms are such a menace to the subsistence economy that international control has started, and across the continent ideas are being exchanged.

In the Kilimanjaro area, local Wachagga farmers say the birds have always been present, and older European and Afrikaaner settlers remember the depredations of African grain crops from their earliest years in the area. It was not until the Northern Wheat Scheme started in 1944, however, that the extent and severity of the damage was actually realized, for until that time all export and estate agriculture on the mountain was in coffee and sisal, respectively, two crops unaffected by the dioch menace. With the beginning of wheat farming on a large-scale commercial basis, the birds soon made their presence felt. One farmer noted "small grain crops . . . are now subjected to such devastating destruction by diochs that unless the Government is prepared to take an immediate and active attitude towards the elimination of this pest, crops . . . can no longer be grown."[5] Another wrote that he was harvesting one bag per acre over devastated fields, as opposed to seven on fields that were lucky enough to escape, while a third had fields yielding fourteen bags per acre completely stripped.

Many farmers believe that heavy dioch attacks occur in cycles of four years, and once again it is interesting that such ideas of regularity should arise in the face of extraordinarily little evidence. It is almost as though the farmers look for regularity in their opponent Nature, and are uncomfortable in the face of randomness and unpredictability. However, one agricultural officer, holding to a less immediately involved perspective, wrote, "[the arrival of dioch birds] appears so far to depend very much upon the luck of the year."[6] Whether the birds attack the wheat fields seems to depend, at least in part, upon the rainfall in the lower plains, which controls, in turn, the birds' supply of wild grass seed. If this is so, then any cyclical pattern of dioch bird swarms is so deteriorated that it may be considered for all practical purposes as a random arrival rate.

The Farmers versus the Environment

The locational and varietal strategies of the farmers, and the rainfall, rust, and dioch strategies of the environment, may be arrayed in all their combinations in a rectangular form called a payoff matrix (fig. 4). In this matrix, each row represents a particular combination of location and wheat variety, while each

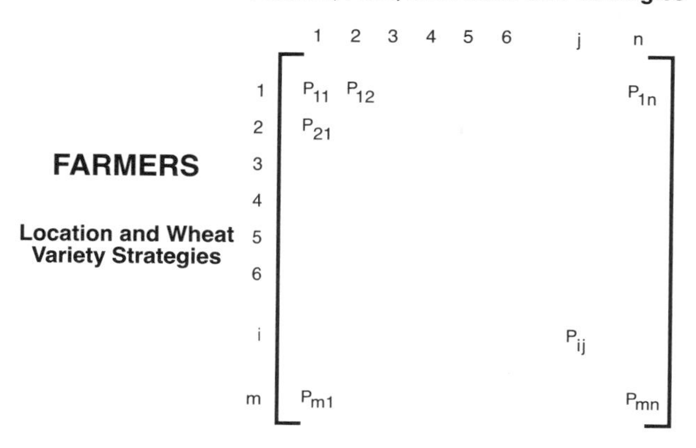

Fig. 4. An illustrative payoff matrix between the wheat farmers choosing locational and wheat variety strategies and the environment "choosing" rainfall, wheat rust, and dioch bird strategies.

column forms one combination of the environment's rainfall, rust, and dioch strategies. The payoff values (p_{11}, p_{12}, and so on) are measured in bags of wheat per acre, which would be equivalent to exact monetary values in a period of guaranteed price supports. Under free market conditions, the payoff values could not be exactly predicted, but might be considered as small random fluctuations around average values or price trend lines. In either case, a two-person-zero-sum game is described, with some payoffs being positive in those areas of the matrix where the environment chooses relatively benign strategies, while others are negative where the environment's rainfall, rust, and dioch strategies force a loss.

In arraying the data in game matrix form, there is the implication that the farmers regard the environment as a vindictive, minimizing player. There is the objection that such a view is unrealistically severe, but when people have to make decisions under conditions of great uncertainty, they may very well choose extremely conservative strategies or, in game theory terms, strategies according to minimax criteria, to maximize their security level, particularly in the initial phases.

In the case of the farmers on Kilimanjaro, there is photographic and cartographic evidence that in the beginning years of wheat growing, the locational strategies were very conservative, being confined almost entirely to the higher slopes of the mountain, where rainfall was heavier (map 9). A few who planted in the early years well down the slopes toward the drier plain, reported by 1951: "a succession of bad years . . . heavy rust attacks and low rainfall have brought financial embarrassment to many."[7]

However, over the years information about rainfall has accumulated, and if systematically analyzed it would be possible to obtain measures of the probabilities with which various amounts of rainfall arrive or, in other words, to obtain some thoughtful notions of the probabilities with which the farmers' opponent will choose certain strategies. These probabilities may be used to weight the payoff matrix, which means that the payoffs accruing to the farmers change relative to one another, so that locations farther down the mountain become more attractive.

In addition, dioch bird control started in earnest in 1955 and focused upon the destruction of the huge breeding sites whenever they could be found. A number of methods were tried, including parathion sprays, firebombs, and gasoline-gelignite mixtures saturating areas of up to seventy acres at a time. Even an itinerant magician was pressed into service—although his results were

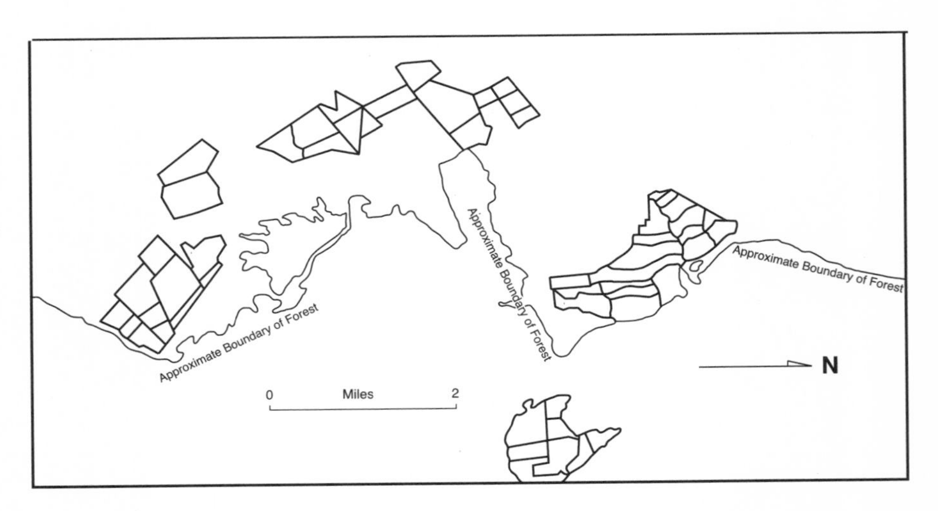

Map 9. The wheat areas under production on northwest Kilimanjaro by the end of 1952. Virtually all the areas planted are confined to the higher, better-watered slopes.

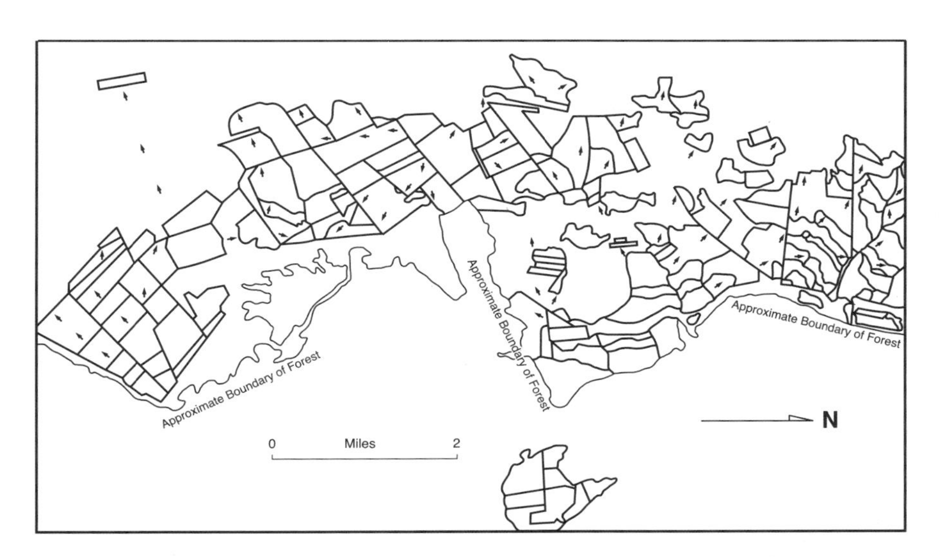

Map 10. Areas under wheat production by 1962. Not only has there been an expansion at the higher levels over the past decade, but the farmers have probed downslope to the drier areas of increased risk.

disappointing. In 1956 over 40 million birds were killed by the control service, rising to over 100 million in 1959, and in the early 1960s swarms had decreased markedly over former years. Here, then, the farmers are changing the strategies open to the environment, and in doing so alter again the values in the payoff matrix as the probability of heavy bird attacks decreases sharply.

As a result, and in the presence of steady prices, locations farther down the mountain become more attractive as the farmers not only learn more about their opposing player, but also alter the strategies open to their opponent, so changing the payoffs accruing to them in the game. Under these changing conditions, one might expect the farmers to mix their locational strategies by extending the wheat-growing areas down the mountain. Photographic and cartographic evidence is available to confirm such expectations (map 10).

The map of wheat growing in 1962 shows a marked extension over the pattern in 1952 and represents a decade of trial-and-error decisions. The farmers have probed both downslope and northwards around the mountain, moving from the most conservative locational strategies to a mixture of the most conservative high locations and the more risky, lower areas. If this mixture represents an equilibrium position, then we should expect further contraction or expansion to depend upon prices and their changes over the long run. This, of course, was said by Heinrich von Thünen 140 years ago.

Yet all is not secure: one totally unknown and unpredictable strategy of the environment remains—wheat rust. Even from the earliest years of wheat growing, new varieties of wheat have been made available to combat rust, but often the farmers have decided to reject the advice of the plant-breeding stations. A visiting plant breeder reported plaintively in 1950: "There were badly rusted stands . . . but it appears that farmers are continuing to grow this variety in spite of advice given in the 1948 report that this variety should be discontinued."[8] Lamented another the following year: "Some farmers still insist in planting varieties which are known to be susceptible."[9] But during these years rust attacked the wheat fields severely, and by 1952 came the report: "It is interesting to note that farmers are increasingly turning to newer, and more rust resistant wheats and abandoning the older susceptible varieties."[10] By 1959 "only a few fields were sown against 'official' advice,"[11] and two years later virtually all farmers were accepting the advice of the plant breeder, and growing only recommended varieties.

Thus, in order to combat the environment's rust strategies, the farmers have *learned* to choose, from all the strategies available to them in the game, those strategies in which the new wheat varieties are a part of the combination, and have shifted away, often late and reluctantly, from the old favorites of earlier years. Such a two-choice situation may be described by a formal stochastic model of learning in which one choice is rewarded, thus reinforcing the behavior or decision, while the other choice goes unrewarded. The situation is

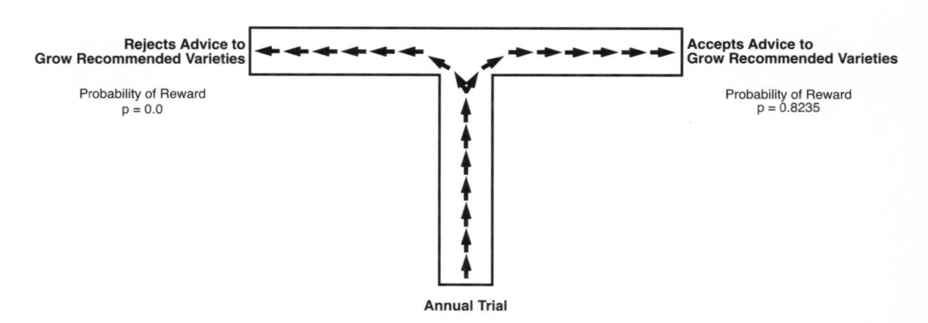

Fig. 5. A formal learning model of partial reinforcement, depending upon whether farmers are willing to follow the advice of plant breeders to grow the newer, rust-resistant varieties of wheat.

analogous to a maze in the shape of a *T* (fig. 5), in which the acceptance of new wheat varieties at the beginning of a season is the equivalent of turning to the right with a certain probability of reward, while a turn to the left represents the rejection of the plant breeder's advice for that year, or trial. In this particular case, a learning model with *partial* reinforcement is required, because while acceptance of new varieties decreases the probability of rust attack, such acceptance does not completely ensure protection. Conversely, a farmer may stick with an old favorite and through sheer luck escape rust for one more year.

If we define "reward" as yields greater than those obtained by the farmers who rejected the advice, then if we know that 14 rusts appeared in 17 years, we may postulate that the behavior will be reinforced approximately 82 percent of the time. To illustrate: over the years the land-use pattern reflects the increasing use of new wheat varieties as more and more farmers come to accept the plant breeder's advice, so that by the 1960s, the curve describing the learning model approaches the expected asymptotic value of 100 percent (fig. 6). This

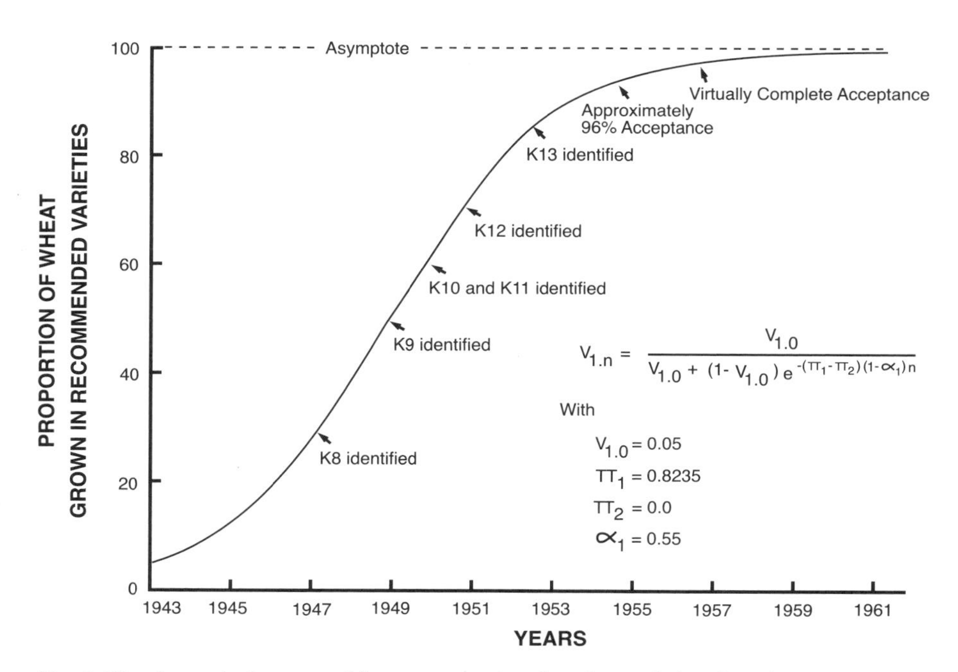

Fig. 6. The theoretical course of farmers accepting the advice of plant breeders to grow rust resistant varieties of wheat. With reasonable, empirically based assumptions, only the parameter α needs to be estimated. The K8 through K13 points are when these mutations of rust were first identified in the area.

smooth curve, representing the theoretical increase (provided the parameters are properly chosen), can only be suggestive of the pattern of acceptances over time. The geographer in the field seldom encounters the sort of rigidly controlled laboratory conditions under which such a model of partial reinforcement has been examined before.

Finally, it is worth noting that the curve has more than passing similarity to the growth of acceptances in other geographic models of diffusion of innovations, and there is some comfort that we may be approaching another, but closely analogous, geographic problem by another route.

Mental Maps

STRANGE how lines of thinking start sometimes. The research on what I called "mental maps" began one day in 1965 while I was fiddling around with a piece of chalk on a blackboard, wondering what would happen if you treated people as the columns of a matrix instead of the rows. It sounds strange, of course, but I have the feeling that many of us who teach in a school or university think of people as rows occasionally, because we all keep lists of students in our courses and record as columns their scores on quizzes, exams, and papers. At a time when factor analysis was still a bit new, I can vaguely remember thinking, "I wonder what would happen if you transposed this and made the people the variables instead. Hmm . . . But if the people are columns now, what would the rows be?" It did not take long to remember that people think about things, and if they thought rather similarly about places, perhaps you could examine how their individual evaluations were related to each other, and so map some sort of collective viewpoint. I called these collective images "mental maps" simply because that is what they seemed to be, and because I had never heard the phrase used before. Later on, some people objected that they were not *really* mental maps at all, but by that time their chagrin was too late. Patient and long-suffering friends at Minnesota, Berkeley, and Alabama (not to mention their long-suffering students) replicated the little "where-would-you-like-to-live-given-your-druthers" map-questionnaire I had given my own students at Penn State, and I found all sorts of interesting things dropping out of such analyses.[1]

Over the next few years, this sort of research was explored in various parts of the world with students and generous colleagues, but it was during the sabbatical year 1969–1970, at the University of Lund in southern Sweden, that I was really able to explore some of the things that seemed to lie behind and shape the hills and valleys of likes and dislikes that people have for places.[2] *Information* (using the word and concept as something meaningful to people, rather than some abstraction of the electrical engineer) seemed terribly important, but it was a devil to define and measure. In the end, I had to use a very simple count of places, an experiment I had tried a few times while teaching students in a first course on spatial analysis at Penn State. Without warning, and in a dramatically timed five minutes, they were asked to write down all the names of towns they could think of in the United States. Furious scribbling for two minutes, then . . . oh, yes, Green Bay (Packers), Baltimore (Orioles), Oakland (Raiders), . . . um . . . aha, Shy Beaver, Cherrytown, Clover Creek [you knew where that student's home was!], . . . until only a few pencils were moving when time was up.

When all the individual student lists were added together it turned out that these counts of simple geographic information were highly predictable from the populations of the states generating the information and how far they were from Pennsylvania, simple gravity model variables used as a first exercise in multiple regression. Most of the students liked the idea of themselves generating a variable used in an exercise, and they were surprised at how predictable they were as a class of eighty to a hundred people. You could also use the exercise to raise some thought-provoking questions with them. After all, if their location in geographic space seemed to determine to a high degree the information they had, what about their location in "ethnic space," "religious space," and "gender space"? And how did information form images, stereotypes, and prejudices? How free really were they when they were trapped by their locations in these other spaces? Never mind the exercise and its technical calibrations: in a university it is an experiment worth repeating every time you teach.

Maps have always fascinated me, and one of the greatest professional regrets in my life is that I never met Brian Harley face-to-face before his tragically early death in 1993. He taught us all to look at maps with fresh eyes, and used a variety of genres of writing to get his various messages across.[3] His joyous, sad, and beautifully written essay, "The Map as Biography,"[4] is one of my favorites, and I do not find it strange that his friend Ed Dahl felt compelled to rearrange some of his prose as poetry.[5] I would not dream of trying to emulate him in this so intensely personal vignette of a part of his life, but it has certainly been the inspiration to return to a place I loved as a teenager through the medium of some of its maps, both old and new. All sorts of boyhood memories came flooding back as my eyes lovingly caressed those sheets, and so I called it "Maps as Memory." What else?

Acquiring Spatial Information

I only ask for information.
—Miss Rosa Dartle
In Charles Dickens, *David Copperfield*

NOT LONG AGO Herbert Simon dusted off, and refurbished in an imaginative allegorical way, an old idea.[1] He noted that it may be very fruitful to assume that human behavior is relatively simple, and that the complexity we observe is due to the complexity of the environment within which the behavior takes place. Such a thoughtful allegory was illustrated by a small ant on a wave-ribbed beach trying to get back to her nest after a foraging expedition. The ant had a clear goal, to get back as quickly as possible along a straight line, but if we were able to observe only her path, perhaps by projecting it on a piece of paper as a map, we might be very puzzled by the complexity of all her twists and turns. Yet the complexity is obviously caused by the structure of her environment, the space through which she has to pass. If we could discover the space of the wave-ribbed beach, we would understand why her behavior is so apparently complex. This image should raise for the geographer questions about the environments that have been of traditional concern, and perhaps lead to an expanded and possibly less limited and more relevant definition.

We may be able to speak of an *information environment*, one of those "invisible landscapes," to adopt the phrase of the psychologist Stea,[2] that are becomingly increasingly important. Perhaps the very traditional idea of the geographer as an explorer and mapper can be extended to the discovery and delimitation of the topography of these newer spaces and landscapes. Though less familiar and more abstract than those to which we have become accustomed, they may, paradoxically, represent descriptions of a more pertinent and relevant reality.

When we talk about information, what we might legitimately call *real* information (the stuff acquired and used by human beings, as opposed to the binary-coded pulses of the electronic engineer that form the rather mislabeled

Somewhat revised from my article "Acquiring Spatial Information," *Economic Geography* 51 (1975): 87–99 area of "information theory"), we enter one of those siren areas of conceptual quicksand that so frequently hold an unhappy attraction for the geographer and other human scientists. After all, everyone *knows* that information is important: the quaternary sectors of postindustrial societies manufacture, distribute, and live off the stuff; people, or rather decision makers, use it, presumably for making decisions; its market value often decays over time; conferences and meetings are organized at great cost to disseminate and exchange it; and social scientists hunt around with the mislabeled theory, desperately trying to find something that will fit it.

The results and insights of the social scientists are usually meager, and sometimes silly, although a few have recognized immediately that information theory has nothing to do with information as it is usually defined and discussed. For example, the Oxford Dictionary defines information as "knowledge, items of knowledge," so it has everything to do with sapient beings and little to do with machines. Some geographers, it is quite true, have used it effectively to generate measures of chaos;[3] but most have hung onto the information theory bandwagon as it has rolled through each of the social and behavioral sciences in turn, even though it is genuinely difficult to see that our insights about information in a meaningful and human sense are intellectually richer once the dust has cleared. The problem is that information is one of those common, "intuitively obvious," but actually will-o'-the-wisp terms that become very difficult to pin down for analytical purposes, and even more difficult to measure and so relate to other things.

When we have great difficulty in defining and measuring something we feel is genuinely important to human affairs, it may be necessary to tolerate very simple, perhaps even naïve measures applied to simple problems and situations. If we are fortunate and perceptive, such initial simplicities may help us define and explore more complicated measures and aspects later, which are more precise, useful, and richer in understanding. But for the moment, the definitions we must employ are simplistic, while the measures are going to be little more than simple counts.

We shall focus, first, upon a rather restricted set of *geographic* information, that is, information consisting of items of knowledge about places. The sets of information are all formed by aggregating individual responses of groups of schoolchildren in Sweden. The samples consist of children of different ages at many locations scattered all over the country. We also need to explore some of the ways in which adults obtain geographic information, and how such knowledge appears to affect the degree of agreement within a group about their shared mental map of residential desirability.

The Information Surfaces of Children

It is quite impossible to monitor the acquisition of geographic information by children. Information of this sort is acquired in many formal and informal ways—travel experience, radio, television, school, travel posters, newspapers, magazines, and so on—and it is retained in varying degrees, at various levels of awareness, by different children. To investigate such information, we require a very simple and inexpensive technique that can be applied quickly without trespassing unduly upon their time. Accordingly, and during the course of a larger investigation into children's mental maps, children were asked, without prior warning, to write down in five minutes all the names of towns in Sweden they could recall. In a sense, the task was to "unload" the names of places as quickly as possible under time pressure. It is difficult to think of a simpler test.

The responses of individual children are, by definition, unique and not particularly meaningful, except perhaps at that highly restricted, personal level. But when many are aggregated on the map, distinct, interpretable, and predictable patterns appear as information surfaces.

In the maps that follow notice two points: first, that the isarithmic intervals are geometric rather than arithmetic. Information appears to be exponentially peaked in geographic space, following many other types of human phenomena. Secondly, the surfaces are interpolated from 70 point values, one for each of the 70 A-level, functional regions in Sweden. In the south, the points are fairly close, and interpolated values are probably quite accurate. In the north, points are far apart and some license has been employed in locating the lines.

To take one example only, nine-and-a-half-year-olds at Skövde, a town lying between the large lakes Väner and Vätter in south-central Sweden (map 11), have a distinct peak of information immediately around their local area, and the surface falls off in all directions, rising only at major towns such as Stockholm, Göteborg, and Malmö. The gradients around all these information nodes are very steep, particularly those east and west of Skövde. The lakes seem to reduce drastically the information coming from these areas, and the information surface is at zero level to the southeast. Other zero-level areas lie immediately to the north of the town, (a region I have named the Västmanland Trough, because it appears as such a consistent depression of information on so many surfaces), as well as in the sparsely settled far-north. In contrast, there appears to be some directional compensation along the corridor between Stockholm and Göteborg. The barrier effects of lakes have often been noted in studies of diffusion in Sweden, and here they produce some of the sharpest gradients on the entire surface.

In eleven-and-a-half-year-olds (map 12), the information surface has risen, but the same peaks and troughs appear. The directional bias is especially marked, indicating that travel patterns are having a strong effect upon the acquisition of information. The gradients across the lakes are still steep, and the Ljungby region is one of only two on the entire surface still at zero level. At age sixteen-and-a-half (map 13), the same configuration appears in a more devel-

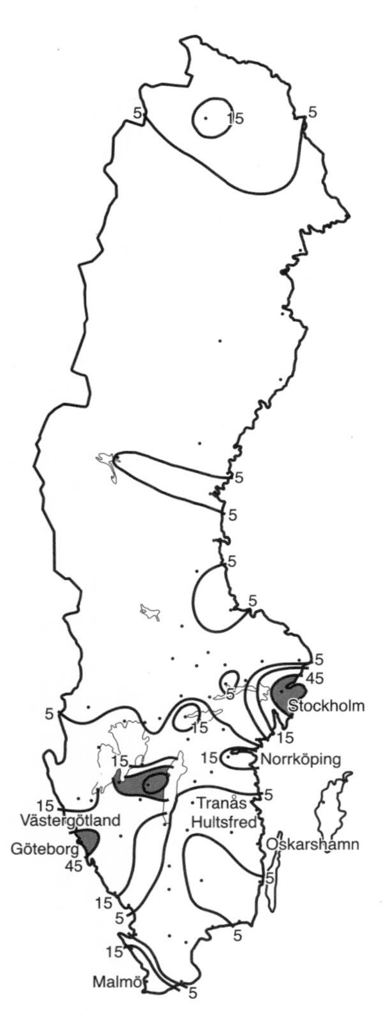

Map 11. The information surface for nine-and-a-half-year-olds at Skövde.

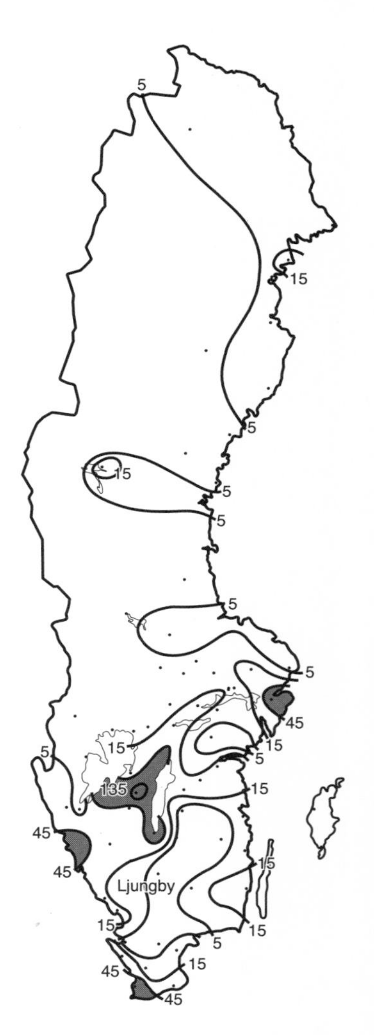

Map 12. The information surface for eleven-and-a-half-year-olds at Skövde.

oped form, and two years later (map 14) it remains reasonably stable for the gymnasium, or senior high school students. On both these maps, for the older teenagers and young adults, the local peaks appear to be saturated (373 and 342, respectively), and the Västergötland-Närke information corridor is still marked, bounded to the east by the steep gradient to Östergötland and Småland.

This particular sequence of information surfaces, and many others we could have taken as illustrations, indicates that geographic information—or at least crude, surrogate measures of it—develops in a quite regular way with age. The

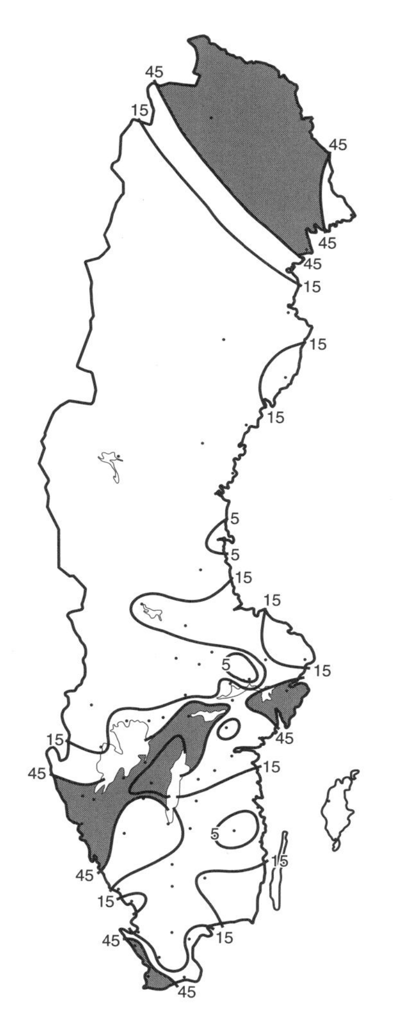

Map 13. The information surface for sixteen-and-a-half-year-olds at Skövde.

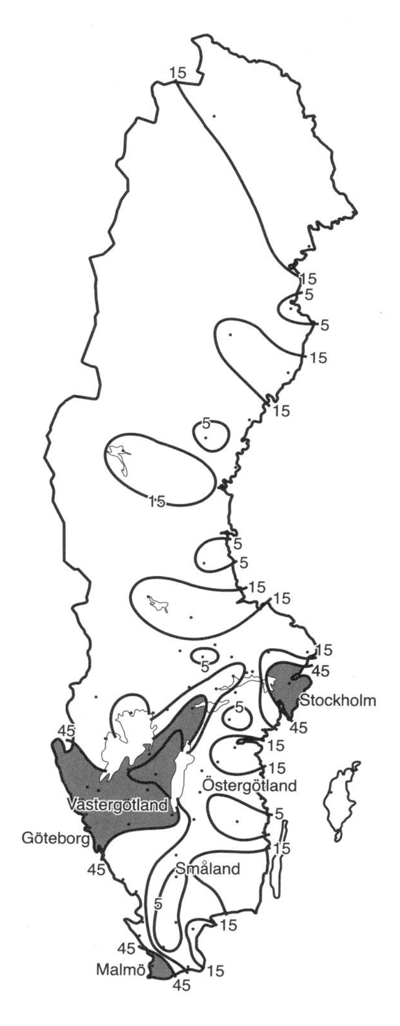

Map 14. The information surface for eighteen-and-a-half-year-olds at Skövde.

information surfaces of the older children are anticipated in the more embryonic forms of the younger ones. It is as though we had some final, saturated information surface slowly rising above a datum plane as the children get older: first, the local region and the three major urban nodes generating high peaks of information appear, and then, with this major nodal framework exposed, the areas between emerge and are slowly filled in.

But the surfaces are predictable in a more formal sense. Except for relatively rare environmental events, it is people who usually generate information, and we might expect regions with many people to generate more information than those with only a few. Conversely, we might expect regions close to a group of children to transmit stronger information "signals" than those farther away. In brief, information may be a function of the traditional gravity model variables, population and distance. For any particular location, this simple model can be expressed as

$$\ln (I + 1) = \ln K + \beta_1 \cdot \ln P - \beta_2 \cdot \ln D,$$

where I is the information generated by a region, P is its population, and D is the distance away. The parameters K, β_1, and β_2 can be estimated by standard methods of calibration. If we do this for 101 samples, involving many thousands of children from 36 regions all over Sweden, this simple and quite traditional gravity model predicts the information surfaces at a reasonably high and consistent level.[4]

At the same time, such formal analyses raise a series of disturbing questions. If we look at a typical sample of parameters (table 1), the value of β_1 is reasonably stable, implying that a region of Sweden generates information in proportion to its population. Taking all the 101 regions and age levels together, the average value of β_1 was 0.85, without much variation from one place to another. But in marked contrast, β_2, the parameter controlling the relationship between information and distance, varies widely. Unfortunately, such variation does not appear to be related in any way to the ages of the children, because the effect of distance on the acquisition of spatial information does not change in any systematic way as children grow up. However, when we examine the parameters based on the samples of the oldest children at a particular location, they appear to vary with considerable regularity in geographic space (map 15). The behavioral implications are so disturbing as to be virtually untenable. Does it mean that the behavior of children in this reasonably egalitarian society—namely, the way they acquire geographic information—varies in a systematic fashion with location? Look how the effect of distance varies from −0.19 at Örebro in central Sweden to −1.03 at Örnsköldsvik in the north, and the way

Table 1
Gravity Model Values for a Selection of Regions and Age Levels

Regional Number	Regional Name	Age Level	Population Parameter β_1	Distance Parameter β_2	Coefficient of Multiple Correlation $R_{1.23}$
2	Norrtälje	16.5	0.80	-0.64	0.72
7	Eskilstuna	11.5	0.83	-0.53	0.69
13	Nässjö	18.5	0.88	-0.68	0.76
20	Kalmar	16.5	0.81	-0.59	0.76
39	Skövde	16.5	0.78	-0.38	0.69
45	Örebro	18.5	0.86	-0.19	0.69
56	Gävle	16.5	0.86	-0.51	0.69
60	Härnösand	11.5	0.95	-0.82	0.74
62	Örnsköldsvik	18.5	0.86	-1.03	0.83
65	Skellefteå	18.5	0.89	-0.95	0.88

the effect appears to increase in a very systematic way both to the north and to the south. But for what possible behavioral reasons?

We enter here a very difficult area of inquiry and interpretation, for the recent theoretical work of Leslie Curry has indicated the extreme and still insuperable difficulties we face in trying to interpret and compare parameter values that purport to measure the frictional effect of distance—in this case, the effect of distance upon the acquisition of information.[5] The problem is that in calibrating some form of the gravity model to a particular set of information, we are convoluting both the effects of human behavior and the effects of map pattern or arrangement. The "Curry Effect" can be easily seen when we write the autocovariance function of a population map such as:

$$R(s) = \overline{P(x) \cdot P(x + s)}.$$

What does this mean? To see what is involved, imagine a map covered with a lattice of square cells, each one containing the number of people living there. Pin this map down on a light table, and now imagine you have an identical map as a transparent overlay. If the overlay were lying directly on top of the map pinned to the table, we could correlate the numbers of people in the cells of the overlay with those in the cells underneath. Of course, this would be an idiotic thing to do, because obviously a map is perfectly correlated ($r = 1.00$) with itself. But now imagine shifting the transparent overlay some distance s, say one, two, three or more cells in any direction, and *then* correlating the over-

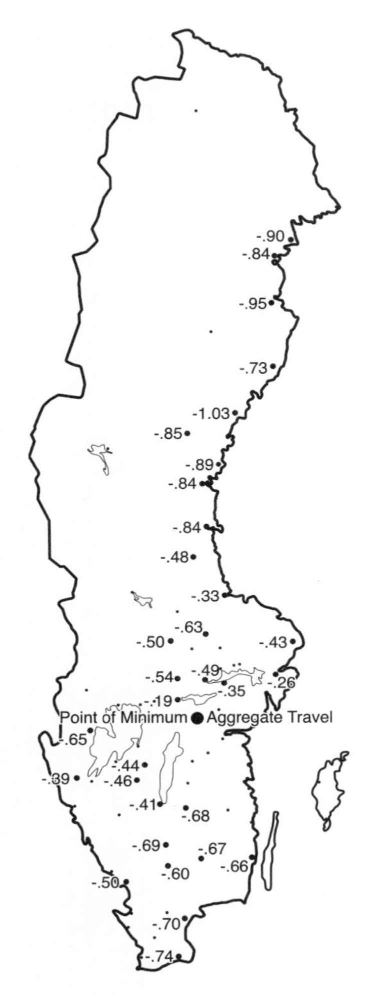

Map 15. The spatial variation in the parameters (β_2) relating information to distance from the home region.

lapping values on the top and bottom maps. Intuitively, you can see that if the population values were just random numbers scattered across the cells of the lattice, then the correlation, the *auto*correlation of the map with itself, would immediately become close to zero for any number of lags in any direction. However, if the population values were not random, if there were some *pattern* to the map, then at various lags or distances of the map with itself the correlation would not be zero.[6]

Now notice that the gravity model could be written as:

$$R(s) = \frac{P(x) \cdot P(x + s)}{f(s)}$$

If the distance function $f(s)$ is D^{β_2}, where $\beta_2 = 0$, the denominator of the expression becomes one, and the gravity model is the same as the autocovariance function, which is a measure of the map pattern. This means that any empirical estimate of β_2 is going to be convoluted with the particular pattern or configuration of population on the map. Such an effect should make us even more hesitant to say that children in Central Sweden form a new, "spatially free" generation, while those of Ystad and Luleå (the extreme southern and northern locations on the map (map 15) are still bound by the old chains of distance. Rather, the distance parameters appear to index the *relative accessibility* of a location within the information space of Sweden. If we take the point of minimum aggregate travel for the Swedish population as our most accessible location, then the estimated exponents of distance (β_2) are a simple linear function of the square root of the distances ($r = 0.79$). Thus we have a strong indication that the parameter values, supposedly measuring the friction of distance, may not be doing so at all, but rather simply indexing the relative location of a group of children in the space. The supposed effect of distance is not simply a distance effect in any traditional or behavioral sense, but rather a matter of where a group is located within the set of information-generating regions.

Information and the Strength of Space Preference Signals

Geographers have often shown that a group of people at a particular location usually display a high level of agreement about the relative residential desirability of other parts of their country, and that a "mental map," or residential desirability surface, can be generated, showing the way in which a most general, shared viewpoint varies. A typical mental map for southern Sweden is that of adults at Ystad (map 16). A high ridge of perceived desirability runs along the south coast (an area with a high proportion of retired people, and, in summer, of vacationers), and the surface falls to highly undesirable areas to the north, with a marked depression, the "Stockholm Sinkhole," around the capital city.

A mental map is often constructed from a principal components analysis, which yields successive eigenvalues from the correlation matrix measuring the degree of agreement in perceived residential desirability between the people in the sample. The ratio of the first two eigenvalues (λ_1 andλ_2) may be used as a measure of group agreement. If the group agrees strongly about the desirability and undesirability of places, the first component associated with λ_1 will account for a large portion of the variance, particularly when it is compared to the second, λ_2. Conversely, if there is considerable disagreement within a group of people, the ratio of the two eigen-

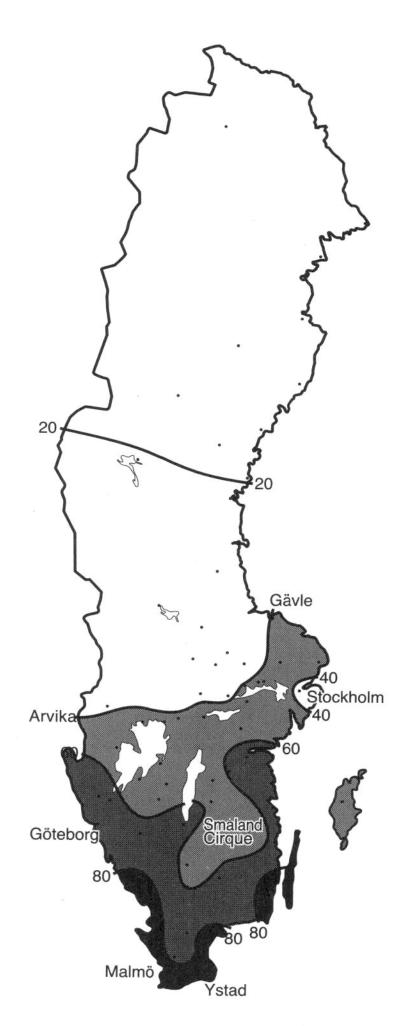

Map 16. The adult mental map of residential desirability from Ystad.

values will be low. We might expect that as children get older and approach adulthood they acquire more and more information and tend to agree more. But if we plot the agreement indices for 284 samples taken all over Sweden (fig. 7), the relation is extremely weak: there is a slight rise in the average agreement levels with age, but the variation is so huge that any generalization about a simple learning model is banal. However, if we plot the agreement levels for the adult samples on the map, an extraordinarily regular pattern results (map 17). The indices vary from 10.53 at Ystad in the south, indicating a high degree of agreement, to 1.20 and 1.28 at Mora and Östersund in north-central Sweden, demonstrating a high degree of disagreement as many of the individual maps conflict with one another. Moreover, the trend is extremely regular.

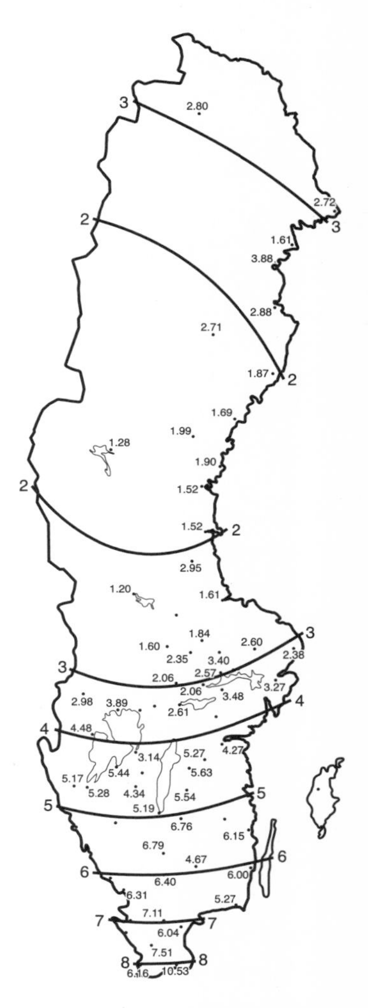

Map 17. Adult agreement indices measuring the homogeneity of perceived residential desirability. A quadratic trend surface, shown simply for illustrative purposes, fits well (0.89 on a scale of 0.00–1.00).

Space Sitters and Space Searchers

Why should levels of agreement in adult groups about residential desirability vary in such an extraordinarily regular fashion with location? It is difficult to marshal absolutely clear-cut evidence at this point, but there is a strong possibility that the agreement levels are related to the degree of informational homogeneity within a group. For example, a very parochial group, sharing the same set of information and displaying the same levels of knowledge and ignorance about other places, would tend to agree very highly. In contrast, people who had very different sets of information might have a very low level of group agreement.

Yet this only begs the question, because we have shifted the burden of explanation for this highly regular pattern to the degree of informational homogeneity within a group. If mental map signals, or agreement ratios, vary in

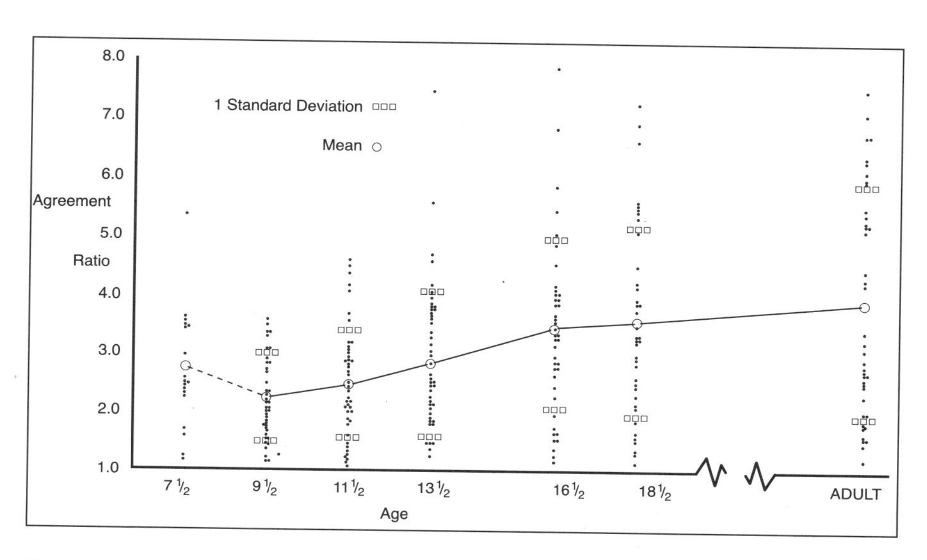

Fig. 7. Agreement levels (λ_1/λ_2) and ages of 284 sample groups, ranging from young children (7.5 years) to adult groups with average ages in the early forties.

such a regular fashion over geographic space, is it also possible to show that homogeneity levels of information also vary in the same way?

Let us consider the sources of spatial information. Thinking of the regions as transmitters of information, and the people within a group as receivers, we know we can predict information surfaces reliably from population and distance values, indicating that people in different parts of the country will have different sets of information. At any particular location, we would expect the information received by a group of people to be relatively homogeneous. But some people do not just acquire spatial information by sitting at one location and receiving it. Some also travel widely for business, recreation, and family visits, and, intentionally or not, they acquire and store information about the space they move in.

But the basic question has only been pushed back still further: can it be demonstrated that the space-searching travel behavior of the Swedish people varies from place to place? Do people in the south tend to be sitters and receivers, while those in the north tend to be movers and searchers? There is a long history of out-migration from northern Sweden, which means that a large proportion of families there have relatives and friends in the south today. Add to such a propensity to visit friends and relatives the fact that the large cities with a great variety of things to buy are all in the south, and that the south coast has the famous summer vacation areas, and we have a distinct possibility of a strong directional bias in movement. It is as though the people of Sweden were small particles doing random walks in a space with elastic barriers whose strength is dependent upon location. That is, the probability of a man or woman from Skåne in the far south going to the north might be very small compared to the chance of a northerner traveling south.

If the analogy to random walks in a space of elastic barriers holds, even crudely, then travel behavior should be spatially more restricted in the south. Groups of southern adults will contain high proportions of passive receivers, and even the space searchers among them will have their movements fairly well restricted to the southern part of the country. They will share very similar sets of transmitted or received information, and such homogeneity will be reflected in their high agreement levels. In the north, however, groups will have relatively low proportions of sitters and high proportions of searchers, whose spatial information will be richer, relatively heterogeneous, and more varied than the southern groups. Such a lack of homogeneity in the information set will be reflected, in turn, in low levels of agreement.

The evidence for such an idea is crude, consisting of simple counts delimiting the travel fields for many adult groups in various parts of the country. These are constructed by recording those regions in which people had stayed for a week or more, an arbitrary time span designed to eliminate casual passings through or over, perhaps at night by train or plane. The travel field of people at Ystad (map 18) is highly restricted to the southern half of the country, and if we draw an arbitrary line between the Bollnäs and Hudiksvall regions, only 6.1 percent of all visits have been made north of this line. Even at Karlstad (map 19), adult travel is still highly restricted to the southern half of the country, and it is not until the line is crossed to Östersund (map 20) that a totally different travel pattern appears. In this region of the lowest agreement index, fully 62 percent of the travel is to the south. Even far to the north at Luleå (map 21), the travel is equally divided between the north and the south, a remarkable and cosmopolitan contrast with the restricted and parochial travel behavior at Ystad and other southern regions. With high proportions of space searchers in their groups, the information sets of Östersund and Luleå are highly heterogenous, so depressing their overall indices of agreement.

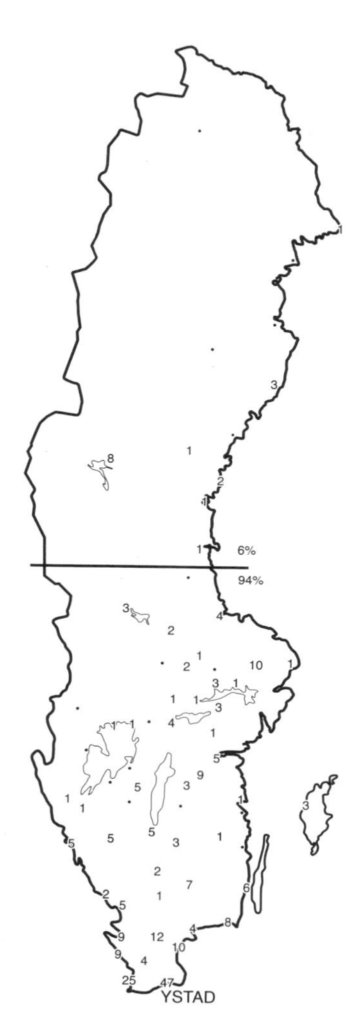

Map 18. The travel field of adults at Ystad.

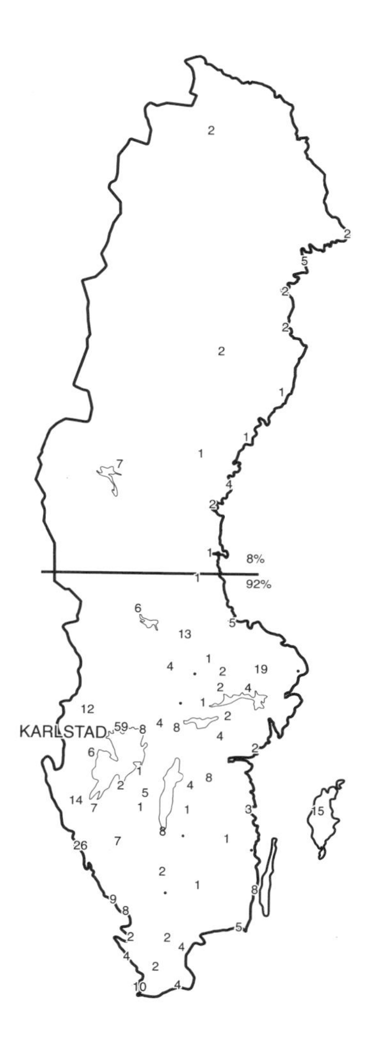

Map 19. The travel field of adults at Karlstad.

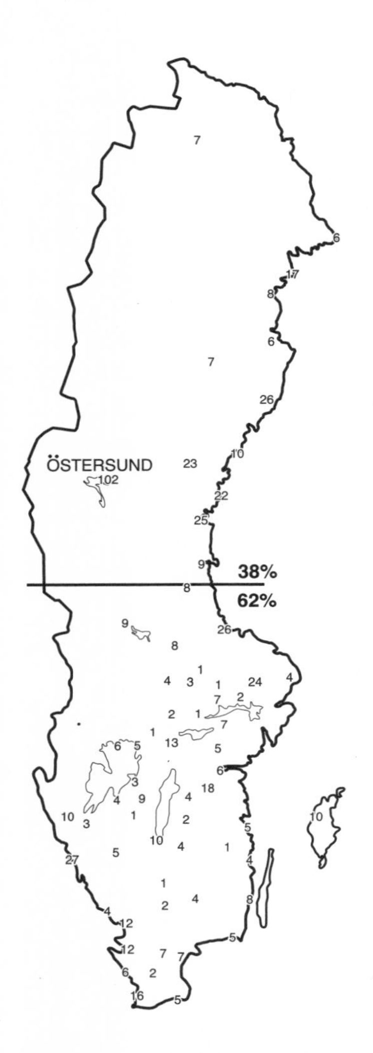

Map 20. The travel field of adults at Östersund.

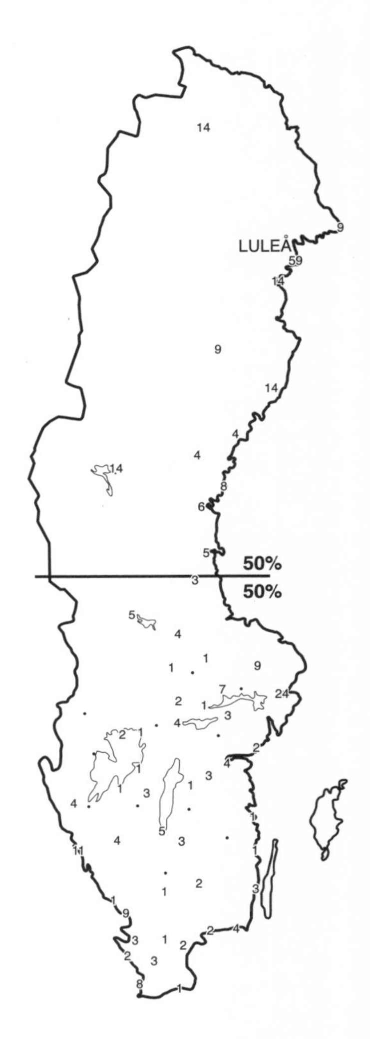

Map 21. The travel field of adults at Luleå.

Conclusion: Location in Information Space

Information is difficult to define, and still more difficult to measure. But by restricting the idea to the relatively small set of spatial information, and using simple counts as aggregate measures, we have seen that information generates very regular and highly predictable patterns in geographic space. *Where* groups of people are located appears to determine, to a very large degree, what sort of information surface they will generate. Moreover, their levels of group agreement about certain, essentially geographic questions are also highly predictable from their locations. These locations, relative to the broad patterns of information-generating population in the country, also shape the aggregate travel fields that represent possibilities of acquiring firsthand, as opposed to received, information.

It is difficult not to conclude that it is *location,* location in information space, that emerges as a crucial variable in any consideration of the amount, composition, and acquisition of spatial information—and perhaps other forms of knowledge as well, particularly if we conceive of information spaces in much broader terms than just the geographic.

Maps as Memory

> In such a way, the map has become a graphic autobiography; it restores time to memory and it recreates for the inner eye the fabric and seasons of a former life.
>
> —Brian Harley
> "The Map as Biography"

BRIAN'S memorable, joyful-sad essay is one of my favorites. And if you should feel, from some sense of trespass, that it cannot be emulated, allow it at least to inspire other memories of other places. For me, four folds of the Ordnance Survey's Sheet 202, Torbay and South Dartmoor, contain my teenage world (plate 1). From the mouth of Devon's River Avon in the west to Start Bay in the east, it was a world circumscribed by a solid Raleigh bicycle, with its three-speed gears helping young legs to pedal over the hills of a south Devonshire landscape. Its center was Salcombe, and if you left on a summer's morning, with the larks singing high above you, you could visit almost any part of it and still be back before dark. Lamp generators never did work well when one was pushing a heavy bike up those hills, and at the end of a long day more than the ingenious gears of Messrs. Sturmey and Archer were needed to get home.

Yet even bicycles had their limits, for this was also a world of footpaths and cliffs, of woods and fields, of sand and water, places where no machine could go. From Prawle Point to Bolt Tail, I knew every path and rock from the many hours and days spent on the wild cliffs watching birds, their first agitated cries calming to disapproving mutters as I lay immobile in the bracken. I had no need for company, and after the crowded, chivvying life of a boarding school it was marvelous to be alone, clambering on those cliffs, doing what you liked, going where you would, reveling in the freedom of vacation time without a care in the world.

An essay accepted by *The Map Collector*, just before it was taken over and incorporated by *Mercator's World*. A truncated version appeared in *The International Map Collectors' Society Journal* 4 (1996): 1–4.

But not always alone: often Robin and I would meet at South Sands and walk the cliff paths to Bolt Tail and Hope. Once you were out of the woods below Overbecks the wind could batter you, and if the tide were low great waves would crash on the bar—Tennyson's bar of:[1]

> Sunset and evening star,
> And one clear call for me.
> And may there be no moaning of the bar,
> When I put out to sea,

The bar never moaned for us, or if it did we never heard it. Perhaps it was drowned by the ghostly screams of the Excisemen murdered by the smugglers? Or was it just the howl of the wind around Sharpitor? For that is what Robin and all the local people called it, and what do map makers really know? From there you looked down into Sterile Bottom, already Londonized "naicely" on the Devonshire Sheet CXXXVIII. NW 1905 into Starehole Bottom, and today removed entirely from any prurient gaze. It took longer to sanitize Sewer in all its Upper, Lower, East, West, Middle, and Mill varieties, but today most of them have gone, and Sewer is an ethereal and more hygienic "Soar." But Robin said "Sewer," and so do I.

Below, in the small bay, lay the great shadowy hulk of the *Herzogin Cecilie,* one of the last great China clipper ships, wrecked while trying to find haven in a 1936 gale. For weeks after, the estuary beaches were covered with South Australian wheat rotting in the sun, and at low tide you could still see the great trunk of one of her broken masts sticking out of the water. As for the huge cave, it was, Robin assured me, the entrance to a tunnel that came up under Malborough church, used by the smugglers to stash their wares. "And one day," said Robin, "a black bull went into the cave, and it came out pure white." But even at low tide we could never find the entrance. "It must have fallen in," said Robin, "just over there, by the rock pile." Even in the depths of winter, without another soul around, there was always a small café on the way back to Bolberry whose kindly lady was prepared to produce scones and jam and cream before we walked on to Malborough. Sometimes if we were lucky, and had a sixpence left over from tea, we caught the bus on its infrequent winter schedule. Often we legged it home, never thinking of hitching a ride. Anyway, there was little traffic then.

Other days we would carry the heavy bikes down the ferry steps, the ferryman never charging two boys for their loads. And then up the steep hill to East Portlemouth, past St. Winwaloe's church, and on to Slapton Ley. Hiding the bicycles behind bushes, we spent hours watching for bearded reedlings, and to this day my much tattered *Birds of the Wayside and Woodland* records "Rare, Slapton Lea [*sic*], 1949" in my rather florid immature hand. Little did we know at the time that in the months just before the Normandy Invasion the entire area for miles around had been cleared of all its people to provide the secrecy necessary for large-scale landing exercises on Slapton Sands. When we were there the people must have come back to their homes, but I have no recollection of seeing anyone on those long and quiet afternoons.

As for Salcombe harbor itself, it was always alive with small craft, except on the stormiest of days. For this was a water world just made for "messing about in boats," whether seriously, to make a living with lobster and crab pots, or for the sheer pleasure of being on the water. When the tide was right, we would putt-putt our small, blunt-nosed pram all the way from South Sands to a mooring by "Ma Gasson's steps," known on Sheet CXXXVI SE 1905 rather primly as Victoria Place. When the tide turned, you had to get out of there in time, or you would be stranded on the mud for another six hours. Always, coming and going, we dropped a line astern with a spinner for pollock or mackerel, and usually had our supper by the time we landed. Once, messing around in the mouth of the estuary, I came across a huge basking shark lazily minding his own business, and with the stupidity and bravado of youth I poked him several times with an oar. Twice as long as my pram, he could have turned me over with a flip of his tail, but he took no notice and wallowed off in disdain.

On almost any fine day of the year the sailboats would be out, most of them small, clinker-built dinghies made slowly and carefully in the local boatyard from steamed and bent strakes held by copper rivets. But sometimes there were special boats. I can still see the great russet sail of the old Channel Islander leaving the harbor for the last time, with many of the local people along the shore waving goodbye. In 1940, ahead of the Germans, he had loaded his entire family into his open fishing boat, pulled up that same patched sail, and found haven across the Channel. One day, when the war was over, and judging a fair wind, he loaded them again in his boat and quietly sailed back. Another time, a small seagoing junk made in Hong Kong finished its long journey through the Pacific, Indian, and Atlantic Oceans to make landfall in the harbor. Little did they know that an upturned hull of a sailing boat, in a meadow visible from the water, its planks rotten, and its white paint peeled away, was all that was left of Joshua Slocum's *Spray* that he had sailed alone around the world.[2]

Yet behind those folded panels of Sheet 202 today, my own map of memory, lie still other maps, ancestors of this modern upstart, reflecting different judgments, as well as older scales less conforming to kilometers and other foreign devices. John Speed, though probably overwhelmed by the detail, would still be at home in this part of the world after four hundred years. Making allowances for some orthographic niceties (surely his Chilton must be our Chivelstone?), all his place names are still familiar.[3] Strangely, this is not true of many of the names of less distant ancestors from the early nineteenth century (plate 2), for this part of England was one of the very earliest to be surveyed by the great Ordnance Survey after Essex and Kent.[4] Indeed, the surveyors' drawings all around the Plymouth Estuary date from 1784, reflecting the strategic importance of the area, and the devastating impact it would have had on Britain's defenses if it had been seized by a French expeditionary force. If you take a map of Devon and Cornwell, divide it into its 47 drawing sheets, and plot the dates at which the initial drawings were finished, you can sketch in rough isochrones showing the expansion of cartographic detail captured in the great net of the trigonometric survey. And for all the local vagaries of trigonometric necessity, caused by wrestling the huge 200-pound theodolite to the top of Dartmoor's tors, perhaps we can legitimately interpret these isolines as perceived "isodangers." In 1779 there were 88 enemy vessels off Plymouth, and in 1797 the French burned farmhouses at Ilfracombe. And then recall the landing barges assembled by Napoleon for his great invasion, like another one that never came nearly a century and a half later.

Although the original (1809) Ordnance Survey Sheets 23 and 24 are at 1:31,680 (two inches to the mile), a bit more detailed than the present Sheet 202 at 1:50,000, a comparison of the two seems legitimate. In this part of my teenage world, a careful count shows that of the 263 names indicating some form of settlement, 64, or 24 percent of them, are no longer to be found on the modern sheet. With a little visual license, they seem to lie in a circle around the fingers of the estuary, although in the center the Salcombe "octopus" itself has swallowed Shabicombe, Horsecombe, Hangar Mill, and Churchill. Many of the old names, in which you can still hear the soft-slow Devon pronunciation, now have a harder bite. This is hardly the imperialistic imposition of names by the Ordnance Survey, recorded so movingly by Brian Friel in his marvelous play *Translations*,[5] which should be read by every cartographer and geographer. Nevertheless, Oaten becomes Anton, Friscombe has been prettied up to Frittiscombe, Yatson is now imperfectly French as Yetsonais, and Hantsknowle is Hansel. (Alas, lonely without his Gretel!) Scores of hamlets—Nutcombe, Tor, Pool, Pitt, and many others—have now been reduced to "Farm," but changes are not just one way. On the current map of 230 places, 31, or 13 percent, are new, some with a suburban feel to them—Homelands, Courtpark, Court Barton, and Icy Park (in Devon?).

And yet, ironically missing on both maps, there is no Great Western Railway line from Kingsbridge to Brent. In the interval 1809–1976, the Railway Age has come and gone, leaving only a thin hatched line on the modern map as a ghostly reminder of the Age of Steam. I used to take it to school: by ferry to Salcombe, bus to Kingsbridge, and then the enchanting ride along the gentle curves cut by the Avon River into a deep gorge. I can still feel the resignation on the outward journey, and remember the anticipation and excitement when I returned to another window of freedom, to the cliffs and birds, the water and the boats. For even today, fifty years later, you could drop me anywhere in my earlier world, and I could direct you on your way. For I need no map, not now. Elizabeth Jennings, in her lovely poem "Map Makers," was quite right:[6]

After the journey we can fill the map
We shall not need; that map can only show
The journey that we need no longer go.

Against the Grain

SOMETIMES you feel that things are just not right—that things are said, claims are made, or people are hurt—and your sense of fairness wells to the surface. When this happens, you feel a need to shove back, to give a different view, to flip the coin over and see what the other side says. When you do this, it often means moving against the grain in various ways: rocking the boat a bit; feeling uncomfortable as a "team player" when you do not like the goals of the team; not being an acquiescent wait-and-see "committeeman"; and sometimes even loosening the ropes on the tightly constrained cannon. One of my favorite geographers once sent me a card that has long been posted on the wall outside of my office. In a large meeting of assenting cows there is one horse looking slightly abashed saying "Neigh." The caption reads "Just because you're in a minority, doesn't mean that you're wrong." Of course, all of the above, and some that follows, may be dismissed as self-serving and self-justifying nonsense, the apologetics of a self-righteous pr(I)g. There is probably something to this, but you can judge for yourself.

In the fall of 1977, two years before the seventy-fifth anniversary of the Association of American Geographers, I was approached by the editor of the *Annals* to write an essay of personal observations about the previous two decades. From my perspective, one shared by many of my peers, it had not been a particularly easy twenty years. At several distinguished universities geography had been dismissed, and too frequently the reaction of the "Old Boys" ("Old Girls" were still a decade or two in the future) was to draw the wagon train of geography into an ever-tighter circle, symbolic of the defensive laager-like attitudes so distressingly prevalent in those days. The same sort of smug traditionalism made sure that the new methodologies and conceptual frameworks were kept well tamped down, even if this often meant tamping down the influence of young colleagues who taught a token course or two on the often dismissed "quantitative techniques."[1]

So accepting the invitation, I wrote what I felt (and still feel today) was an open and honest account of those difficult times. Over the next couple of years following publication I pieced together some of the reaction, a highly polarized pro and con reaction with little in between. Some of the Establishment were so outraged that the essay had seen the light of day at all that they made sure that nothing like it would ever appear again. Anyone comparing the *Annals* of March 1980 with the next issue in June will see that within a year an editorial board of nine was put in position to "inform" the editor. Two years later, when he stepped down, the board was expanded to twenty-three to make sure that any such accounts in the future were replaced by soothing and self-congratulatory anodynes.[2] In California, the heart and soul of Carl Sauer's "Berkeley School," there was a stamping of naughty little feet that someone should dare to point out that *Maître* had packed almost all the universities on the West Coast with his own students. Why this came as such a revelation I do not know: all you had to do was browse the current *Guide to Graduate Departments* to document the inbreeding.[3]

Fortunately, the reactions were not all negative. I received a score of letters saying "Well done! About time!"—often from those who had gone through the same sort of thing. "At least," said one, "the whole miserable business won't be forgotten entirely," echoing almost exactly the words of the original invitation ". . . an account for future generations."[4] I still have a form letter to the twenty-seven contributors of the anniversary issue saying "I wish I could share with each of you the growing stack of mail . . . The response has been strongly positive . . . This may be the first issue of the *Annals* in years that so many people claim to have read from cover to cover without putting it down," together with a personal handwritten note to me saying "Your item was the star of the issue!"[5] Of course, the times and the reactions are long past now, but only the other day a graduate student said to me, "You know, I never realized how strong the feelings were against you guys in those days!" So perhaps a faint memory of those times will tend to thwart any future rise of Guruism, something that seemed to be a congenital defect of geography "in those days."

And speaking of Guruism, I have never been very good at genuflecting to gurus surrounded by adoring acolytes. I have the gut feeling that we are all mortals stumbling around the lower slopes of Mount Olympus, and claims to godhood should be viewed with proper suspicion. Which is not to say that I do not have my own and quite private intellectual Pantheon, an always-open assembly to which I add members from time to time. Genuine attempts to illuminate our complex world always command respect.

I am afraid that my critique of dissipative structures is by far the most difficult essay in this book, and you may wish to skip over it altogether. But I have included it because it constitutes an attempt to think against claims made out of the physical and mathematical realms toward the human world. The claims were put forward by Ilya Prigogine, who was awarded the Nobel Prize for chemistry in 1977. He is not the first of those so elevated to be tempted to expound in areas that have nothing to do with highly limited regions of scientific expertise.

The critique was originally given at a conference in 1985 about modeling complex systems, held in a huge auditorium at the Ilya Prigogine Center for Statistical Mechanics and Thermodynamics at the University of Texas. Maître was present, surrounded by adoring acolytes from many fields far removed from physical chemistry. Maître obviously did not like the idea of critique and soon interrupted my presentation from the floor. I had assumed questions would be posed at the end, but I answered politely and continued. A few paragraphs later Maître interrupted again, and this time I asked him if he would be so kind as to hold his questions until the end. There was "a breathless hush in the Close to-night," Sprint's dropped pin would have echoed from wall to wall, and all the acolytes turned chalk white, while Maître turned pink and walked deliberately up the aisle and out. I think there were some questions at the end. Frankly, I cannot remember. In the coffee break, no one dared to approach me, a questioning pariah in a sea of truth and certainty.

The revised and final papers were to be collected together and published as an issue of the *European Journal of Operational Research.* It did not appear until two and a half years later. I was told privately that in the meantime my paper had been sent to an unusually large number of people in mathematics and philosophy, indicating a most gratifyingly thorough review process, and not to be interpreted as a search for critics who would reject it. However, while I took total responsibility for the paper, it had been carefully read and reviewed by two colleagues, a fine mathematician and an equally fine philosopher. Apparently the arguments were sound, and no official reviewers could be found to reject them. So it was published, but at the very end of the journal, with a petulant little so-there-I'll-have-the-last-word-after-all note of "commentary" by an acolyte.[6] But the *éminence grise* was not very well hidden in the shadows. As for dissipative thinking and bifurcation in the human realm, we have not heard very much about them for the past ten years. Who knows, perhaps they bifurcated and dissipated?

"Sharing a Tradition" had a rather similar initial presentation, this time as the plenary address to the Canadian Geographical Association in Ottawa in 1993. As a courtesy to Canada, and especially to Ottawa lying on the boundary between Ontario and Quebec, the first and last quarters were given in

French, the middle half in English. A number of geographers, working in fairly new directions, were calling for forthright criticism of deplorably incorrect past practices, something I welcomed since there had been times when the distinction between critique and criticism had become blurred in my own writing. So I decided to take them at their word and turn the eye of critical reflection back upon some of the writings in radical feminism, postcolonialism, and so-called postmodernism. Many of these writings evoked the century of Enlightenment, insisting it was a time of rapacious Reason with a capital *R*, a precursor of hateful modernity from which *post*modern insights would finally release us. Unfortunately, most invocations of the Age of Enlightenment only displayed their author's unbounded ignorance of everything except clichés taken from secondary and tertiary sources. It is always nice to believe that you are at the advancing front of an eschatology of your own making.

Again reactions were highly polarized; many, both men and women, came down to the podium afterwards to say "Thanks, it's about time," while others walked swiftly away to complain to the Association that a white European male chauvinist pig with a hegemonic phallocentric gaze should never have been invited in the first place. Perhaps they forgot that the century of Enlightenment was marked by critical reflection that brought previously unthinkable ideas out into the light of reasoned debate. Three people were chosen to comment on the published address, and I was allowed to reply. You, from your "situated hermeneutic stance of positionality" may judge this "Other" yourself.

The various genres of writing emerging out of an excitingly refurbished social and cultural geography are frequently enlightening, thought-provoking, and intellectually satisfying, in that they help us to see something in a different way, from a perspective or viewpoint we had not taken before. But some throw up (the image is deliberate) writing that is so totally biased, shrill, and predictable that it eventually becomes boring. To give a parallel example, it is all the difference between original Marxist insight—Marx writing his marvelous *Braumaire,*[7] or his wonderful series of essays on the 1848 revolution,[8] or David Harvey on *Kapital* or nineteenth century Paris,[9] etc.—and the parrot-like repetition of jargony phrases about dictatorship of the proletariat, hegemonic capitalism, forces of production, and so on. Repetitious jargon nearly always means genuine thinking has been replaced by clichés.

So after I had been threatening to write *Cathartic Geography* for a couple of years, the final catalyst was the invitation to give a geographical address at St. Andrew's in Scotland during March 1997. Getting close to the third millennium, I wanted to make this as forward looking and as positive as possible. But before being able to give a "Forward to the bright sunlit spatial uplands of the twenty-first century" exhortation, in other words to provide the setting for an address "*Beyond* Cathartic Geography," I obviously had to write "Cathartic Geography" as a sort of response to all the cathartic nonsense presently splattering some of our journals. In one sense, it was not difficult to write, with the plethora of awful examples available. In another sense, it was very difficult to write, for I needed to persuade even the antagonistic reader that in the various subgenres of contemporary human geography there are also many genuinely illuminating and thought-provoking pieces of scholarship. Others may wish to separate the sheep from the goats in different ways, but that there *are* both sheep and goats out there presently browsing the meadows of geography, few would dispute.

Geography 1957–1977
The Augean Period

> And documented batterings
> Are better left alone;
> What shocks her isn't crawling things
> But lifting up the stone.
>
> —Roger Woddis
> "Too Bloody True"

BITING BULLETS leaves a nasty taste in the mouth, leads to mutiny, even madness, and is not to be recommended as a steady diet. But given the specific charge of duties and responsibilities by the editor, I really have no alternative.[1] The two decades 1957–1977 were the "best of times [and] the worst of times," and Dickens's phrase captures exactly the ambivalence and ambiguity of any truly revolutionary period. It was a time of great intellectual excitement, the sort of excitement that can only come from seeing new paths opening up, new connections being made, and real challenges to be met. There was a sense of discovery, and forging, of breaking out of the banal, factual boxes erected by the old men, and a sense of reaching out to scholars in fields to which we had never been properly introduced, but which seemed friendly enough if you were prepared to learn. It was a precious and memorable time, one that was probably impossible to sustain at such a high pitch for too long, even as it is difficult to recapture today all those feelings of past excitement. But I hope they come again, and that what we did is replaced by something better.

The Best of Times . . .

Looking back on my early graduate days in the fall of 1956 at Northwestern, I feel incredibly lucky and privileged. Challenges were there, not the least the

---

Somewhat shortened, particularly in the 76 original footnotes, from my essay in the *Annals of the Association of American Geographers* 69 (1979): 139–151.

presence of Ed Thomas, a third-year doctoral student, who traveled down to Chicago for a weekly liaison with . . . a *computer!* Multiple regression . . . *four* variables . . . urban relationships . . . explained variance.

"Ah yes, . . . variance" we repeated, furrowing our brows with the profundity of the concept. And then, when he was out of earshot, "What the hell is *variance?*" for we would have impaled ourselves on our soil augers before admitting that we did not know. Hurried consultations that night; frantic scanning of Croxton and Cowden; what on earth did Snedecor say?[2] But the next day we casually dropped a few beta coefficients of our own.

And then there was Ned Taaffe, inexhaustible patience after long days and long commuting drives through dawn and evening rush-hour traffic, learning one page ahead of us, his four-colored pencil patterning the map, sharing his own delight and excitement with all who wanted to learn with him. During my first term, when I had calculated the parameters of my first quadratic regression line, I proudly showed him the result. "Why not plot the residuals?" he asked, having learned about them the week before. But not knowing what they were, and being too stupidly proud to ask, I said in a rather chagrined voice, "Oh, I don't think I will, it's hardly worth it, is it?"—and walked quickly out of the door before he could question me further.

And when we all got stuck, as we often did, there was Bill Krumbein in geology one floor below, a scholar and teacher of the first rank, with that rich Middle Western drawl that can still gather an audience of hundreds spellbound into the palm of his hand. Taking nothing for granted (he, too, had taught himself), he would patiently show us how easy it really was, instead of how difficult the books had made it. And Charlie Tiebout in economics, one of the best teachers I ever had,[3] with that rare and precious gift for simplifying the complex with a magical mixture of analogy and chalk, spiced by language of the purest and most exquisite obscenity, accompanied by draws on his ever-burning cigarette that reduced Humphrey Bogart to an understudy. Very personal memories, as they can only be of those starting days; but after exchanging many notes and memories I know now that others at Iowa and Washington shared the same sorts of things and more.

. . . *And the Worst of Times*

Unfortunately, but perhaps inevitably, there was another side to these years, for they were sometimes ugly and vicious times, characterized by a meanness of the human spirit and the threatened postures of the traditionalist lashing out in defense of his comfortable, Establishment world. This side, too, must be recorded, for we need "an account of this for future generations," and you, who have picked up this essay in the twenty-first century and beyond, read closely. Like the McCarthyism that immediately preceded these decades, the ugliness of these times must be recorded and remembered, for those who cannot remember cannot learn. I would like to think that the things recorded here are as honest and as true as I can make them, but I am acutely aware of the way a particular viewpoint can shape historical material, and as you read, remember the Whig historians.

These times have been called the "Quantitative Revolution," but in some ways the term was a disastrous misnomer. Certainly, and with relatively few exceptions, the late 1950s were characterized by the application of statistical methods and simple optimization techniques to geographic problems, and obviously quantities, in the sense of raw numbers, were involved. But the threatened traditionalists misread the signs and pinned on the wrong labels. It was not the numbers that were important, but a whole new way of looking at things geographic that can be summed up in Whitehead's definition of scientific thought, "To see what is general in what is particular and what is permanent in what is transitory.[4] How shabby, parochial, and unintelligent are the contrasting words of Sauer at the time, "The geographer is any competent amateur. . . . [W]e may leave most enumeration to census takers. . . . [T]o my mind we are concerned with processes that are largely non-recurrent and involve time spans beyond the short runs available to enumeration."[5] But he was faced with a new generation, one that was both sick and ashamed of the bumbling amateurism and antiquarianism that had spent nearly half a century of opportunity in the university piling up a slagheap of unstructured factual accounts. It was a generation that, without exception, was completely conventionally trained, but one that knew in its bones that there was something better, more challenging, more demanding of the human intellect than riffling through the factual middens and learning the proper genuflections from *The Nature of Geography*.[6]

For we must remember that these years were marked by events that seem almost unbelievable today. With the exception of one or two works of scholarship in historical geography, it was practically impossible to find a book in the field that one could put in the hands of a scholar in another discipline without feeling ashamed. Only when Bill Bunge's doctoral dissertation and Peter Haggett's synthesis were published was there something that one could offer without apologies.[7] The latter was published in England, and the former never did appear in the United States. Rejected savagely and repeatedly by the same reviewers, the book was finally published in the distinguished Lund Series of Human Geography in Sweden. "I didn't agree with all of it, it was a bit too geometrical for me," the editor of the series once told me, "but I knew something about the way it was being stopped, and I thought it should be published." Such perspicacious generosity was not to be found on this side of the Atlantic River in those days. Bill Bunge had walked out of Wisconsin, after telling his doctoral committee that their questions were inappropriate—or words to that effect—and had found an intellectual haven with William Garrison at the University of Washington. There, in the company of some young and unknown graduate students like Brian Berry, Waldo Tobler, John Nystuen, Michael Dacey, Arthur Getis, Duane Marble, and Richard Morrill, together with a Swedish visitor that nobody had even heard of called Torsten Hägerstrand, he wrote one of the first real attempts "to see what is general in what is particular." Not that things were easy at Washington either, for, with one or two exceptions, conservatism ruled there too. But as some of those former graduate students have told me, "It was Bill Garrison who put his foot in the door and got us through."

Unbelievable though it may seem today, those were times when not one geographer in a hundred knew the names of von Thünen, Christaller, and Lösch, men whose basic principles of spatial organization appear today in beginning university courses. Richard Hartshorne, for example, gave von Thünen a "one-liner" in his *Nature of Geography*, but condescendingly dismissed him with the comment, "Whatever validity this analysis may have had in earlier periods has largely been destroyed," and then added a footnote to his own work, which must have been the *coup de grâce*. Even as Berry and Garrison were building on Christaller's work of twenty-three years before, confirming the principles of central place theory from the dense core of Chicago to the rangelands of Montana, an older generation smirked because they could not actually see the hexagons. Except perhaps in Iowa: "But," they cautioned, "not every place is Iowa—every place is unique, no other place exists in the same space and time as any other!" And with these warnings to a foolish younger generation they bumbled off to oblivion. A whole mind-set was involved here: geography was working down a checklist—geology, vegetation, soils, climate, population, etc.—and when you got to the bottom of the list you were finished. So, nearly, was the field as an intellectual discipline, for these were also the days when universities like Harvard, Yale, and Stanford threw geography out on its ear, and many other major universities, with proper intellectual standards, refused to invite it in. They were the days when the Social Science Research Council politely, but firmly, rejected a bid to give geography full status, after members of the committee examined some of the recent issues of the "major journals in the field." It was all done terribly politely, with sufficient circumlocutions so that loss of face was minimized, but the message was clear—and well deserved.

These were also the gatekeeping days, when the journals of the field closed ranks and stopped up the chinks to prevent the winds of new ideas from getting through. Some of the letters of rejection were extraordinary for their arrogance, sarcasm, and sheer venom, and we used to circulate the prize examples among ourselves as paper witnesses to the reaction of the current order. Eventually, the reactions from both east and west became so predictable that few bothered to submit their writing thereafter, and in general this pattern has continued, so that even when a less conservative editor takes over, the traditional reputation of a journal lingers on. The result is that most of the demanding and worthwhile geographical writing now appears in the newer geographic journals (eight appeared in the years 1958–1977), or in the journals of other, more open, disciplines. Even today the distortions of the scholarly process can be large. One editor, of a well-known geographic journal, has suffered so greatly from the pressures from both the right and the left that he has been forced to return papers containing mathematical notation to the authors without even sending them out for review. He apologized, poor fellow, but said he had no choice given those around him.

But two exceptions stand out in this early period. Robert Platt of the *Annals* was a tall man in all senses of the word, and even from his flat Middle Western plain he could see farther than those who walked about the western hills or climbed the eastern monadnocks. A number of us can still remember him coming up after a meeting to ask, in his open and generous way, about a paper and the new approaches it contained. And Wilma Fairchild, too, with her incredibly difficult job of walking a tightrope between the professional and the layman, but willing always to consider new ideas if they were clearly expressed, embodied true editorship in a highly professional sense.

Yet beneath all the excitement of new ideas, and despite the increasing dimensionality of the intellectual space of geography, the conservative reactions were often sharp—if they were not merely lethargic. ("It's no use being idealistic; you've just got to think of a big, blind elephant," one of my teachers used to say. "All you can hope to do is prod it, and then hope it blunders off in a different direction.") It was the time when the California-Oregon system began packing itself with its own products, so that even today over 42 percent of the faculties are their own graduates, ranging from 75 percent and 70 percent at

Riverside and Oregon, to 12 percent at Santa Barbara. At the latter, geography was finally put out of its misery by the administration, and then reinstated with a completely different cast. Today it stands as one of the most intellectually demanding and exciting departments in the country.[8]

Fragments from a Kaleidoscope

The dark side of those days now appears as a set of fragments that somehow form a coherent pattern, although I realize that with a different eye and a shake of the kaleidoscope a different configuration might appear.

Fragment One: when an old guard relaxed its grip for a moment, and allowed a newer generation to honor Warren Thornthwaite as president of the Association of American Geographers, after years of exclusion. In the 1930s he had criticized much that purported to be research, and they never forgave him.

Fragment Two: a graduate student, now a member of the National Academy of Sciences, writing a highly creative dissertation using linear programming in a geographic and behavioral context. Of course, it was unacceptable, and had to be completely rewritten to conform to accepted geographic standards.

Fragment Three: Reino Ajo, who never received a permanent university position among the folklorists of his native Finland, supporting his family in a small provincial town as an inspector of automobiles, working in the evenings at home in order to publish some of the most innovative geographic analysis of his day in German, Swedish, Finnish, French, and English, sometimes with Latin abstracts. The Association of American Geographers finally honored him in 1972.

Fragment Four: the attempt to honor Walter Christaller, still living frail and tired in his native Germany. The attempt failed: a visa was refused because as a young man in the 1920s, Christaller had been a member of the Communist party for a few months. It remained for Sweden to honor him with her gold medal.

Fragment Five: the decline of the American Geographical Society, the despair of its later directors, who tried so hard to move it out of its white ties and annual banquet halls into the geographical reality of the twentieth century. Under pressure from the academics in the 1920s, it virtually relinquished the honorable role of public education that was its founding purpose, and it never seemed to sense the vitality of the new geography until it was too late. With creative and imaginative direction, it could have transformed itself into one of the great centers of urban research, with direct inputs to the thousand and one problems of New York, the world's most contradictory city. But it preferred to catalog its library, draw its maps, and hand out gold medals. After living on its endowments for years, it finally seems to have fallen asleep.

Fragment Six: field camp, the apex of methodological training, in which we spent over a thousand student hours compiling a land-use map with fractional coding. Unfortunately, nobody knew what to do with it when it was finished. And then more hundreds of hours mapping the land use of a small Wisconsin town for the purpose of delimiting that most profound of geographic concepts, the CBD—something that any inhabitant over five years old could have done with his eyes shut. It put me off urban geography for twenty years.

Fragment Seven: the director of the large, interdisciplinary study on the behavioral and social sciences meeting with the geography committee for the first time: "Now let's see, you people deal with topographic maps, or something like that, don't you?" We all cringed, because here was a highly intelligent, perceptive man, with a broad overview of the social sciences. I know he had both a deeper and brighter view of the field when we had finished our work, but what had been going on during the past four decades to leave such an intellectually flabby impression? The answer was very little; that was the problem, and that was what a new generation was reacting against.

Reaching Out

Virtually severed from other fields in the human sciences, geography had stagnated for decades, without tools, without methodological insight and development, without principles and constructs that gave coherence and order to observations, and without concepts to guide their gathering. It is hardly surprising that the first reactions against such intellectual stagnation were methodological and conceptual. Explanation, even good and worthwhile description, involved relating things to things, and we had to have tools. With only a few, imaginative, and pioneering exceptions, we nearly all turned to statistics, for here was a body of well-developed methodology that seemed both useful and well tested in the world of real problems. Now, with twenty years of hindsight, we can see that many of those early borrowings were naïve and inappropriate, but they represented a tremendous change from what had gone before, and eventually many of their inadequacies exposed large and unexplored areas within the field of statistics itself. Today it is geographers who have been responsible for significantly extending the bounds of statistical theory in the plane.

There were, however, some remarkable exceptions to the general trend of applying statistics to geographic research. First, and through the seminal influence of Walter Isard, via William Garrison at Washington, we had the first uses of optimizing methods in geography. The very idea of efficient spatial arrangement and partition was raised in the humanitarian context of medical care, and later applications were to come in the realm of the polity at the request of Supreme Courts as geographers were asked to solve problems of reapportionment. Secondly, and in a remarkable burst of individual creative scholarship, without discernible geographic precedent, Leslie Curry started his passionate search for spatial theory by taking the probabilistic road. Starting from queuing processes, whose acquaintance he made at Johns Hopkins, he worked with random arrival rates of rain, grass, and lambs in New Zealand, climatic change, landscape, settlement and economy, spatial structure and interaction, and in the process he changed the way we were able to look at the natural and man-made world. Finally, the enormous contribution of Torsten Hägerstrand must be acknowledged. In an age when every high school student can draw random numbers and Monte Carlo her diffusion map exercise, we tend to forget what an exciting innovation his work represented. In a direct and short line from a mimeographed paper by John von Neumann, carried from Texas by a visiting Swedish computer scientist, Karl-Erik Froberg, the work of Hägerstrand in the late 1940s and early 1950s represents, to the best of my knowledge, the first true Monte Carlo simulation in the social and behavioral sciences. The fundamental idea of trying to write "the rules of the geographic game" was his, and today computer simulation is a standard approach in the conceptual armory of the contemporary geographer.

It was from these examples, and the often overbold applications of statistical methodology, that a new excitement began to be generated. It appeared that geographers could actually solve some problems, or at least take a reasonably sophisticated cut at them. If nothing else, statistical approaches raised the question of proper experimental design, as well as the problems of marshaling reasonably secure evidence. Geography was no longer just making maps of land use that no one knew what to do with or compiling large, and numbingly boring, inventories. Some geographers, those with growing methodological competence, even began to be consulted about practical problems both at home and abroad. The hexagons may have been difficult for one generation to see, but it seems that those trained in central place theory had much to contribute to practical problems of urban and regional development. After all, central places are the articulation points of most human activity, and a concern for their arrangement, size, and relationships was fundamental to any thought about spatial organization in the future. The theoretical and empirical work of Berry and Garrison led to practical investigations of urban and regional structure in many countries of the developed and Third Worlds, and a similar concern at Iowa led to the involvement of Gerry Rushton in the practical work of articulating central place networks in India, and the specification of health care delivery systems in Iowa. The long and continuing expertise of Gilbert White in water resources rubbed off on his students, and with a blend of the new methods and earlier substantive knowledge, they forged ahead to make contributions of their own. One of the outstanding characteristics of those who were students in the late 1950s and early 1960s is that they are often called upon to consult and advise, and many of their students, who have shared and extended the methodological competence of those days, are also often involved. The list is a very long one, ranging from remote sensing, census planning, the care of the mentally less capable, to the hydrology of catastrophic floods, urban growth, and many more highly practical topics. Indeed, it is obvious that the range of geographic inquiry has expanded greatly in the last twenty years, mainly because methodological developments and extensions have given geographers an intellectual ability to move much more freely in the problem space. The structure, the geometry, of the intellectual space called geography has changed and sharply increased in dimensionality.

Four Strands to the Present

This is not the place to write a history; the charge is for a shorter, more immediate and personal view. But four strands, which sometimes cross and weave in a tangle of impressions, seem to stand out today, and if we follow them back into the labyrinth of those years we find certain people on the way. Later on, when we have greater perspective on these times, with all the loss of immediacy, and all the gain of hindsight, we can place another, but still and always contingent, structure on the data.

It seems to me that for the first time in nearly two hundred years a genuine concern for philosophical issues has arisen in geography. Oh, I know, we can parade our superficial catalog of Ritters and Ratzels, but I am talking about something of quite a different order. We have evidence today of much more profound questions being asked, questions concerning the implications of difficult inquiries in contemporary philosophy for our own work. There is not space to explore these developments here, and they appear as a strong component of another strand. A second aspect is the concern for the philosophy of science, a concern that itself reflects a marked change in the field, for it says that geography shares the questions of the sciences, something no field defined by inquiry into the particular and unique could tolerate. We see ourselves in an enlarged paradigm, and this is what I implied when I said that the dimensionality of geography has increased. We are not flip-flopping around from one paradigm to another, but enlarging our perspective by reaching out. The tools we have acquired over the last twenty years—statistics, optimization, systems analysis, operations research, even certain general areas of mathematics—have enabled us to see the general in the particular, the specific instance as a special, constrained case of a more general statement and proposition. Finally we seem to realize, and accept, that our structures are contingent, to be replaced without too much regret when a more powerful ordering construct comes along. This, too, is a large step forward.

But most of all, we seem to have generated a deep concern for language, for what can be said, what can be captured in our net of words, and what, perhaps, lies beyond the "bubble of discourse." It is, of course, the concern of Wittgenstein, and his influence is most directly seen in the work of Gunnar Olsson and that of some of his former students. It is an influence that has been quietly at work for several decades, appearing in places and people where you least expect it. Peter Haggett, I am sure, would acknowledge the subtle force of his influence, and I keep coming across people who were at Cambridge in the last years of Wittgenstein's life who seem to have absorbed an ever-present concern for matters of language. From this thread, and the people we meet on the way, we can acknowledge an ethical concern that is much more explicit and open today. What does urban, regional, and national planning mean in human terms? Can we, should we, freeze the spatial status quo with top-down administrative impositions? Or can we move to more carefully specified goals and steer toward them in the future? In the 1950s, for example, I honestly do not think geographers ever consciously thought about such matters. I certainly did not, and I do not recall any seminar, article, or book that did.

Close to this philosophical thread, and often woven tightly with it, is the Marxist concern. Of course, giving a blanket label to something as subtle, complex, and varied as the acknowledgment of Marxist thought in geography is asking for trouble, but we have to call these important innovations something. One thing is certain: we hear today expressions of human concern that were not heard before, and we have gained perspectives on past research to which we were totally oblivious. There is a sense of social outrage today, expressed in the annual meetings, only to be met with conservative counterreactions. The books of David Harvey, and a growing number of other geographers, express pointedly ethical questions, and increasingly the journals are reflecting the growing concern for problems of social justice in a spatial setting. As a result, we can now pose questions from a very different perspective, and much of the research on the spatial aspects of modernization appears naïve in the harsh light of such criticism. Even though we were highly constrained by the data, the fact remains that we did not think to pose the deeper questions, questions about the structure underlying and shaping the superficial forms and changes we were describing. In contrast, those of us who work and consult in the Third World today are obliged to ask whether we can give our intellectual skills to Gulf and Western in the Dominican Republic, or whether we should reserve the insights we have to help agricultural development in Guyana.

Closer to home, literally in the streets and alleyways around us, Bill Bunge was forming and leading his Detroit, and later Toronto, Geographical Expeditions. We owe him an enormous debt for heightening our sensibilities, for showing us that simple observation and mapping, focused by sharp, pertinent questions of humane concern, can amplify the local environment to the point of excoriation. His maps of road deaths of black children, of the density of child-maiming debris, of infant death rates, have been nailed to the doors of our conscience. Those who have worked with him testify by their own research to his remarkable influence, whether it informs an analysis of municipal chicanery, or a combinatorial analysis of alternative spatial partitions.[9]

It seems to me that the real problem facing this particuar tradition of geographic enquiry is whether it can move on to enlarge the perspective set forth with such extraordinary skill and clarity over a century ago.[10] Even in the last twenty years, the bounds of biological knowledge have exploded outwards, but there is little evidence that these equally radical insights are being digested and incorporated into current thinking. There is the constant danger that a framework, which at first liberates human thought, later becomes transformed into dogma and myth. The word *grille* captures the dilemma nicely: it can stand for a framework in the sense of a conceptual and intellectual setting (grid), as well as the bars over a window that imprison and constrain (grill). As others have also noted, we must move on and enlarge the base of insight with the new knowledge we have today. Surely Marx would be the first to approve?

A third major strand of these twenty years has been the development of true cartographic research, genuine intellectual inquiry that has extended and enlarged our definition and our understanding of maps and the act of mapping. It has been the work of virtually one man—others simply do not come to mind. From his doctoral dissertation at Washington onward, Waldo Tobler has so extended the bounds of cartography that an innocent ignorance of his work now constitutes a constraint on the geographic imagination. That is to say, if a graduate student is not aware of certain pieces of Tobler's research, her own research abilities are jeopardized because she cannot gain a new and crucial perspective, and so evaluate its potential for her own area of interest.

Just two or three highlights from a very rich body of insight: from his inquiry into ancient map projections, and the extensions he has made to traditional cartographic transformations, Tobler's development of two-dimensional regression gives us residuals as vector fields, fields that can indicate the warpings of geographic space under various mental or physical influences. From his research on nonmetric transformations and trilateration in archaeology and botany has come his imaginative research on "winds of influence," cartographic statements and models that summarize and clarify enormous quantities of numerical information. His work on iterative methods of transformation are in the direct tradition of d'Arcy Thompson, and I sometimes wonder if he keeps that maroon, two-volume work by his bedside.[11] In one short research note, he told traditional cartographers exactly where they could put their choropleths (arbitrary values chosen for mapping data), and in doing so pointed up the triviality of the question upon which so much time, effort, and money had been expended. Such hubris on Tobler's part is difficult to forgive. Despite its imaginative comments on cognitive mapping, a recent book on cartography completely ignores the entire body of Tobler's work over the past twenty years.[12] A student reading that book would never know the man existed.

The final thread we must follow back is a mathematical one, for the use of mathematics as a "language," as a means of geographic expression, constitutes a marked departure from the writings of twenty years ago. Just look at the journals if you do not believe me. I think I am right when I say that none of us in those early years came to graduate work with any mathematical competence beyond a hazy recollection of some integral and differential calculus. What we know today, and speaking for myself it is not much, we taught ourselves; and when we became responsible for teaching others, we desperately tried to stay a few pages ahead of our students. The good ones, wallowing in that golden and unconstrained time of graduate education, when the only responsibility was the privilege of learning, overtook us very quickly and pulled us along in the vortex that they left behind. God knows we needed every bit of help we could get! A few, with true gifts and tenacity, went on to make original contributions to mathematics. The work of Michael Dacey, and a number of his former students, represents genuine extensions not only to geographic theory, but to mathematics itself. Andrew Cliff, another student of Dacey's, and Keith Ord have made a number of contributions to statistical theory that came straight out of such fundamental geographic problems as comparing spatial patterns and distributions on maps. Others have applied some of the deeper areas of mathematics to traditional geographic questions. Bernard Marchand, after coming across Haggett's *Locational Analysis*, sat down in the middle of Venezuela and taught himeself doctoral level topology from Dugundji, building straight up from a base of traditional French schoolboy mathematics.[13] Later on, I remember that when he took the three-term graduate course here at Penn State, the "gatekeeper" sequence for the doctoral program in mathematics, he found that the first two terms were only "useful review." The tradition of translating geographic observation and hypothesis from the verbal to the more manipulatable language of mathematics was also helped by the example of Michael Webber. His extraordinarily clear translations from words to symbols, his exploration of the consequences of stated assumptions, and his testing of the consequences with historical and contemporary data stand at the beginning of a tradition of geographic inquiry unknown a decade before.

And we were all lucky when Alan Wilson finally recognized his true calling and joined us as professor of geography and regional planning. His work in entropy maximization models has given us another way to look at the world, whether it is the description of large and complex urban regions, or the way a landscape represents a most likely state of a system developing under human and physical constraints. Blending his concern for theory with a socialist's passion for real data about the human condition in the old parts of England's cities, he, and those who work with him on the large research team, have operationalized areas of statistical mechanics for geographic understanding. Perhaps more than any body of work since Hägerstrand's simulations, Wilson's approaches have opened up the ideas of systems analysis. Geographers are no strangers today to these adjacent areas of cybernetics, automata theory, and general systems analysis, and they share the concern for the troubling questions of stability and equilibrium that arise in large and highly connected systems. The description of complexity is certainly not a new problem in geographic research, but with the prosthetic help of large computers newer languages of algebraic topology are now being applied to harden up some of the notoriously soft and ill-defined concepts and observations that must inevitably characterize the initial phases of any geographic inquiry.

Janus in a Geographical Setting

From the perspective of 1978, my own graduate days of the late 1950s seem a long time ago.[14] There is no question about it—we have made some considerable strides in the last two decades, and the field has been drastically transformed in many respects. New books and new journals making genuine demands upon the intellect are now abundant, and we can put them in the hands of others without apology. On the other hand, we must not exaggerate the progress we have made. Twenty years ago there was nowhere to go but up, and despite the real gains I am worried and anxious, for I have the uncomfortable feeling that some of the momentum has gone. Many intellectually capable people seem to be growing out of geography, although this has its positive side too. It means that geography may be a good base to grow from, rather than a permanent base to reside in. The growing *from* seems to be a growing *up*, in that we are connecting with the more general, more powerful disciplines, such as philosophy, operations research, general systems theory, and the higher forms of mathematics. These are all areas of very general inquiry that give us deeper insights into the fundamental structures of geographic and other problems, and they allow us to move with greater freedom, creativity, and imagination in the problem space.

But two urgent tasks face us, as they always will. First, we must extend the bounds and deepen our understanding: the obvious, crucial, and traditional tasks of the scholar. As always, such extensions will depend upon a few really creative, original, and imaginative people, with others just behind them ready to exploit the new holes punched in the walls of ignorance. We know so little about creativity that there does not seem to be too much we can do except nurture it wherever we find it, and increase the range and sensitivity of our own intellectual antennae to pick up the signals of new and imaginative insights. This is so obvious it hardly needs saying.

But the second task, that of teaching, is equally urgent, and it is here that my feelings of disquiet have their origin. So much geographic teaching still seems to be characterized by the factual compendiums that were rife in the earlier days, rather than the newer conceptual frameworks that allow more general insights. It is here that the contrast between American and British universities is most vivid: fifteen, even ten years ago we were ahead; the slow, but more flexible movement of shared responsibility for curriculum in this country had moved past the impacted forms of Britain, where the professorial power of one man could kill a department for decades on end. But get the old out, and the young in, and British departments can change with a speed impossible to match under the constraints faced by many departments on this side of the Atlantic River. Generally speaking, the undergraduate curricula at many British universities today are superior; that is they make greater demands upon the intellect and sensibilities, compared to most over here.

The causes are two: first, and external to the discipline, is the general financial problem, and the increasing allocation of budgets on the basis of student credit hours. What department in the university cannot afford to water down its big beginning courses, since this means bigger enrollments of bored students who are forced to fill slots in distribution requirement sheets? Such requirements, originally designed to broaden a student's education, have been perverted into rules that give every department a "cut" of the pie. If you think this is too harsh a judgment, just look at the passion created whenever a suggestion is made to change the rules, so that some department might increase its total—always, and by definition, at the expense of others. Or see what howls about academic freedom and integrity are involved when one suggests that mandatory limitations are observed on grade distributions in beginning courses—so that no department can acquire student credit hours by easy grading. Such pressures could not have come at a worse time for geography, for we were just beginning to upgrade the intellectual content in the 1960s before the new pressures began.

But the second cause is an internal one: the revision of undergraduate curricula has been agonizingly slow, generally neglected in light of the problems of graduate revisions, and equally traumatic where it has occurred at all. Let me tell a story against myself. In the late 1960s three of us wrote an undergraduate text that departed from the traditional introduction.[15] We wrote it for university sophomores, although it did not make nearly the demands of a junior level text in high school physics or chemistry, and we hoped it would "trickle down" to the freshman level in a few years. Abroad it has done quite well: apparently Swedish, Dutch, British, and other universities use it in some beginning courses, and it has been used as supplementary reading in the British sixth forms (juniors and seniors in high school). Over here it has hardly sold at all; we have been told that it is generally regarded as too difficult, and that at a number of places it has been used as a base for a graduate seminar. This is absurd. Today, a beginning course in meteorology starts from a base of two years of physics and a working knowledge of the elementary differential and integral calculus. I know, because my son, a senior in high school, is taking one now, but I wonder how many graduate students in geography could get through it—or how many faculty? Some simple, and now classic, texts in geography have had to be "revised" in light of declining standards, and the editors of most commercial publishers edit down in order to sell a text at all. Prerequisites on upper level courses are usually death, and even when a geography major is required to take basic university mathematics, the student complains, usually quite rightly, that he never uses it thereafter.

There still seems to be a reluctance and fear of using mathematics in undergraduate teaching, a feeling that somehow it is inappropriate, that undergraduates must be protected from the dangers of these abstract, manipulatable languages lest they be forever perverted. But as Bernard Marchand pointed out, these same attitudes were prevalent hundreds of years ago when printing first made its appearance. What will the lower classes do when they can read the vulgar tongue? The answer, of course, is not to suppress literacy, but make it universal. We face the same sort of problem today. As a result, few undergraduate majors ever face demanding intellectual experiences. Lots of busywork, lots of coloring in the maps, yes: but genuinely difficult work, when thinking is really stretched, no. Ask the graduate students today, anonymously if necessary, if they were ever really challenged by an undergraduate course in geography, and they will tell you.

So we have a long way to go, and there can be no room for complacency. Yet let us also recognize the progress that has been made. Things are very different today, for we have solid and growing bodies of insight, tools, concepts, and methodology from which to construct our teaching programs. This was not true before. Despite a recent, and to my mind ridiculous, review of a popular exposition, it is a whole new ballgame.[16] It is no good pulling a sentence or two out of context, and giving it a new twist in light of contemporary developments to claim that there is nothing new under the sun and "we were doing it all along." The trouble is that they were not doing it all along, and that is precisely what the years 1957–1977 were all about.

A Critique of Dissipative Structures in the Human Realm

> Turning and turning in the widening gyre
> The falcon cannot hear the falconer;
> Things fall apart, the center cannot hold;
> —William Butler Yeats
> "The Second Coming"

Critique and Criticism

In each formal field of inquiry, within which we dwell professionally as beings who seek to illuminate some chosen portion of our physical, biological, or human realm, we attach particular meanings to words. Within the close and familiar world shared with our co-workers and co-thinkers, such meanings are precise, familiar and useful. But language is slippery and ambiguous: outside of a particular discipline or line of research, other, unintended meanings may be placed upon the same words, and may create confusion, which sometimes is only a short step from resentment. Since resentment is not a mood conducive to clear and open thinking, clarifying the intended meaning of words becomes essential. Otherwise we shall not be able to stand open so readily to new possibilities as they are granted to us.

It is important, therefore, to clarify the word *critique*, which, in philosophical discourse, is wholly separate and distinct from *criticism*. The latter holds out to us a harsh connotation, a finger pointing to error. In marked contrast, *critique* holds no such connotation, and at its best it is saturated with caring and concern—a caring and concern to lay out fairly and appropriately claims that have been made, and to focus on the conditions of the possibility of those claims. To be sure, one does not engage in the difficult and troublesome task

I have modified this somewhat from the original to make it a bit more accessible to the nonspecialist. The original with the same title appeared in the *European Journal of Operational Research* 30 (1987): 211–221. Reprinted with permission from Elsevier Science.

of critique if no sense of doubt or disquiet has passed over the area of concern. Doubt is like the falcon's shadow that momentarily comes between the ground and the illuminating sun. Sometimes it passes on, leaving no disturbance behind, and the sun continues to shine and illuminate as before. But occasionally something catches the attention of the falcon of doubt, and then it hovers, and finally stoops to examine more closely that upon which the shadow initially fell. To engage in critique is to doubt for the sake of the truth, a phrase familiar to an apostle, and yet a maxim embedded at the heart of the best tradition of scientific inquiry. Perhaps it is one worth reflecting upon in this age of terrible certainty. It is wholly within this spirit that the following reflections are offered.

These reflections may be roughly characterized as lying in two domains of concern: the mathematical, in both its pure and applied aspects, and the philosophical. The former is relatively straightforward, if that degree of total certainty which arises from pure analytical solutions is not the fundamental goal of inquiry. For those who are not familiar with some basic aspects of mathematics, such a statement may require some explication, starting with the idea of an *algebra,* which is nothing more than a well-defined set of "things" (numbers of various sorts, abstract symbols, etc.), together with some equally well-defined operations that "connect" the things in logical and consistent ways. Formally, the operations are called "binary operations" precisely because they can operate on two things at a time, like $2+2 = 4$, or $x-y = z$. The trick in creating an algebra lies in that adjective "well-defined," qualifying both the set of things and the operations you are allowed to do on them.

Take the familiar algebraic structure of our everyday arithmetic on the set of the positive integers, 1, 2, 3, . . . What binary operations are well defined? Certainly addition (+), because you can take *any* two positive integers, perform the binary operation of addition, and get another positive integer lying within the set. Obviously, the operation of multiplication (×) is equally well defined for the same reason. But, and equally obviously, the operation of subtraction (−) is not, because $4-4 = 0$, which is not a positive integer, and $3-7 = -4$, which is also not in our set. Children, with perfect logical consistency, say that "minus 4 apples is just silly." So we have to enlarge our notion of number, to the set of all integers plus zero, until we want to divide (÷), at which point some ratios, say $-8/4$ or 88263308/1123, are still in the set, but others, like 88263308/1122 or 22/7 are not.

And so we have to enlarge our notion of what is a number once again. In one, very limited sense, much of the history of mathematics up to about 1850 is contained in this "expansionist tale" about enlarging the very *idea* of what counts as a number. Some numbers, like $\sqrt{2}$, are algebraic; others, like π and e, are transcendental;[1] both sorts "go on forever" if you try to compute them. And here is the rub: as long as you stick to those "pure analytical solutions" I noted previously, you can manipulate these symbols any way you like with the binary operations allowed in your algebra. But as soon as you start using your pure algebraic structure as a possible description of the way other "things" (atoms, amino acids, heavenly bodies, human bodies, etc.) are connected by such binary operations, then these sorts of going-on-forever numbers cannot "go on forever," but have to stop somewhere, even if it is the sixty-fourth decimal place of a computer. And as soon as you start to compute with the binary arithmetic of the computer, rounding error is going to creep up those decimal places until it overwhelms any useful results we might be trying to compute. Our finite abilities to measure prevent us from assuming a vatic role, and so becoming gods. It also means that any computational mathematics used to describe concrete things, rather than the purely abstract fantasies posited by mathematicians, needs only the integers, for 7.215689 kilometers can always be scaled to 7,215,689 millimeters. So that to the degree that numerical and computational approximations can be accepted (as, indeed, they are often, and quite properly, accepted in many applied areas), then the task becomes one of describing carefully what takes place in that shift from analytical to computational procedures. If the descriptions are faithful, these are unlikely to be controversial. However, in the philosophical domain, critique begins to focus upon the conditions of possibility that give rise to particular "ways of looking," and so, inevitably, raises questions about the appropriateness of particular frameworks, especially when these are characterized by certain mathematical structures. It is here that controversy may arise, but to the degree that critique is undertaken in a mode of concern for the truth, and results in clearer statements of scientific intention in the future, it may have a clarifying, and so beneficial, effect.

Over the past fifteen years, we have seen an old and familiar mode of mathematical thinking illuminate a realm of physical chemistry. From this starting point, it spread first to the world of living things, and later to the human realm. Since the mode is mathematical, it shapes our seeing by requiring the human realm to conform to the mathematical structures within which the descriptions are cast. It seems appropriate, therefore, to lay out, as fairly and as truthfully as possible, what *actually* and *operationally* happens when mathematical thinking about dissipative structures is transferred to the human realm and used there as a framework upon which certain aspects of the human world are projected. Such an attempt constitutes a phenomenological description of what actually has been achieved in quite concrete, operational terms in a limited domain of application. Within the context of such a description, it will then be possible to examine a concept and a claim that are carried over as an integral part of such a framework that was initially employed to describe the physical world of things.

Delimiting the Domain of Concern

The domain of concern is deliberately restrictive: I disqualify myself immediately from comment on the appropriateness of such mathematical structures as descriptions within the physical and biological realms. It is clearly for chemists and biologists to judge the illuminating power of this framework for their particular areas of concern. Nor, quite obviously, is the mathematical framework itself in doubt. Nonlinear concern reaches back to Apollonius, and even to the Pythagoreans;[2] and the *way of seeing* that explicitly brought to the fore the possibility of the one-to-many mapping is grounded in the highly original work of the Italian algebraic geometers of the late nineteenth century, and possibly reaches back to Riemann. This is an important but rather technical point that may be difficult to get across to those unfamiliar with basic mathematics, while I suspect any simplification will make mathematicians uncomfortable.

Most people are familiar with functions ($Y = a + bX$, $F = ma$, $e = mc^2$, Fat = No Exercise + Too Much Food, etc.), in which you compute something on the right-hand side of an equation, and get a single number on the left. But in the nineteenth century, German and Italian mathematicians began to enlarge the idea of a function to something called a *mapping*, a process somewhat analogous to the idea of number itself being enlarged when it became too restrictive. In these new terms, a function is now a one-to-one or many-to-one mapping, because whatever you combine on the right-hand side is transformed to a single value on the left. People like Riemann and Levi-Civitas literally created new conditions of possibility for thinking in mathematical ways, because now one-to-many, or many-to-many mappings can send things on the right-hand side to more than one thing on the left. What this means is that sometimes you can compute "on the right-hand side," the way a function "will go," perhaps by plotting the corresponding left-hand values on a piece of graph paper, but at some, highly critical points, your *computation*, with only a smidgen of rounding error, could make your function branch, sending it one way or another, rather than to a single value. It is these *bifurcation points* that cause problems in any real, as opposed to abstract, situation and application.

However, when the mathematical descriptions of dissipative structures make that sideways shift into the human realm, particularly that region of inquiry called human geography, then it is appropriate that we reflect carefully upon the power of illumination that they give us. The basic outline of the extensive research in this area is well known at both the *inter*urban, and *intra*urban scales.[3] The former would be labeled by the geographer as the "dynamics of central place theory," the latter as the "geography of urban growth and change." In both cases, a set of simple, straightforward nested equations, describing the balance of positive and negative feedback effects encountered in a birth-death process, operate upon or within a lattice structure, allowing numerical values at those points to change, so generating what we might call a "spatial history" of our abstract, lattice-based region.

The outputs, consisting of a series of maps in "generational time," are enormously seductive in any intellectual sense. No geographer can resist telling a story about such a sequence, bringing that hermeneutic perspective to bear upon the spatial configurations created out of technical concern, impressing upon the graphic and text-like sequence of numbers a human meaning. As a simulated spatiotemporal "geohistory" unfolds, we see a town suddenly take hold, and begin to structure a region; and we may recall the way in which, all over America, the initial advantage of a town during pioneer settlement days appeared to confer upon it what we might call the tenacity of historical inertia. Once a town begins to get a grip on a region, drawing large amounts of capital investment to it, it appears to create its own momentum that carries it through time, and up and down the roller coaster of economic boom and bust. Towns, like old soldiers, never die; like the mill towns of New England, they only fade away.

Later on, a burgeoning, locally space-organizing town may produce repulsive forces at its center, so spilling over into surrounding dormitory towns, nearby points of the lattice that receive the people from the center to produce that collapsed cone of the urban volcano that reminds us of bleak inner cities and deserted downtowns at night—New York, Baltimore, Philadelphia, Detroit, and so on. Far away, where the distance decay effect sufficiently weakens the urban field, another town begins to command, organize and structure another part of the region. But now, perhaps, it has a competitor, and we observe the push and shove until one emerges victorious. Locally, in my part of Pennsylvania, I think of the old county seat of Bellefonte gradually succumbing to the upstart State College, where, after decades of millions of dollars pouring into the Penn State University, the rival overcomes the older center, turning it into a dormitory village for itself. Anyone can easily fill in her own example of a local Rome declining and falling to a new, not to say nouveau, rival.

All of these stories that we are capable of telling about the map sequences, generated out of those birth-death, positive-negative feedback equations, are familiar to us, but notice how our interpretations are saturated with the conceptual images of late seventeenth- and early eighteenth-century physics: the "inertia" of a town, the "momentum" it generates, the production of repulsive "forces," urban "fields" whose strengths decline with distance. Newton, too, never dies: indeed, there is little evidence that he is even fading away. And how could it be otherwise? If we choose the mathematical structures devised to describe the physical world, as it was thought in careful and reverent reflection in the late seventeenth century to describe the human world of today, then that human world can only look like, and be interpreted in terms of, a piece of celestial mechanics, in which momentum, inertia, forces, fields, and trajectories are an integral part of the closed, tautological system of definitions that we call today (following Eddington) "classical physics." That initial choice, to take the mathematical structures, grounded in binary operations on the real numbers, as our framework or "language" to describe the human world does not allow us any other choice. In projecting the human world onto a mathematical framework devised three hundred years ago to describe faithfully the physical world of things, the human world must appear as, cannot appear as anything else but, a piece of seventeenth-century celestial mechanics.

The Newtonian Trap

The problem is that where we have no other choice but to describe and interpret within a framework constructed with a concern for the truth of Newton's physical world, we must also face the fact that we—our thinking that is us—is trapped. To be trapped is to have no choice, and to have no choice is to admit that we no longer have any freedom to inquire, to stand open to other possibilities and ways of seeing. Even when the condition of seeing changes, and brings out of concealment the one-to-many mapping as an enlargement of possibility, we are still trapped within the somewhat more capacious, but still essentially mathematical, framework that allows thinking to move in certain directions, but not others, for the task of human description. Once again, it is of crucial importance to emphasize that this is a point of critique, not criticism. *Any* framework upon which the human (or physical or biological) world is projected has the capacity to trap thinking, unless we are constantly at a very high state of awareness, and are prepared to stand open to other possibilities even as we go about our daily research tasks, whether in or out of the scientific mode of thinking. This point of trapping also applies to more readily recognizable "ideological" positions, such as a Marxist viewpoint that brings certain variables and relations into the realm of concern and projects them onto a particular set of expectations, but not others.

Now it may be contended that this is not really a faithful account of the use of dissipative structures as descriptive frameworks in the human realm. It is true that we write nested sets of differential equations, but in concrete, practical, and operational terms we evaluate these in finite increments as difference equations, and at the end of each "round" of scanning and recalculating the values at our lattice points, we end up doing something that Newton would have found intolerable on theological grounds. For let us never forget that Newton thought and computed for the glory of God. To Newton was granted the awesome privilege of announcing that which God, in his omnipotence, and in his perfection, had created. Moreover, God, in his omnipotence, knew that which was to come, and had granted the insight that from the description of conditions now, that which is to come is determined by that which is now. What was awesome, in the very literal sense of the word, was that given the mathematical definitions, then this wholly mathematical structure described with extraordinary fidelity the deterministic world that was God's perfect handiwork.

And yet, in the area of thinking that we call dissipative structures, what do we do operationally to such Newtonian perfection? Into this perfection we deliberately and knowingly inject handfuls of ignorance, calling these handfuls of chaos "stochastic perturbations," while philosophers of a certain propensity announce, like a Greek chorus, that God is dead. This, surely, must appear as an extraordinary state of affairs, and one to be examined carefully.

Bifurcations in the Human Realm

We start by considering a typical set of nested equations that constitute the transformational rule that sends the integer numbers at lattice points in the regional network to new values at the end of each round of computation or "time period." The system so described is nonlinear, and so contains within it the possibility of bifurcation. As an aid to thinking, we shall represent the successive states of such a system as points in a much reduced "state space" (fig. 8), in which the axes x_i and x_j in the diagram stand for the set of values $X = \{x_1, \ldots, x_n\}$ characterizing each of the n points in the regional lattice, while the axis λ stands for a set of parameters $\Lambda = \{\lambda_1, \ldots, \lambda_k\}$ characterizing those in the system of equations capable of reaching numerical values specifying bifurcation points.[4] Thus, any actual state space will have $n + k$ dimensions, so that initial conditions will be specified as a point, say t_0, whose coordinates are some combination of the Cartesian product of the union of the sets X and Λ. It is important to note in passing that traditional analytic concern for the behavior of such many-valued functions (containing one-to-many mapping possibilities), was grounded on the continuum of the real numbers. However, conceptually, as well as in practical and concrete computational terms, the numerical values of the regional lattice points (the $x_i \in X$) are constrained to the integers, while the parameter values (the $\lambda_i \in \Lambda$) are constrained to the rationals. Thus, the state space is anisotropic and discrete, and so forms a mul-

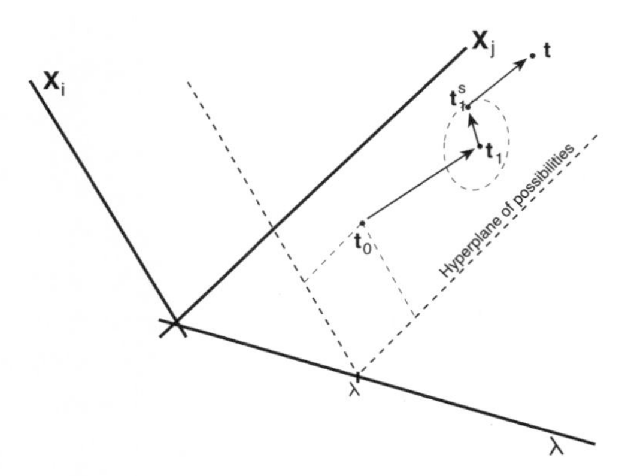

Fig. 8. The hyperplane (lattice) of possibilities defined by a fixed set of parameters (λ), within which a system trajectory is represented by a sequence of ordered points generated by both the deterministic structural equations and the stochastic shock on the regional points.

tidimensional lattice upon which any particular trajectory of the system will constitute a series of discrete points. This also means, within the conceptual and computational context of practical, rather than analytic, concern, that bifurcation points can exist only at parameter values in the set of rationals.

In investigating the behavior of such a system, four alternative ways of proceeding appear as possibilities. The first provides for stochastic shock to be applied to the numerical values at each point in the region after each round of computation. A typical run consists of specifying the initial conditions as some combination of regional and parameter values to locate t_0 (fig. 8). Notice that if the parameter values (λ) are constant, and are not recalculated endogenously within the model (as they are not), and also are not stochastically shocked exogenously by a pseudorandom number generator (as they are not in this first alternative), then they define a hyperplane within the state space, a subspace containing all the permissible states to which the regional system is confined. If such an a priori specification of the parameters does not happen to include a value at which a bifurcation point exists (and the probability of hitting such a value, while not infinitely small, is close to the impossible in the finite context in which actual research takes place), then, tautologically, bifurcation will not occur.

Consider a typical trajectory under such conditions (fig. 8). The system starts at t_0, and the noncontinuous jump to t_1 is computed from the wholly deterministic structural equations. At this point the "clock" is stopped, and the stochastic shock is applied to the regional points $\{x_1, \ldots, x_n\}$. This allows the system state to be dislocated to a new point, say $t^s{}_1$, lying on the hyperplane, since the parameter values are held constant. The process repeats, sending the system on a trajectory that is a result of a convolution of the deterministic structural equations and the probability distribution from which the stochastic inputs were generated. In essence, any actual trajectory is determined in an elementary and classical fashion by the structural equations, the initial conditions, and perturbations that constitute small "random walks." Nothing that can be observed of a single trajectory, or even a large set of trajectories, provides evidence that bifurcation has taken place, or that it is possible. Initial values of the regional points (the x's) may be changed, but neither marked changes in a trajectory from previous runs nor sudden alterations in trajectory direction provide the slightest indication of bifurcation so long as the parameter values remain constant.

The second alternative consists of applying stochastic shock to the parameter values (the λ's) at the end of each round of computation. As before, the choice of *x*'s and λ's defines the hyperplane of initial possibilities and the starting state t_0 (fig. 9), and a straightforward deterministic trajectory for t_1 is computed from the structural equations. However, at t_1 stochastic shock is now applied to one or more of the parameters, and this discontinuously jumps the system state t_1 parallel to the λ axes, and out of the subspace initially defined, into another subspace of possibilities containing $t^s{}_1$. By definition, the new subspace will contain states not contained within, and not reachable from, the first. Again, by definition, and so tautologically, any trajectory generated by successively shocking the parameters must appear different from those generated by constant parametric values, since new possibilities (new state space points) appear and disappear at the end of each round of computation, whether the stochastic shock sending λ_0 to λ_1 has forced a parameter value to pass through a bifurcation point or not. Under these conditions, where the hyperplane subspace is doing a random walk along the parameter axes, no system trajectory, or even a set of trajectories, can provide us with certain knowledge that bifurcation has occurred, even though trajectories generated from successive runs may depart markedly from one another.

Of course, certainty, which can come only from analytical, logically closed and tautological solutions to the structural equations, is really an unfair criterion, and not to be expected in the practical world of computational investigation. Such assurance can come only from extremely careful and patient investigation of system trajectories characterized by parameter values at, or

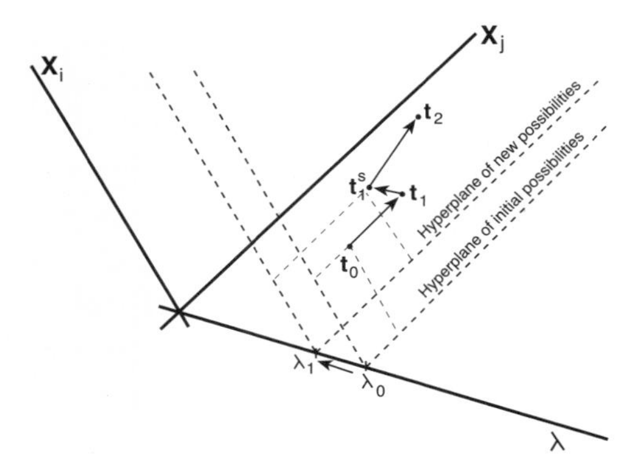

Fig. 9. The hyperplane of initial possibilities (λ_0) containing the deterministically generated trajectory, and the hyperplane of new possibilities (λ_1) to which the system is dislocated by stochastic shock to the parameter values.

near, suspected bifurcation points, and then following up such experimental investigations by equally careful examinations of system convergence and divergence over a sufficiently long sequence to marshal evidence that bifurcation may, in fact, have taken place. No statements about such experiments appear to date in the literature, and it must be presumed that they are still possibilities for future research.[5]

The third alternative is to change parameter values deliberately, either to specify new initial conditions or to intervene and alter such values during the course of a model run. Clearly, a change in trajectory will be observed with new initial parameters, or with parameters deliberately changed during a run, but with or without stochastic shock (the fourth alternative or variation on the basic themes), trajectory change, in and of itself, cannot provide evidence for bifurcation. For example, by changing a parameter β in the model (interpreted as the "friction of distance," or the average effect of distance on human behavior), either globally for the entire system, or locally (for example, when some points or towns on the regional lattice are connected by a new road), the regional system will certainly veer from its prior trajectory and reach a new succession of system states. But some such change will appear whether the change from β_1 to β_2 has passed through a bifurcation value or not. A simple presentation of the new trajectory and system state as a map configuration cannot constitute evidence of bifurcation, since *any* change in a parameter value will always and immediately produce change (sometimes quite marked change), that may still allow the system ultimately to converge back to its governing steady state or cycle.

Bifurcation is a concept traditionally well-defined over the set of real numbers. Analytical solutions provide mathematical certainty of the existence and numerical values of bifurcation points. In contrast, always discrete numerical approaches require extremely careful a priori examination of possible bifurcation values, or considerable, and carefully accumulated, experimental evidence that bifurcations can exist at values that are believably characteristic of a particular physical, biological, or human system. This latter point is of crucial importance as thinking moves sideways out of the purely mathematical realm towards those substantive areas that are projected onto a particular mathematical framework. *Mathematically*, parameter values characterizing bifurcation points may range over whatever set of numbers is considered appropriate, and these, presumably, are infinite. Certainly this is what is implied by diagrams indicating successive bifurcation trees at ever-higher parameter values. However, *conceptually*, within the concrete empirical domain of an actual system, such values may be meaningless, since no actual system could achieve them. For example, bifurcation may occur in our interregional model at $\beta = 17.2$, and do so in a mathematically impeccable fashion, but a friction-of-distance effect of this magnitude on human behavior would constitute something close to total geographical paralysis in any society we can conceive. People would die many times over from thirst before they could get to the well located on such a viscous surface. Moreover, the existence of such an *interpretive* problem would be maintained in full acknowledgment of all the difficulties raised over the past ten years, problems of empirical estimation that stem from the realization that map pattern appears to be inseparably convoluted with distance in any empirical estimates of β.[6]

Given the great empirical and conceptual difficulties associated with determining bifurcation points, it might be appropriate to approach this impeccably grounded mathematical concept with great care in the human realm, and invoke such a concept with even greater thoughtfulness. The condition of the possibility of bifurcation in human systems always comes from the a priori methodological decision to employ mathematical structures allowing a one-to-many mapping. If we choose such a mathematical structure as our framework upon which to project the regional system that is, say, Ghana, Portugal, or Iceland, we shall see the possibility of bifurcations—just as in an older social physics we saw the concepts of momentum, inertia, and force of celestial mechanics. But what we observe, as social scientists, is *a* trajectory of Ghana, Portugal, or Iceland, all unique, nonrepeatable history-trajectories, and our observations, rather than the mathematical structures we have chosen and imposed, do not provide any evidence of bifurcation—nor can they. Bifurcation lies wholly as a possibility within the methodology we impose, and not in the spatial histories of the human world we are trying to illuminate. In our model, the regional system is simply jumping along a series of points in a discrete state space, partly as a result of the deterministic equations, partly as a result of the wholly unknowable stochastic noise injected at regular intervals to make that which is determined and knowable *un*predictable and *un*knowable. The *form* of such "quantum thinking" from the realm of physical chemistry has been observed, but the content of repeatable events, upon which such quantum mechanical thinking rests, has evaporated. It might be argued that if a regional model could be calibrated for Portugal, then its parameter values might generate the history of Ghana and Iceland. To establish the plausibility of such a claim requires that we believe that the history of Portugal has been governed by a single set of structural equations since time immemorial. Others have noted that real histories may be characterized by actual structural *change*, not simply parametric shift and bifurcation,[7] but such thinking requires a less constraining, essentially topological, view to be brought to bear.

The Meaningfulness of Statistical Description

The deliberate attempt to make unknowable that which we might know from the very process of mathematical description, generated by the set of nested structural equations, has been achieved by making computationally operational a concept of amplitude fluctuation considered illuminating in the realm of physical chemistry. But such an operational and conceptual requirement also goes a long way to destroying the very ground of mathematization and mathematical description. Literally the raison d'être, the reason for being mathematically described, to illuminate in the only way mathematical description can illuminate—by foreseeing—is destroyed. This applies whether the purpose is to predict a numerical value, or whether it is to acquire a qualitative feel for the always *future* behavior of the system. Both are undertaken mathematically to achieve some foreseeing, some knowledge of what will arrive out of the future. This is a strong claim, and it does not rest upon mathematics itself. Rather, the problem is essentially philosophical. How might it be sustained and made plausible in the realm of human systems?

We must recall the meaning of mathematics itself, and let us do so with a thinker who dwelt upon the meaning of words almost as a daily habit. Martin Heidegger was no stranger to mathematics or to science.[8] As a scholar who had written treatises on logic and served on doctoral committees of mathematics at Freiburg, Heidegger was acknowledged by Heisenberg himself to be among his two or three peers in the opening realm of quantum physics. Thus, Heidegger's thinking on science was not that of the amateur standing idly on the sidelines, but of a deeply engaged thinker thinking towards the meaning of a world increasingly saturated and shaped by science. At the very least, we should listen carefully to him in this realm, and here especially when he writes: "*Ta mathemata* means for the Greeks that which man knows in advance in his observation of whatever is and in his intercourse with things."[9] Newton, and all physical scientists up to our own century, would understand and acknowledge his primordial meaning of mathematics, and so, I suspect, would all those in NASA today who are charged with steering a spacecraft to Jupiter and beyond. The whole point, the fundamental underlying reason for mathematization, for describing something in mathematical terms, is to know in advance of our observation of whatever is in our intercourse with things.[10] Otherwise there is no point in casting a description into the mathematical realm. If by so doing we cannot illuminate by telling in advance, there is no reason for attempting such a description in the first place.

But in projecting the human world upon the nonlinear mathematical structures characteristic of dissipative systems, we deliberately destroy the foundational reason for such mathematical description by injecting into the deterministic order of the difference equations such chaos that we acknowledge that *nothing* can be known in advance of our observations of whatever is. In the words of *Order Out of Chaos*, "We cannot predict the details of temporal evolution."[11] We start in the tradition of foretelling, and then deliberately deny ourselves the possibility of foretelling.

But a further claim is made: ". . . only a statistical description is possible."[12] Suppose, for reasons of emphasis and stark contrast, that a counterclaim were made in this human realm into which we have analogically shifted: a statistical description is probably not possible in anything approaching a concrete, operational sense, and even if achieved by massive amounts of computation it could not be interpreted meaningfully, in the sense that such a description could illuminate an actual and concrete instance of regional dynamics and development in our world.

How can such a counterclaim be supported? Our thought experiment proceeds as follows: we start our regional lattice in a state that is any one of a huge number of combinatorial possibilities, possibilities whose only bounds are delimited by the believability that an actual region could achieve such a starting configuration. We then compute the regional lattice values $\{x_1, \ldots, x_n\}$ according to the structural equations, and at the end of each round perturbate the parameter values with stochastic noise. As we have seen, this moves our regional system onto a hyperplane of new possibilities, although we have, by definition, no knowledge of what these might be. The location of the hyperplane of new possibilities (used for the next deterministically computed point in the state space), is influenced solely by the type and bounding of whatever distribution we happen to draw our stochastic noise from, but in the human realm we have nothing to inform our choice about what this might be, and it is difficult to see how such knowledge might be gained.

Let us suppose that after an arbitrarily chosen number of computational generations we stop and store the spatial, or geographic, configuration of our regional lattice. We return to the beginning, and whether we choose the same initial conditions, or a new set from the combinatorially huge number of believable possibilities, we run our program again and again. Suppose we end up with one million configurations each as unknowable and as unpredictable (by definition) as the other. Any statistical description that is claimed can be achieved only by summarizing in some verbal, graphical, or algebraic way the distribution of maps as three-dimensional arrays. It is conceivable that "spatial modalities" will be discernible, and these might indicate that bifurcation has occurred in the system of transformational rules, although we could not necessarily expect sharp modalities to inform us, particularly if initial conditions changed between runs. Indeed, we must be extremely cautious here in carrying over conventional, and aspatial "histogram" thinking into the geographic realm. The task of statistical description does not simply involve summarizing single values as a frequency distribution, but two-dimensional arrays of n-point lattices whose numerical values are not independent "events" or outcomes, but highly dependent upon adjacent values. Each lattice point of the regional system has a frequency distribution associated with it, but we have no interpretive experience to guide us here, even though the Monte Carlo simulation of map sequences has a respectably long history in geography.[13] Moreover, we must double our caution for two additional reasons: first, the *spatial* distribution on the lattice of the frequency distributions of final outcomes will be strongly affected by the boundary conditions—boundary conditions in the literal geographic sense of the shape of the region. Secondly, any statistical description must reflect the characteristics of the distribution used to generate the random unpredictability we injected to destroy the ground of *ta mathemata* before we ever undertook such an exercise.

To this original counterclaim may be added a corollary. However we choose to express the statistical distribution of outcomes, it remains essentially meaningless in this context, because its characterizations possess not the slightest chance of illuminating anything actual and concrete about a real regional system. The claim that a statistical description has the power to illuminate arises in the quantum world of physical things where events have the appearance of repetition, and where an inequality involving Planck's constant allows a narrowing of location while momentum goes to infinity—or vice versa. I do not think that there is anything even vaguely and analogically equivalent in the human realm. Nor can there be. Once again, it should be noted that this is a philosophical claim, not a mathematical one.

In brief, can any statistical distribution reflecting either the spread of outcomes from a combinatorial selection of initial conditions acting through wholly deterministic structures, or the convolution of the moments of the distribution from which our stochastic ignorance was drawn, illuminate any concrete, actual, and nonrepetitive instance of our human world—a Ghana, a Portugal, or an Iceland? Claims may be made in the social sciences that revolutions are evidence of bifurcations in human systems, but what possible historical evidence could ever be marshaled for such a claim? Is John Locke merely a local perturbation that grows to affect global trajectories? Is the plural form *trajectories* even appropriate here? Or does it arise in the realm of dissipative thinking that once again, as a modern variant of an old social physics, traps, rather than releases, our thinking to describe faithfully the human realm?

To believe that human systems are simply describable into the future as statistical smears generated from systems of nonlinear, and, therefore, bifurcating "functions" (one-to-many mappings), constitutes a belief that human change is predicated only upon the possibilities of bifurcation. In brief, by moving from the condition of possibility predicated by systems in which bifurcation is not thinkable, to the somewhat enlarged condition of possibility in which the one-to-many bifurcation is possible, we feel that the problem of faithfully describing the human world is thereby resolved. But what we find ourselves in is not the condition to describe human systems as they are, even with the newly seen mathematical freedom to compute possible new directions, but simply a new trap, a new mechanism, which allows only those courses or trajectories predicated by certain parameter values. It is precisely here that we see the grave problems that arise when we shift thinking analogically from the nonsentient, nonconscious world of mechanical things, to the conscious realm of self-reflective capacity of the human world. Having released ourselves from the constrained Newtonian trap of functional thinking, we find ourselves only in a larger and more capacious trap.

The human world is not bound by the conditions of possibility engendered by the strict algebraic requirements of parametric change in one-to-many transformational rules. Otherwise, that which is primordially constitutive of humankind, namely, that "its own potentiality-for-Being is an issue,"[14] is no longer something opening freely into the future, but a mechanical supposition bound by parametric change, cast in an inappropriate mathematics that, no matter what bifurcating possibilities it can generate, always remains a framework upon which the human realm is projected. When we accept such a projection of the human world upon a constrained mathematical framework arising from the descriptive, and historically contingent, requirements in the world of physical things, the true requirements for a description of the human world are pushed into concealment.

A concept and a claim constitutive of dissipative structures, the framework of seeing they represent, have been shown to be practically and operationally difficult to interpret meaningfully in the human realm. This is not a cause for rejoicing. All thoughtful attempts to illuminate an aspect of our human world command our respect. When they raise in their turn grave interpretive difficulties, when the claims made for them raise questions of the appropriateness of such claims in the new and radically different realm to which they have been shifted, we are forced to reflect, if we genuinely care, upon such deconstruction.

Thus, and in contingent conclusion, allow me to point to two possibilities. First, I believe the time will come when thoughtful people in the human sciences will look back upon these claims for dissipative structures and see, at this historical moment, one more indication of the immense difficulty of mathematizing in the human world. The difficulty arises because it is a world *defined* by beings with the capacity to reflect upon, and so contradict, any mathematical description made of them. And, it must be added, this possibility of denial exists at all levels of aggregation. Societies, as highly complex aggregations of individuals, also have the capacity to undo any mathematical description we may care to make for human, as opposed to merely physical, reasons.

Secondly, we may have one more indication of the absolute necessity of moving away from those mathematical structures devised to describe the physical world of things, toward new forms of mathematical structures that arise out of the wholly different descriptive requirements generated by the human world. But this must be a mathematics where the tyranny of the binary operation, that algebraic command that perhaps is the kernel of those statements we term *laws*,[15] is weakened to produce more appropriate structures that allow, forbid, but do not require; structures that seem to characterize the human world in which we live and have our Being. Indeed, if doubts about concepts and claims in the human realm are sustained *in any degree*, perhaps these can then form the condition of possibility for a return to the physical realm from which such thinking emerged, a realm where we may examine such claims once more for their truthfulness.

Sharing a Tradition
Geographies from the Enlightenment

> Vigilance and self-criticism are crucial, since there is an inherent danger to oppositional effort becoming institutionalized, marginality turning into separatism, and resistance hardening into dogma. . . . there is always a need to keep community before coercion, criticism before solidarity, and vigilance ahead of assent.
>
> —Edward Said
> *Culture and Imperialism*

Who Is in Disarray?

I WOULD LIKE to touch on a number of questions and confront you with them, in the sense of placing them in front of you for your reflection and consideration, rather than being *confrontational* in any hard or aggressive sense. Of course, in the process, it is not impossible that I may tweak a few noses, but it is not my fault if there are some Pinocchios around. In any case, I want to direct most of my remarks to students—of all ages—because those who are gurus already know the Truth.

In my forty years as a student of geography, I have never seen our field so intellectually vibrant. I envy those entering the discipline of geography toward the end of the twentieth century, providing they can bear the frustration of never having time to learn, read, and think about all the colorful threads that weave together and form the shimmering fabric of geography today. Advice may appear gratuitous, but if I could offer one piece it would be: resist spe-

---

Originally given as the Wiley Lecture to the Canadian Association of Geographers in Ottawa, June 1993. The published essay was dedicated to Dr. Sylvie Adam of Rouen, who helped me make my French in the original address understandable, while being sensitive to my wish not to appear as an educated French geographer who would have phrased many points more elegantly. Dr. Adam's enthusiasm for geography was unbounded. She died, after a long and unsuccessful operation, December 11, 1993.

cialization, despite all the pressures. Try to leave a corner of your professional life open for new perspectives and new influences, and remember that Young Turks can become Old Establishment with a speed that is distressing. Follow your own interests, read eclectically, and your head off: nothing is ever wasted, and you cannot make connections if you have nothing to connect.[1]

In brief, you can see that I do not share the opinions of such diverse figures as David Stoddart, Ron Johnston, Bob Bennett, Ron Abler, David Harvey, Mel Marcus, Michael Dear, and others who see and fear fragmentation everywhere.[2] I find the idea of fragmentation absurd, and too frequently raised by people who long for others to conform to their monolithic ideological positions.

Let me take one, specific example: suppose there is a Specialty Group of only fifty members in the Association of American Geographers interested in Native Americans, an example cited by Johnston. Despite some contrary opinion, this is no evidence of fragmentation, no evidence that they are disconnecting from the rich body of ideas that constitute the geographic "way of looking" today. For their research, they may be using remotely sensed images, geographic information systems, structural-Marxist concepts, hydrological science, socially constructed ideas of the environment, Michel Foucault and the power of surveillance, deconstructions of official government reports, goal programming, a heightened awareness of gender issues—I mean, you name it!—to illuminate the spatiotemporal human condition at a *place*, say northern New Mexico, embedded in a larger regional and national *space*. With all that conceptual and methodological integration, where, literally on earth, is the fragmentation? The important thing is an ability to *illuminate*, a word of crucial importance for me, because it means an ability to bring something out of the darkness and into the light of human understanding.

And there are no privileged positions here, no ideological stances capable of sustaining exclusive claims, no *meta-hodos-logos*, no path-to-knowledge, no methodology, that constitutes an exclusive choice. All choices constitute conditions of possibility to illuminate, to bring out of concealment, but every concrete choice made in the course of a particular investigation may eliminate others, and so close off possibilities for seeing. Which is why I said "don't specialize," why I said "read eclectically." The more choices you open up, for yourself and for others, the more numerous the conditions of possibility to make choices to illuminate. We all know of geographers who have made exclusive choices—of ideology, of methodology, of style, of subject matter—who have defended their sand castles while the seas of geographic inquiry moved on and over them. Sometimes we can still hear their faint theoretical cries above the waves. I hope they are learning how to swim, because even life belts, stenciled *The Good Ship Marx* or . . . *Saussure* or . . . *Freud* or . . . *Foucault* or even . . . *Heidegger* become waterlogged when submerged too long.[3]

A Questioning Tradition

Geography is characterized today by a great number of conditions of possibility to illuminate, to enlighten, which is why I have emphasized in the title "geographies *from* the Enlightenment," and the fact that we *all* stand in that tradition and share it. Over the past year I have had the privilege of reading in and about the Enlightenment, and although I have only scratched the surface, I realize how varied it was over space and time, and how shallow and uninformed are many of the criticisms of its supposedly overriding rationality. But what does seem to bind those years together is a willingness to question. Not necessarily in each individual—people were still quite capable of making exclusive claims just as strong as the statements of religious belief they often opposed—but if one distances oneself just a little, achieving a somewhat detached posture I want to examine later, one can see a constant questioning, doubting, thinking, inquiring, and experimenting as characteristic of that eighteenth-century world, guided by a curiosity that was unleashed by a renewed will to "unthink" the authority of established and institutionalized opinion. Transformed to, and constituting another "world" (using "world" as Heidegger used it) that sense of questioning is a major characteristic of geography today. Our "world" today contains the condition of possibility for questioning that renewed itself out of the Enlightenment tradition. There is no break here, despite all the pretensions of *post*modernism, but rather a vibrant continuation and concatenation of clarifying questions. The geographies available to us today, the perspectives that constitute the conditions of possibility making up our "world," arise out of that questioning tradition, the willingness to confront and challenge claims to a whole succession of essentialist ideologies posited by those who wish to wear the mantle of messianic guruism—or have it thrust around them. Frankly, I would rather have naked emperors—at least you can see that they are not hiding anything.

The close questioning of the past decade has not been easy. To put it more directly, it has been hard, and "hard" in two senses: hard on those who have been subject to critique, and hard intellectually. Questions that are essentially philosophical, though they arise again and again in forms meaningful to successive generations, are not easy, and they demand the very best of patient thinking. They are not resolved by taking a few books out of the library and flipping through them for some quotable quotes. It would be interesting to see when the words *ontology* and *ontological* first appear in our literature. My guess is that they do not appear much before 1980, and even if used clumsily, and sometimes in ways not readily recognizable by professional philosophers, what sort of secular "soif ontologique" (secular ontological thirst), to borrow Massignon's phrase,[4] do they represent? And what reactions do they generate? When such questioning exposes the shallowness of exclusive claims to ideological truths, to "theory" in human affairs that is not simplistic physical analogy, the result is too frequently petulance, strident and sarcastic disclaimers, and occasionally sheer venom. We need, desperately, to develop a tradition in our field of critique that goes for the ideas, not the person. And when such questioning discloses the absence of firm ground anywhere, others react as though moral choice is not possible, and assume that nihilism is the order of the day. To make an ungrounded claim myself, nothing could be further from the truth, as a number of recent papers constituting a "critique of arrogance" have pointed out.[5] As geographers, as "writers of the earth," our writing is, ironically, always ungrounded and, therefore, always temporary in its human finitude.

This means the taken-for-granteds—in other words, the assumptions that remain unthought, and perhaps unthinkable, today—must appear spatiotemporally parochial in the future. This becomes clear as, for the first time in our discipline, a new generation of geographers, well-grounded in contemporary intellectual history and philosophy of science, and without Hettnerian or de la Blachian axes to grind, examines the "unexamined discourses" of earlier times, discourses whose unexamined assumptions are sometimes explicitly stated, but often disclosed unwittingly in language itself.[6] We are shocked by the oblivious arrogance of racism, class attitudes, paternalism, and claims to divine blessing that always make *us* the chosen people rather than *them*. But what are our own unthought assumptions today? And how will they appear to future generations?

A take-nothing-for-granted Enlightenment tradition of questioning characterizes many of the geographical perspectives that form and enrich our conditions of possibility for geographic seeing today. It arises in feminist geography, in concern for the many forms of Third World disintegration, in the rethinking of social and cultural geography, in wholly original views of cartography, in a recasting of geography as a contested enterprise, in the injection of "spatiality" into adjacent areas as varied as archaeology, epidemiology, sociology, and political science. Often the critiques are against the grain of Establishment thought, or even contrary to the shared thinking of small coteries. They challenge, sometimes with deliberate overstatement, using polemic to grab the ear of an established coterie, and rhetoric to persuade it of its folly.

Radical Feminist Geography

Given this larger context of questioning, I want to stand in that tradition of critique and challenge some of the challengers. First, feminist geography, where I am sure a questioning and critical statement from one of Geographia's many lovers will be greeted with less than enthusiasm.[7] I am unaware of any critical statements in print of radical feminist geography, which may mean it has been elevated rapidly to the sacred, where faith and belief rule but reason is not relevant. A more enlightened position would assume that what is sauce for the goose is sauce for the gander. And let me be explicit: nothing I say should be construed as a devaluation of thoughtful feminist critique and the perspective it provides as yet another valuable condition of possibility for seeing. From various stances (for feminist geography is itself complex), it is a perspective that has illuminated a "world" of women in an Italian-Swiss valley, the spatiotemporal constraints on women in a small Swedish town, the effects of economic restructuring on women, the challenges facing a Marxism trying to rethink itself, and many other arenas of feminist concern.

But under the exploding hegemonic power of strident feminist critique, we increasingly see male geographers scurrying around in print, flashing their gendered sensitivity. Their precious posturing, informed by a gloomy morality, obliterates even a historical understanding of the very language in which they write. They would rather destroy the potential for rhythmic structure in an English sentence, with an ugly "she/he" or "his/her," or even "s/he" than use one or another nominative or possessive form in their interchangeable possibilities. Contrary to some claims, possessive adjectives are not a mark of phallocentrism; they arise in the structure of some languages and not others. Emphasizing the "his" of "*his*tory" displays only philological ignorance of the Greek heritage—or should it be *her*itage? As for the constant interjection of *sic* into every quotation from the past that used a masculine possessive adjective, I find the posturing so shallowly flamboyant and historically uninformed and ignorant, that it may fairly be called "sic-ening." Notice that the French are seldom bothered by such trivialities: *lui* can refer to male or female; their possessive adjectives agree with their gendered nouns. Even *fraternité* and *paternité* are feminine. One has to admire the hermaphroditic *jouissance* of the French.[8]

But there is a deeper concern here. For the first time in our literature, I have come across not just anger, which is perfectly understandable, but something close to hate. We have in our literature today such words as "dislike," "envy," "despise," and "hate" to describe academic colleagues.[9] And in the same piece by Gillian Rose, one would think that no woman has ever looked with desire upon a man. This is a theme taken up by other radical feminists, who extol visual transvestitism and label any "gaze" as "phallocentric."[10] Poets like Máighréad Medbh, Rita Ann Higgens, and Edna St. Vincent Millay know better than make such silly, and so often bitter, claims. Frankly, I stand with them, for why should my very being, and the "world" into which I was thrown, and about which I had no choice, be used epithetically by clitorocentrists of any gender? I *am* a "white European male." It is true, I am not dead yet, but give me a chance. I would not dream of using "black African woman," or any other racist phrase, as a term of rejection and denigration, and I refuse to offer an unprotesting license to any human being who judges me on hateful racist, national, or gender grounds. The last thing we need is yet another division, another scarifying line across which the human family is divided into *us* and *them*. Like a Marxist clique's castigation of Nigel Thrift and Gordon Clark for their "betrayal" in founding *Society and Space*,[11] or the accusations hurled repeatedly, but always behind his back, at Peter Gould for being a "traitor" to the quantitative revolution, these feminist epithets are not needed in any caring human discourse, and I suggest that it is time we all grow up.

A Curious Gaze

But such divisions, resulting in a distant viewing of the "Other," raise a further difficult and potentially divisive point at a quite different scale. Thanks to the courage and insight of people like Edward Said, we have become more sensitive and thoughtful about representations of other lands, peoples, cultures, practices, ways of life—in brief, to our particular propensity for "the world as picture" or, in shorthand, our viewing as "the gaze." And thanks to people like Timothy Mitchell and his skillful historical exploration of one colonizing experience,[12] we have acquired a much deeper perspective on the transformations colonialism can bring about. And while I buy all of Gunnar Olsson's warnings about the power of the concrete example,[13] Mitchell's spatial exposé demonstrates that abstraction is nothing if it cannot illuminate the concrete instance. As geographers, we have another condition of possibility for seeing, and, more importantly, for thinking.

But these and other sensitizing perspectives are not beyond the reach of critique. In the act of sensitizing, they also challenge, and what they challenge is a Western world in which intense curiosity has been a driving force on and off for at least 2,500 years. Other cultures have traditions of theater—one thinks of Japanese Noh and Indonesian shadow plays—but none so far-reaching as those of Aeschylus, Sophocles, Euripides, and Aristophanes, on through the medieval morality play to Racine and Shakespeare, to our own modern-day theater. And let us remind ourselves that the *thea* of *theater* and *theoria* means "the outward look," of bringing a representation of the world and its human affairs into the public gaze. "All the world's a stage," wrote Shakespeare, but that metaphor goes back to Horace and Plato.

The Arab world, of which Said and Mitchell write, appears to have no such tradition. Of dramatic, ritualized storytelling, of poetic recitation, of making the word of God a spoken, living presence—yes. Of theater, no. Mitchell records the astonishment of Tahtawi, an Egyptian student visiting Paris in 1820, at the very *idea* of theater, of a stage, of actors and actresses dressed to imitate others, speaking words not their own, all placed in the detached gaze of a knowing public.

Pace the pre-Socratics, the condition of possibility for the detached and curious gaze has been present in the West for millennia, and once the authority of the gods has been given up or denied in post-Enlightenment times, there is something approaching a necessity that it transform itself into "world as picture," world as representation. *Pace* Kant, there are no "things in themselves": only *we* give them meaning, only in us do things "come to presence"; and in sharing such presencing, in "coming with presents" to others, in *communus*, in communicating, we can *only* re-present. In a making public, even if to just one another, a curiosity-driven presencing can only be a re-presentation. This does not deny the private, literally privileged, experience of the devoutly religious person, and surely lovers know that linguistic representation is not required to share a present.

So any experience of the Western Enlightenment tradition must constitute a standing apart, a positing of how things are, will be, ought to be. Which is why Heidegger remarked "We are all positivists today," and all geographic writing posits because it can do no other. Once the god who posited has fled, there remains only us to posit a meaning on the world. How can it be emphasized sufficiently, that in a post-Enlightenment world, meaning is posited by the centered human presence, not the authority of the god as recorded in holy writ? This is why *all* posited perspectives constituting conditions of possibility hold the potential for creating geographies *from* the Enlightenment. Whether this appears currently congenial or not, the feminist critique, in particular, is drenched in the rationalistic and scientific viewpoint, denying the patriarchal authority embodied in the Judeo-Christian tradition, and positing either an egalitarian world, or one that is structured by a gender-guarded matriarchal authority that is as exclusive and discriminatory as that previously revealed to the monotheistic faithful. The fact is that we do not teach geography in textual stages as the Koran and its commentaries are still read. Neither does an Edward Said follow such a tradition, for he stands squarely in the questioning tradition of the Enlightenment. In human positing there is a condition of possibility of re-positing, of re-examining, of changing a perspective and a meaning; in brief, of enlarging the horizon. And it is in this context that I want to think about colonialism and the colonial experience, for it is very much a taken-for-granted today, one that shapes, almost to the exclusion of any other perspective, how the Third World is approached by geographers. Too many approach the once-colonized Third World knowing what the answer will be before they have even looked.

Reapproaching Colonialism

Rank heresy though it may sound to some, I want to challenge that singular perspective, for there is more than a cliché here; namely, that it is important to understand the past as a guide to the future. We are going to see a remarkable reexamination of the colonial experience, its geography, and its history, as it catches the attention of those who never knew it. Like the Enlightenment itself, colonialism, in its late-nineteenth-early-twentieth-century forms, becomes enormously complex and diverse as soon as one examines it at all closely. This will not please the makers of ideological umbrella-phrases, those singular perspectives claiming to explain everything, but it will confirm the issue of diversity for a newer generation whose more tolerant and self-aware thinking can sometimes approach the phenomenological. That appreciation and reasoned tolerance of diversity also come straight from the Enlightenment tradition—despite numerous individual cases of intolerance being tolerated. I said the Enlightenment was complex. What we lump together as "colonialism" ranges from the viciously cruel rule of Portugal across three continents—a rule that produced one Ph.D. in four hundred years (my fellow student, colleague, and friend Eduardo Mondlane, leader of FRELIMO until he was murdered)—to forty years of trusteeship in Togo by France and Britain for the United Nations. It ranges from the deliberate nineteenth-century penetration by the French of most of West and Central Africa, to the reluctant foot-dragging of the British Foreign Office at the turn of the century to prevent takeovers of hinterlands in Sierra Leone, Ghana, and Nigeria. Let us recall that for four hundred years after the Portuguese built Elmina in 1482, there was virtually no penetration of Africa by Europeans from the small trading enclaves dotted along the coast, and that almost every man, woman, and child placed in slavery was so placed by fellow Africans. Slavery was an indigenous institution, just as it was in the silver mines of Athens two millennia before. These were *pre*-Enlightenment days, and they still shape the human geography of the world today.

As for administrations, they ranged from direct impositions from the Metropole, to the "indirect rule" of a Lugard. In Niger today, young politicians complain that the French destroyed their indigenous political institutions; two kilometers to the south in Nigeria, they complain that traditional powers were left intact to become conservative anchors against modernizing movements. Colonialists are damned if they do, and damned if they don't. Perhaps rightly so, although some regimes also displayed a capacity to give much; not the least, scrupulous honesty in all matters of accounting, and for many a sense of service to their fellow human beings—in medicine, in engineering, in education, in law, in administrations that ensured that people who held to different religions, languages, and ways of life could live in peace side by side with others. Compare this to the literally billions stashed away in Swiss bank accounts by the Mobutus and Bokassas; to the warlords of Somalia; to Uganda under Idi Amin; to the ethnoreligious cleansing in southern Sudan, and so on. This is not an apology for colonialism: it is a plea for fair and honest geographical scholarship to seize the conditions of possibility newly released by the continuing tradition of the Enlightenment to question, a tradition in which all of us share today.

And I raise these questions, and challenge you with them, not simply as an academic exercise—too frequently a euphemism for irrelevancy—but because we stand in a world today in which men and women of the United Nations are on peacekeeping, humanitarian missions in countries like Somalia, Mozambique, Rwanda, Cambodia, and Liberia, and perhaps one day in other places like the Bronx, Los Angeles, Belfast, and Londonderry. It is a world in which *serious* consideration is being given to United Nations takeovers in countries in such chaos, in places of such human, political, economic, and environmental disorder, that a sense of what is decent and right demands that we collectively, and wearing light-blue berets, consider interfering deliberately in the internal affairs of sovereign nations—often to the relief of a large majority of ordinary people. From a detached view, a gaze on the other, we are going to invoke Enlightenment ideas that this need not go on, that something can be done, that there is still hope, that the hungry can be fed, the sick helped, that we can stop the slaughter of the innocents.

I suggest that such new, humanely inspired "colonizing" comes straight from the Enlightenment, that Reason can, or at least *ought*, to prevail. And if you shy away from the implications, think of the alternatives: Amin's ethnic cleansing of the Baganda, the killing fields of Pol Pot, the racist-religious genocide of the Sudan, the slavery of the Amazon, the tribal slaughter of Liberia, the population-environmental catastrophe of Haiti, the forest fire of AIDS in Thailand, . . . The list seems tragically endless, and the consequences will touch generations to come. As a new, internationally sanctioned "colonialism" comes to the fore, with its distanced gaze of Reason, it will result in a radical reappraisal of prior colonial experiences, a radical resorting into the positive and effective, and the selfish and exploitative. In this reappraisal, geographers will have an important part to play, in the academy as well as in agencies dealing with immediate human and environmental problems in specific places, and, perhaps increasingly, in both arenas of activity. Canada is bound to be involved: it has never appeared as an a threat to others, it has itself bicultural and bilingual traditions, and its men and women are already serving around the world in United Nations programs.

This willingness to question, to reappraise, to reexamine presents us with an enormous challenge—it is the old yet ever-new question of what do we teach, what is worth teaching? All planning, all reconstruction, is inherently spatial and geographic. How do we open the horizons of thinking so that the spatiotemporality of human existence can illuminate such complexity? How do we give students—and that always means ourselves—perspectives, tools, sensitivities? How do we open the horizon of the technical specialist to the possibility that an answer may not be technical? How do we give the concerned cultural geographer a sense of the potential for cross-cultural dialogue in goal programming to incorporate local desires incommensurable with the reduced monetary units employed in banal cost-benefit analysis? We need to teach a diversity of geographic perspectives to think *with*, diverse and Enlightened geographies as conditions of possibility to lighten the darkness.

An Interlude of Commentaries

As I fully expected, the address produced highly mixed reactions. A number of women and men came down to the podium afterwards and thanked me for being willing to take the less-than-orthodox view, and so provide an alternative perspective from which to reflect upon some current trends, and increasing orthodoxies, in the field. Others, it became clear later, were absolutely incensed that anyone should have the temerity to question, let alone open up a critique, of their absolute and unshakable certainties. At the president's reception later that afternoon, few approached me or wanted to be seen in my company, and one radical feminist, who was unfortunate enough to be "caught" by being introduced to me, made it clear to others around her that she had no wish whatsoever to shake my hand. I know I should not be, but I am always surprised that those who call most stridently for critique are inevitably the most shrill when their call is taken up and turned back upon their own sureties. Their outrage always reminds me of Derek Gregory's insightful phrase that ideologies are "unexamined discourses."[14] After all, I was only asking that certain discourses in the field be examined.

Three commentaries followed initial publication of the address, the two by women geographers difficult to respond to because it seemed they were unloading their own frustrated agendas rather than responding to the address. For one of them, Janet Momsen, I was a Super Mario gobbling up threats, an image that meant nothing to me at the time because I have never played computer games. After delivering that electronic image, she wandered off into Brazilian Ph.D.'s, and an obscure and unpublished paper delivered more than a decade before. The other, Linda Peake, objected to my love for the goddess Geographia, and declared my interest "prurient," before noting that a former Young Turk had developed into an Old Turkey—a splendid phrase of transformation, and one I wish I had thought of myself.

In marked contrast, the third commentary, by Derek Gregory, was a thoughtful "open letter," which, more than anything else, showed that we had been sharing a fascinating literature both in and beyond our own field. It was a response employing a number of rhetorical devices quite skillfully, all of which

I appreciated with a chuckle, even as I recognized the effect they might have on readers who were less familiar with his sociologically tinged style. But his concern was real, and put forward in good faith to raise questions about the questions I had raised.

And so followed my reply.

A Reply to My Commentators

> To me, this idea that you can't dispute the territory, that criticism or debate is not really tolerated outside of a certain set of very circumscribed parameters is dangerous. In order to be a vital movement, feminism should be able to sustain some kind of critique, voices that don't all say just one thing.
>
> —Katie Roiphe, author of *The Morning After,* interview by Art Carey

In these postmodern days of different perspectives and hermeneutic stances, anyone who thinks that a publication, a "making public," will come to reflect the author's intention is either ignorant or naïve. We launch our little caravels into the stormy seas of critique where the waves buffet them as they will. Still, it is a more vigorous life than rotting in some quiet haven—as Tennyson's Ulysses knew well.[15] So I thank my commentators, Janet Momsen, Linda Peake, and Derek Gregory, for the courtesy of their responding. Two of them, however, do not make it easy to respond in turn, since their commentaries barely touch, let alone engage, the questions that the text provided. Let me take these commentaries up first, starting with Janet Momsen, who sees me as a politically correct Super Mario—a character I have never heard of before, but who must be, from the context, both heinous and electronic.

Where in the text have I written anything that could be even faintly construed as an apologetic look at "the Quantitative Revolution of the 60s"? I mention neither it nor the years. Where does this literally extraneous thought come from? I certainly do not feel in the least "a 'traitor' to quantitative geography"—this was not my term, but that of others—and I have regularly taught courses in quantitative methodology for thirty-four years, including our Spatial Analysis II this very semester (Spring 1994).

In marked contrast, my claims are based on explicit texts: whatever Ron Johnston said over a decade ago, at a meeting at which I was not present, he *wrote,* in 1991, a chapter of thirty-seven pages in *A Question of Place* entitled "A Fragmented Discipline," in which he did his utmost to support his contention that "Fragmentation is seen as a problem because it fails to acknowledge the integration necessary for a proper understanding of much of the subject matter of geography. . . ."[16] Whether he actually "fears this fragmentation as Gould suggests" I cannot tell, but I can *suggest* that after thirty-seven pages he does give the distinct impression that he is desperately worried about it. Why? Because at the end of the chapter he declares he is going to write the rest of the book to remedy it by arguing for "an integrated human geography focusing on the nature of places." Different interpretations are possible in these days of postmodern scholarship but we do not have *carte blanche*.

Feminism and imperialism are hardly "windmills," but issues to be thought about with the utmost seriousness, and then reflected upon critically. My quoted clause, employing the subjunctive of doubt and possibility, was preceded by "I am unaware of any critical statements in print of radical feminist geography . . ."; if any reader can point to a counterexample in the literature, I would be most grateful for the correction.[17] I am also grateful that the example I cited "is regretted and disowned by most feminist geographers," although I would find no need for the qualifying adjective. I think most caring people would find such language offensive.

The comment about Brazilian doctorates is simply bizarre: Brazil has been an independent nation since 1822; my context is clearly African; the overlap between Portuguese "colonialism and the doctorate as a relatively common practice" is about one hundred years, surely enough time for Portugal's "civilizing mission" in Africa to produce more than one, and that by an American university. Equally bizarre is the claim that I reject difference and the ungroundedness of human finitude that some perversely interpret as "relativism," or that I am searching for an alternative to postmodernism. On the contrary, *amor fati*, both generally and geographically: I rejoice in the opening of horizons to the increased conditions of possibility they promise. By recalling the questioning tradition of the Enlightenment, rather than "[arguing] the metanarrative," I place the postmodern squarely in its realm. Where the gods give the Truth, and may not be questioned themselves, enlightened questions are not permitted. Ask the women of Isfahan, Tehran, Riyadh, or Khartoum. My text extols the "appreciation and 'reasoned tolerance' of diversity," and ends with a plea that the teaching task will honor the diversity of perspectives in our field. But surely an ex-Brit Super Mario cannot possibly mean what he writes?

It is equally difficult to respond to what can only be termed the tirade of Linda Peake, who makes it clear that she had formed her opinion before even walking into the lecture hall. She did not like my book *The Geographer at Work*, and refers to "the furore that greeted [its] publication . . . in 1985." But the "furore" came not from the thousands of readers, women and men, who found the depictions of the goddess Geographia inoffensive, amusing, and symbolic, but a handful of radical feminists who had formed themselves into the Geographical Perspectives on Women Specialty Group, and represented themselves as "writing on behalf of the Board of Directors of the Association of American Geographers. . . ."[18] I replied courteously and at length,[19] pointing out that the first two depictions of Geographia were already in the literature, and were comments on the abduction of our goddess by uncouth "quantifiers" and "Marxists." The third, commissioned for the book, showed a complete reversal, a beautiful Geographia fully in command, carrying a bunch of men in total disarray back to geographical reality. I would maintain that anyone whose sense of humor has not been totally lobotomized sees this as symbolic of the rise of the feminist perspective in our field, portrayed sympathetically.

My reply, to which I never received the courtesy of a response, was labeled "rather brave" by the chair of yet another AAG committee, this time on the Status of Women in Geography.[20] Why "brave" I do not know: are we to roll over and play dead every time an AAG committee makes public a challenge or critique? I had simply asked the editors of the *Professional Geographer* and the AAG *Newsletter* to publish the original letter from the group together with my reply.[21] Both publications declined to do so, the former courteously,[22] the latter after three previous letters written over eleven weeks had gone unanswered.[23] I then began to get telephone calls and letters from former women graduate students saying that they were being approached "off the record" by members of the AAG Women's Specialty Group. They were rightly disgusted with such witch-hunting methods and informed me immediately, with one of them writing, "I got irate letters from members . . . asking if I have any scabrous evidence or opinions concerning your alleged sexist pigism. I have none, and have made that clear."[24] I called this to the attention of the AAG, noting that I would seek legal counsel if it continued.[25] I received replies from both the executive director[26] and the president, the latter indicating that he had recently attended a session on liability insurance.[27] Both indicated that such practices were not sanctioned by the AAG, and that they would cease immediately.

I respond in this detailed way to indicate that once a hermeneutic stance has been taken to an author, it may be difficult to approach a specific text openly and fairly. Inquisitors seven years ago noted that they objected to the nakedness of Geographia, referring to it as "prurient,"[28] the same objection, and even the same language, employed by Linda Peake. Another labeled the half-naked man being abducted by Geographia as "disgraceful."[29] In our Western world, the naked human form, male or female, is generally regarded as beautiful, and has been a subject for sculptures and painters for over 2,500 years. I have lived in cultures from Sweden to Ghana, from Japan to Jamaica, in which viewing nakedness as "prurient" or "disgraceful" would be considered very strange indeed. I have no problem with it, especially for emperors, but since I am not a trained psychoanalyst I can do little to help those who do.

A hermeneutic stance is one thing: having words put in your mouth, and others "quoted" from your text that were never there, is something else. Neither in the address, the first and last quarters of which were in French and understood perfectly by French and Quebecois colleagues, nor in the English text do I criticize "certain radical feminist geographers on the ground that they have recreated men." I assume that even the most radical would not aspire to such *hubris.* Man is born of woman, of course; but once he is out of the womb, she can keep her sticky little recreative fingers off him. Nowhere in the text are there the purported quotations about "fine scholars," "fine prose," or "fine mind"; indeed, I never use the word *mind,* because I do not know what it is, and have never met anyone who can tell me. Nowhere do I even suggest writing history "as it happened." We are all naïve in different ways, but only the totally ignorant in these postmodern days could think it possible to write some unique and authoritatively true account of past events in such a way.

Which raises the matter of my appeal to reflect upon the way in which the history and geography of colonialism are approached, to recognize the diversity and complexity, and to attempt an open and fair approach to such materials and events of the past. I make explicit the contemporary context in which we might rethink such events and experiences, even as I state explicitly "this is not an apology for colonialism." Nowhere do I imply that fiscal honesty and a sense of service should be "held up as sufficient redemption (or even as poor excuses)." As a finite mortal, I have no concern to redeem or excuse events over which I had no control. I might as well try to redeem or excuse Genghis Khan.

As for the phrase "black African woman," I juxtaposed it deliberately with "white European male," a phrase used frequently, and always epithetically, by radical feminists, rather than out of descriptive necessity, noting that I would not dream of using it in such a way. Both phrases used epithetically are hateful precisely because they prejudge a fellow human being on racist (no slip of my pen), national, or gender grounds. I stand by it: I do not wish to divide the human family yet again for hateful purposes. I will not indulge in name-calling in a scholarly professional journal, and I suggest it is time that some people grow up.

It is, therefore, with a sense of relief that I turn to the thoughtful and concerned critique of Derek Gregory, a distinction I am quite sure will be misinterpreted. And in keeping with his caring style, let me change mine too, writing now in the more personal way of a shared and open letter, for I would be the first to agree that the conventions of commentaries are often too constraining. We really need an extended conversation so that gestures, tones of voice, wry grins, pursed lips, and eyes narrowed in worry or opened wide in feigned astonishment can inform our communication, that "sharing of gifts." So let me work my way slowly, and as thoughtfully as I can in the space allowed, through your "letter."

"Resist specialization": I think we both recognize immediately that it is impossible to resist it, yet somehow it must be done. Somehow we must try to live with this contradiction and leave a corner in an often frenetically stressed, overcommitted, sometimes absolutely mad, professional life for those new perspectives, those ways of seeing and understanding, we did not suspect or think about before. And yes, let them pry open those closures of compartmentalization: for decades I have thought it absurd to teach courses on "——— (fill in the blank) Geography." We need to teach elementary, intermediate, and advanced geography, not disconnected bits and pieces to satisfy tidy administrators and so confirm them in their nineteenth-century partitions.

But don't stick me too hard with those definite and indefinite articles; after all, the spectrum of approaches, methods, perspectives, and concepts I cite in my example should convince you that I believe there are many possibilities within *the* geographic way of looking. Does that definite article imply a totality? No. A project to be fulfilled, some attainable *telos* that one day will be within our grasp? No. Does that particular way of looking, wondering, delighting, thinking, and thanking constitute a closed compartment? Not on your life! And thirty years of graduate students, quite a few of whom have held interdisciplinary fellowships in the humanities, will tell you that I have encouraged them to inquire in fields as diverse as philosophy, computer science, intellectual history, operations research, epidemiology, and many others. And despite the obduracy of some (I am not a favorite with many epidemiologists today), I rejoice in what I see as the spatiality, the sheer "geographicness" of human life, being thoughtfully appreciated more and more by many around us. Not the least because *your* generation and *your* students refuse the defensive, laager-like attitudes of the Age of Adoration we have just passed through—and barely survived.

But do I believe there is *a*, rather particular, way of geographic looking? A spacious house of many chambers, yet a mansion nevertheless? Know what (wry grin!)? I think I really do—well, sorta. Moreover, I think we should share our intellectual wealth, just as we should be open to any and all ways of looking that help us understand ourselves, all of us, the whole human family, and what it is to be human. In brief, I'm in total sympathy with your first two paragraphs, so let us get to that "But"!

Have you noticed that we both use "enlightening" words, phrases, and metaphors—illumination, making visible, optics, visuality, gaze, and so on? But in the midst of this terribly complex matter of thinking through the visual, with all its attendant strengths and dangers, let me pose what I see as *a* (you have made me chary of definite articles!) crucially important question, perhaps literally the *crux*. It is simply whether we have any alternative today. And by *we* I mean all of us educated in a Western mode of university education, itself rooted in disputations of the sacred from the thirteenth century on, and deeply influenced by the enlarged possibilities for secular challenge radically *(radix)* informed by the questioning tradition of the Enlightenment. Nor, in invoking the university as a Western institution, am I forgetting our immense debt to Arab scholarship and Byzantine guardianship. Both, however, were still saturated with an unchallenged acceptance of a world given by the god. In our "world" the god has been displaced, and once the world is made, not given, the human possibility of curiosity is unleashed—for good and ill. Pandora's box, that compartmentalized closure, was prised ajar by the Greeks, and has been gaping wide open since the eighteenth century. We will question even unto death. We know this, fear this, and still we do it.

Once we find ourselves in a world we must make, rather than accept as given by sacred text and theological authority, once curiosity has been unleashed, what can pull us back? And what alternative do we have but to gaze, sometimes with open astonishment, at something that was not simply within our "world," but beyond the horizons of our imagination? From those fifteenth-century explorers bumping into huge unknown landmasses, to those Pacific expeditions of Cook and many others, we have been gazing, learning, classifying, ordering the plethora of visualization, *making* our world. That which comes to presence in us we have been re-presenting to others ever since. You do it constantly, oh child of the Western world and its university, and you do it because you cannot do otherwise. Are you thereby determined? Driven without choice? Only you can answer that. Or can you? Is that Enlightenment release to questioning freedom, that heritage in which we all stand, a liberating blessing or a curse toward death?

None of us who think, inquire, question, and write can escape that sense of detachment, no matter how involved, committed, and rooted in a viewpoint we think we are, no matter how sensitive we think we are to our own "situatedness." It requires no Archimedean point: every act of writing constitutes an act of standing apart, even if it is only a short sideways step out of the immediate hubbub of the world to re-present it to others. Can it be otherwise? What alternative is possible? Even in the act of critique *contra* that "world," one finds oneself nevertheless within it—inescapably in it. To be human is to have a "world," that concretization of the conditions of possibility available to us. Are there alternatives? Can we move back to that religious world of faith where we accept the given? Or move forward to the "poetics of experimentation"? Do my adverbs betray me—forward and back? These are private thoughts, not for publication. But even if we were to make such moves, we would still have to acknowledge the *place*, our "world," from which such a move might be made. And don't misunderstand that "poetics of experimentation"; it is no emotionally informed dalliance, but much closer to the politicized aesthetics of Lyotard, which you summarize in a manner I judge to be approving, or at least thought-provoking.[30] But these are matters for an extended conversation at another time.

For the moment, let me emphasize something that in all our contestations none of us should forget. We *all* of us stand in that Western, university, questioning world, and can do no other. Donna Haraway, Edward Said, Walter Benjamin, Henri Lefebvre, Gayatri Spivak, . . . are questioning, challenging children of that Enlightenment, trying to contest and persuade by reasonable argument. These are no intellectual Dadaists. That is why I wrote that even the most radical feminism is drenched in reason, no matter how unreasonable a singular face of it may appear. Reasonableness, like beauty, lies in the eye of the beholder with her hermeneutic gaze. The work of these people, and that of many others, informs, touches, sensitizes, challenges my thinking, just as it does yours. Sometimes I share their sense of problematic, sometimes I don't, sometimes I once did but find it has since dissolved. Is the "corporality of vision" genuinely problematic to you? What else could it be? Is the "social construction of knowledge" problematic to you? Knowledge is human: what else could it be?

But let me finish this "letter" on two things that are problematic. First, do I mean to "feminize feminist geography"? Do you mean by that to limit the sense of caring to women only? Of course not. But on the other hand, do

I question the sincerity of certain men "flashing their gendered sensitivity"? You bet I do! And, as you pointed out, so do many women, a number of whom have told me that they see nothing quite so ridiculous as men trying to be more feminist than the feminists. Take that "*sic*-ening" insertion of *sic:* are those insertions in historical quotations, following the use of the masculine possessive adjective, made in the best tradition of scholarship, to indicate to the reader that a word, expression, or spelling was not a mistake, but was truly in the original text? Or are they simply cute Jack Hornerisms, insertions of plums into the textual pie so that the author can declare "Oh what a good boy am I!"? Let me suggest that the interpretation, like the mathematician's proof, is elementary and left to the reader. People can write what they like, and take responsibility for it, but can we not be sensitive to the fact that "his" and "her" can be used interchangeably? I have been using them this way since the 1970s.

But for all my respect for language, and my belief that in some deep and carefully explicated sense we are it, this problematic is trivial compared to the last I raise, that of colonialism, its geography and history, and what we might learn from an open approach. Not an unsituated approach, not trying to see "everything from nowhere": we know this is not possible. We know we cannot approach, or perhaps even give meaning anymore, to such Kantian phrases such as "the things in the themselves"; we know it is we who give meaning from the multiple and diverse perspectives that constitute the conditions of possibility available to us today. But we also know that fairness does not equate with apologetic; from our own "situatednesses" we both recognize the careful choice of evidence and example that ignores, buries, and obliterates counterexample, those awkward artifacts that prove embarrassing to one's prejudices (prejudgments) and preconceptions. Not for nothing does Heidegger speak of truth as unconcealment, noting how things can also go back into concealment. Not for nothing do I note how a George Trevelyan changed from the prejudiced Whig historian to one willing to acknowledge other views.

But the context within which we examine the colonial record today presents us with truly hideous choices, lying far beyond the bounds of scholarship. And I write that fully sensitive to the manner in which such scholarship is always socially embedded and informed, situated, a child of its time. How could it be otherwise—once you have seen it? Yet you and I live in two of the 184 countries that are signatories of the United Nations Charter. We have a choice: to accept it or to reject it. We can label it idealistic, naïve, impractical, totally unrealistic, in brief an ineffective manifesto of an institution that can bring no hope. Or we can see that document as an enlightened one, a reasonable one, but not invoking "Reason in all of its singular majesty." Human affairs are too complex for reason alone: we are often torn by hideous choices and finite capabilities that reason cannot resolve. I know that; you know that.

And it is not just "connections between Enlightenment projects and public spheres [that] are tense and tangled." The human condition, at all times, and at all geographic scales, is tense and tangled. Which is my point: colonialism is too complex, too varied, too redolent of the possibilities of good and evil, to be approached as many are approaching it today, "knowing what the answer will be before they have even looked." Rome colonized from the walls of Hadrian to those of Alexandria; Arab colonial rule extended from Bokhara to Grenada; the imperial might of Napoleon from Madrid almost to Moscow. These histories and geographies have nothing to teach us today? All dismissed and condemned? What sense of arrogant futility do we display ourselves by refusing to reflect openly on events that happened long ago and over which we had no control whatsoever. But we can reflect and perhaps even learn; we can recall that many rejoiced in *Pax Romana*, that many preferred the religiously tolerant Caliphates of Spain, perceived as just, compared to the brutal Christian regime of Isabel and Ferdinand. Why was Napoleon greeted as a liberator by thousands in northern Italy? Is even mentioning these things to be construed as an apologetic for colonialism? Obviously not: but can we learn, can we make thoughtful and fair judgments? Perhaps the answer is No; perhaps our "situatednesses" produce incommensurabilities.

But I think not. And why? Because we recognize a sense of obligation, no matter how we interpret it in concrete, often hellish and obscene, situations. We face a choice, we always face a choice, to care or not to care, which of necessity means to intervene where we believe we may be able to help or to "pass by on the other side." Where does this sense of obligation come from? Who knows?[31] Some feel it, some don't. Some merely shrug their shoulders like those nineteenth-century Anglican churchmen—after all, the poor will always be with us.

At the international level, we have such choices—choices to intervene, by force if necessary. Or we can find reasons and excuses, always lots of excuses, for not getting involved, for passing by on the other side. Bosnia . . . for two years we have watched the most hideous mass rape, pillage, and slaughter with genocide as the aim. We're good at passing by on the other side. Cowards usually are. And then Somalia, . . . Cambodia, . . . Liberia, . . . why go on? What are we, the collectivity of the United Nations, to do? After all, we can't be the world's police, can we? So to assuage our sense of futility and despair, let's invoke a good old Devil's theory of history and whip those two boys called colonialism and capitalism. It won't bring peace, it won't feed kids, it won't reconstruct anything, and it certainly won't bring hope. But at least we can cut a few more notches on our academic bibliographies musing on "the hegemonic ontology of Hegelianism in the Althusserian hermeneutic on Gramsci." Profound though such contributions to human welfare and understanding may be, I don't think in this context they're good enough for you, and they certainly aren't for me.

But ultimately, intervention, armed or not, means taking on the responsibility of reconstruction, and reconstruction means taking on the responsibility of ruling, as Somalia so clearly shows us. And ruling means force—as little as possible (an iron fist well covered in a velvet glove)—but the possibility always there. And ruling means roads and railways and bridges and administrations and schools and hospitals and . . . all the accoutrements, all the necessities, of *modern* life that people welcome as signs of hope. And ruling means planning, and planning means calling on the geographer's tasks and insights, and those of many others, tasks that are never done.

Not nice, is it? The thought of practical, real, caring intervention. Well, we don't have to gaze on it; we can always look the other way and walk by on the other side. But in your situatedness, that's not good enough for you. And in my situatedness it's not good enough for me.

Let me leave you with one thought, one to think about together another time, in another conversation. What is hope, and can we give it—at least to some? What does it imply? Does it imply that we are human? Is it human to hope? Hope for what? Surely to move from where we are now, where we see little justice, to a place and time where things are "decent and right." Where is the light, a star *(astron, astrum)* to follow that will lead us out of our disaster? But does such a possibility of hope mean that we shall have to believe in progress?

That's what the people of the Enlightenment believed.

Cathartic Geography

> Instead of constructive social policy suggestions . . . we have endless testimonials, diatribes, and spurious science from people who imagine that their personal experiences, the dynamics of their particular family, sexual taste, childhood trauma, and personal inclination constitute universals.
>
> —Diane Johnson
> "What Do Women Want?"

A NEW GENRE of writing, purporting to be geographic, has arisen in the past ten or fifteen years. It has various, fairly easily recognizable strands—postmodern, homosexual, radical feminist, postcolonialist, etc.—sometimes loosely braided together in various presentations. All, however, are marked by a sense of emotional release on the part of the author, rather than the traditional aim of illuminating some aspect of our world. Aristotle, a dead white European male, whose opinion can therefore be safely dismissed, called such public emotional release "catharsis,"[1] although its original meaning was "laxative," a meaning not necessarily inconsistent with some of the examples encountered here.

Cathartic writing, claiming geographic status, can frequently be recognized by its vocabulary and style of writing, mainly because this trope of appropriating instanciations, arising out of a subliminal genre of counterhegemonic and totalizing megadiscourses of ontologically privileging fetishes of liminal disempowerment, fractures, *tout court*, the elided hybridity of the socially constructed and phallocentrically situated gaze of the subaltern "other" to produce a narrative multiplex of cinematic scope. Alternatively, the cinematic scope of a multiplex narrative produces an "other" subaltern, whose gaze is situated phallocentrically and constructed socially so that the hybridity elided *court tout* fractures those disempowered liminal fetishes privileging ontologically the megadiscourses totalizing the genre subliminally to produce instanciations ap-

---

Written for this volume, as a cathartic experience, which then provided the more positive opportunity to present "Beyond Cathartic Geography" at a symposium on geography close to the turn of the millennium, held at St. Andrews, Scotland, in March 1996.

propriating this trope. It does not really matter which way one reads this *parc*, for the purpose is not illumination of anything of interest, but simply the pure experience of cathartic release by the author. It is a great pity that Alan Sokal was not a geographer:[2] we have all missed a marvelous opportunity.

It should be made absolutely clear, although this strong caveat will be ignored by those who heed only examples of their own esoteric writing, that this is not an unnuanced reflection. Within each of the subgenres mentioned, there are fine examples of genuinely illuminating writing, in which one feels caring, concern, curiosity, and even moral outrage underpinning a particular analysis and narrative. Genuine complaint must not be confused with an excuse for cathartic excess. David Harvey's warning of research commodification and the ephemeralization of ideas[3] and Jane Wills's astute marshaling of evidence for sometimes brutally overstressed and exploited academics[4] are two examples among many pointing to conditions that desperately need attention. But such examples are marked by a concern for general conditions, even if the bite is felt personally. Nor must we confuse morally touched and informed narratives of geographically more focused situations with sheer cathartic excess. For all the dangers of a case study,[5] one way to bring a general point home is to help a reader reflect out of the specific toward the more general. One thinks of Michael Watts on Gambia,[6] Allen Pred on Stockholm,[7] David Harvey on Paris,[8] Anne Godlewska and Timothy Mitchell (in their different ways) on Egypt,[9] and Lucy Jarosz on Madagascar,[10] among many others. One Barings Bank incident, deftly dealt with by an Adam Tickell, is more provocative of thoughtful reflection than all the caterwaulings over a reified "Capitalist System."[11]

These sorts of concerned writing, and many others that could be cited, make an important point. If you use esoteric jargon, in the sense of clichéd phrases couched within a particular ideological stance, you simply put off the very people you want to convince. In other words, not those who are already converted to righteousness, but those whose eyes have yet to be opened. Rather than use publication for a mutual cathartic rant, it is much more effective and less arrogant to assume that readers can be morally touched by straightforward narratives of events seen to be searing by the author. I think of David Harvey telling me about a study of "redlining" by financial institutions in Baltimore.[12] A straightforward report on the practice, and the human distress it caused, was accepted, "although few of them realized that they were nodding their heads in agreement over something that could have come straight out of Engels and nineteenth-century Manchester."

"Postmodern" Critical Eyes

Most claims to the geographic postmodern are so uninformed about the rise of modernity, about intellectual history, about the peculiarly Western heritage rooted in the Greek and Middle Eastern "worlds" informing the Judeo-Christian tradition, that one faces an almost impossible task trying to refute them. The tradition of questioning out of doubt has ancient roots in the worlds of Plato and Aristotle, and these rhizomes of thinking seek the light in a medieval Paris of Aquinas, before they become quintessentially modern in the century of Enlightenment, a century in whose light we still stand. Even if one were to question a capitalized Reason reasonably, or rail against it unreasonably, such questioning itself would be the mark of someone standing squarely in the modern tradition. Ontological insecurity and angst for the moral condition are hardly *post*modern, and what a truly postmodern world would look like, what will appear upon the birth of the rough beast now slouching towards Bethlehem,[13] I have not the faintest idea.[14] In my own mortal finitude I find as little to sustain eschatologies as I do teleologies, all the while hoping for the best in an institution devoted to ignorance precisely because it may be redeemable.

Not so the critical theorists, hovering like hungry jackals around the latest gobbet to be thrown their way. Unable to illuminate anything very much themselves, and avoiding the careful distinction between concerned critique and career-promoting criticism, they rush into print knowing that the rapid turnover and ephemeralization of ideas will shield them with short memories if they are shown to fall flat on their faces. But one can hear the voices now: "Do you mean to say that you would bar all criticism of ideas from our journals?," a rhetorical ploy used with increasing frequency in those same journals to divert attention from the main, and inevitably awkward, issue at hand. It hardly needs saying that this essay itself refutes the rhetorical question without further elaboration. Examples are many, but three are especially meritorious in their ability to illuminate this destructive, as opposed to concernful deconstructive, tradition.

In 1993, Reginald Golledge published an extensively documented review essay calling upon geographers to use their skills, intellectual insights, and practical imaginations to help people disabled by blindness.[15] He pointed to extensive geographic research on the poor and the elderly, but noted how very little had focused on the physically disabled, for whom potential research held out the possibility of helping blind people navigate and lead less frustrating lives in complex physical and human environments. Such concern for caring and applied research blew right past two champions of political correction and the tortured vocabulary, not to say the editors and reviewers who brought their cathartic posturings to print. It is almost impossible to précis or do justice to the overall tone of the commentaries by Rob Imbrie and B. Gleeson,[16] and I can only offer an invitation to read them in full to see two postmods in full sail without rudders. For Imbrie, leaning constantly on Micheal Oliver,[17] a blind person is not disabled at all compared to a person with sight, but simply someone marginalized and oppressed by sociopolitical processes. Blindness is not, apparently, a relatively rare physiological condition, and any attempt to deal with it, medically or otherwise, would destroy "Any sense of celebrating . . . the vitality of difference."[18] He continues, with some of the most insensitive rhetoric to be found in a geographic journal, "Yet, how do we know . . . visually impaired people inhabit 'impoverished environments'? . . . And how is it possible to deduce that the experiences of disabled populations are 'less complex'?" Words and their conventional meanings, such as "normal," "ordinary," "conventional," and "handicapped" are simply disallowed, drummed out of the politically correct vocabulary, while "ableism" is made into a dirty word denoting "social oppression," although to any fair and thoughtful person it can only mean a perspective on blindness that helps people so inflicted to lead more able lives.

Even this is pale stuff compared to Gleeson's catharsis, which would deny the existence of any disability arising from physiological cause. On the contrary, via critical discourse, critical engagement, and sociopolitical emancipation, we see that blindness is not physiologically rooted at all, as Golledge claims quite erroneously, but is an ontologically limited, socially constructed experience arising out of an oppressive socialization of difference. Historical materialism (the latter word anointed by a reference to David Harvey,[19] although he would be the first to deny his paper says anything about disability) rides to the rescue, redefining "dis-ability" in the vocabulary of common usage, and substituting the euphemistically correct "impairment." Capitalism is also a major part of the problem: disablement was not, according to Gleeson, referencing himself, a feature of the feudal European world; only impairment was around as a "prosaic feature." Will those who extracted confessions and saved souls by prosaic blinding please copy. Any alternative view only betrays "Golledge's Whiggish sympathies . . . revealed through both a positivistic ideology of disability and a faith in the liberating potential for impaired people of environmental technology."[20] "Surely," Gleeson goes on, "the inhabitants of these so-called 'distorted spaces,' rather than [non-disabled] geographers [Imbrie and Gleeson?] are the best authors for an authentic description of their own experiences?"[21] At the end, the sense of catharsis is palpable.

What could one say to such people? Golledge's own reply is devastating,[22] pointing squarely to the cathartic nature of both commentaries ("Criticism for the sake of being critical . . ."), but I doubt that it was read with anything but condescension at Golledge's pathetically inadequate grasp of their own politics of certitude. Were they aware that Golledge, in the prime of life, and reveling in its visual beauty, woke up one morning to find that the capillaries feeding the optic nerves had closed down and destroyed them? So our heroes, quoting Oliver, assert that "the materiality of disability is one of its sociopolitical production,"[23] and then tell Golledge that he ought to experience blindness himself before writing about it. Do they know that he has *devoted* (the emphasis is deliberate) his professional life to helping visually and mentally handicapped people navigate complex environments? Would they be sensitive to the fact that when they trash his concern for technological enabling (an extraordinary denigration if they have the faintest idea of what "materialism" means), that his remarkable knowledge and research rests in large part today on an extraordinary machine that can transform standard text into spoken words? Would they preach political consciousness and verbal correctness if they knew that he teaches, does research, arranges conferences, and chairs university committees with great professional skill, without an ounce of self-pity, and is deeply admired by all who have the privilege of contact with him as student and colleague? No wonder that after all the "implies" and "infers," which place in Golledge's paper things that were never there, he himself writes that he finds it "hard . . . to believe that Imbrie and Gleeson read the same paper I wrote."[24]

At the same time, nothing written here should be taken as denying that different cultures and times construe—that is to say, interpret and give meaning to—disease, infirmity, or disability in different ways. Of course such meanings are socially constructed. Where else, since the century of Enlightenment, could such interpretations have come from? They are either given to us by the gods—our optic nerves are starved of blood by Pallas Athena, who works in her mysterious, and therefore humanly unfathomable, ways—or such meaning is a product of human society at a particular place and time. Nor should such qualification be taken as a genuflection to the current shibboleth of "positionality," a word about to overtake "phallocentric" in postmodern popularity. To say, often with more than a hint of intellectual condescension, that something is "socially constructed" is banal. Unless all our postmodern geographers are closet deists, we can only rejoice that they have finally got the point after two hundred years.

But the art of accusation is not confined to abusing those concerned to help others directly, and is often employed to enhance images of precious political correctness in other realms. In 1993, Gunnar Olsson published what was one more step along a long and rather lonely road to understand the complexities of people in societies.[25] Like a number of his other essays, "Chiasm of Thought-and-Action" employs some mild sexual images to exemplify a larger human desire, and to point to "an erotic in every truth."[26] These are images that might be viewed sympathetically by those who let their own sexuality hang out of a Lacano-Freudian matrix, but the images employ the sexuality not of men to men, or women to women, thus allowing another cheer for diversity, but the sexuality of both men and women paired together. This is an unfortunate biological necessity for the propagation of our species, leading to an awkward contretemps for the politically correct postmodernists of whatever gender.

Such a contretemps did not stop Matthew Sparke from jumping in with both feet to produce what must be one of the most authoritarian tracts on political acceptability in a genre marked by such authoritarianism.[27] For in an accusation tortuously written in the latest jargon, and armored in all the Lacano-Freudian panoply of the unsatisfied radical feminist, he proceeds to entangle himself in the lime of ambiguity smeared so artfully to catch the noisy and unwary sparrow. Despite his angry and angst-ridden posturings, his complaints that he cannot find a single unambiguous reading, his railing about brutish patriarchal politics, and so, wearily, on, he insists on seeing what he wants to see, injecting what is not there, and misreading passages even when warning signs have been placed both before and after. The reply by Olsson takes the tirade of Sparke apart piece by piece, but Olsson's four page reply is worth reading in full.[28] As he notes, "feminism has more to fear from its hired hatchet-men than from me."[29]

Authoritarian Homosexuality

Sexuality, in a variety of ways, appears to give problems to self-elevated social theorists and critics. It is, of course, one of the sharpest of two-edged swords characterizing the human condition, capable of cutting a path to the sublime, even as it can hack a way to horror, unspeakable cruelty, and degradation. It is an attribute of humanity that seems to create great conceptual difficulties for homosexual communities, difficulties that then appear as quite extraordinary and aggressive claims. For example, those writing out of the homosexual experience and community assert that the AIDS epidemic is essentially theirs, that only "they" (such *Us-Them*ism is their own characterization) can say anything meaningful about it, that only "they" understand the human destruction and desolation going on around them. While any humane person acknowledges the tragedy, such claims, or even the slightest move toward them, should be firmly rejected. They are claims to exclusive understanding that are dangerous on many counts, not the least of which is that they focus attention on homosexual transmission, when on a global basis male-to-male transmissions are about 1 percent, about as rare as homosexuality itself. Such a counterclaim will not please those within these communities pushing, quite unsuccessfully, the claim of 10 percent, but it is important to try to be truthful rather than wishful. Notice that from the homosexual perspective, the bodies of the other 99 percent of the human race, children, women, and men, simply appear as candidates for "erasure"—to turn a favorite expression back on those writing in this vein.[30]

Does it ever occur to those writing out their catharsis that most people today could not care less what sexual proclivities are displayed by consenting adults in private? Yet the vast majority of society is still seen as some hegemonic enemy suffering from heterosexism. One Canadian colleague, who wanted to investigate the HIV/AIDS epidemic with a view to helping by education, asked representatives of the homosexual community in his city for their help and advice. He never really got past their first question, "Are you gay?" When he said he was not, further inquiry was useless, and the responses became increasingly denigrating and aggressive. Finally, in a mood of mystified exasperation, he asked whether he, as a citizen and taxpayer, also had some rights, since the larger society of which he is a member was bearing an enormous and increasing financial burden in caring for those converting to the opportunistic diseases of AIDS. That produced such an expletive uproar that the interview was ended.

Homosexual authoritarianism appears in many guises, but two examples will have to suffice as an introduction to the genre. Unfortunately, it means breaking a golden rule of mine never to reply to a review, but, for all the potential for misinterpretation, it cannot be helped. Michael Brown did not like my book *The Slow Plague*, charging me with eliding the topic, erasing the experiences of homosexuals, polishing the book to an antiseptic sheen, as well as displaying homophobia in general, a final charge a number of my friends would find amusing.[31] "Indeed," he writes, "the term 'homosexual' is persistently used across this book,"[32] apparently oblivious to the fact that he uses the term "heterosexual" twice in the same, castigating paragraph without feeling a twinge of contradiction. He is particularly cross that a chapter focuses on the Bronx, where I draw on the great wealth of research by Rodrick and Deborah Wallace,[33] rather than Greenwich Village, where I knew of no *geographic* studies to draw upon at the time of writing. I also pay "little attention to gays' experiences in places," and I conspire "to exclude their . . . geographies."[34] The problem is that chanting "Place, Place, Place" like a mantra is not enough to make a study geographical. It is like a historian saying that dating an account to a specific time somehow qualifies it as historical. At this point, *any* topic capable of being specified in space and time becomes "geographic" and "historic," but only in the most puerile and banal sense.

Brown's own essay on the AIDS epidemic in Vancouver is an excellent example of geographic claims going unfulfilled.[35] Assuming he has mislabeled his histograms comparing *rates* across Canada,[36] Vancouver is unexceptional, with an actual rate roughly comparable to any North American city of the same size. This says nothing, of course, of the agony of those afflicted, but this is the subject for autobiographical accounts or for poets, because the geographer, as *geographer*, can illuminate nothing here except his own cathartic requirements. In the last analysis, we always die alone.[37] Brown wants geographers to focus on homosexual bodies, bodies erased by any spatial or epidemiological analysis, but as I have pointed out repeatedly elsewhere, there is no geography in a person's body, except in a metaphorical sense, and despite his shrill accusations about geographers ignoring the homosexual communities, the reader at the end has not the faintest idea how these were formed or structured, or even *where* they might be. Talk about "erasure": there is not a map in sight. Brown accuses everyone else of "[sacrificing] geographic knowledge,"[38] while supplying not a shred of genuine geographic insight with his "ethnography," and he goes on to complain that "these renderings . . . [make homosexual bodies], their contacts, their travels, their behavior . . . important only to the extent that they enable the virus to move across space."[39] Just so: the virus clearly has the capacity to be transmitted from one male body to another, and in this sense the body is a vector. What euphemism would Brown like to substitute? The fact is that individual behavior is essentially unknowable, or at the very least incredibly difficult to reconstruct in any reliable way when some have several hundred sexual partners each year. As I have noted elsewhere, there is almost nothing of *general* value in an individual medical portfolio.[40]

In contrast, scientific insight into this terrible pandemic is not without merit. Unable to say anything at the individual case level, we should nonetheless point out that AIDS rates are *essentially* density dependent,[41] which implies that *all* forms of transmission, not just male-to-male transmissions, are roughly distributed equally. Moreover, we know that an air passenger operator probabilistically smears the HIV across the urban hierarchy to predict AIDS *rates*, 0.80;[42] and that if New York is transformed into "commuter space," local accessibility indices predict AIDS rates over six years about 0.90.[43] Bodies, needles, blood supplies, and surgical "sticks" *are* the vectors carrying the virus from one person to another. Any conflating of the AIDS epidemic and small homosexual communities in a few cities is largely the result of strident claims made by these communities that this is *their* epidemic—and the rest might as well be erased.

In contrast, Scandinavia has virtually shut down the transmission of HIV by treating it as any other sexually transmitted disease, openly, forthrightly, and with the full cooperation of the homosexual and drug-injecting communities. Brown does not like mandatory reporting in Finland, and implies that ethical questions are not considered. One can turn this around, and ask what are the ethical questions arising for the other 99 percent of society by not reporting, i.e. hiding, a disease that is clearly lethal. The United States has now accumulated close to half a million AIDS cases, with perhaps another million HIV infected. Control of the pandemic will not be achieved by writing "lesbian geographies"—whatever these might conceivably be. Brown continues in this vein for twenty-four pages, but at the end he succeeds only in erasing a geographic perspective he does not wish to adopt himself. Despite wallowing in an intellectual milieu of "situated polyphonic perspectivalism," he requires only *his* single perspective to have any value or credence. Yet not one life will be saved, not one transmission will be stopped, by this cathartic exercise. The result is, quite literally, use-less—except for catharsis and academic advancement. What are the ethical questions that should be posed now?

Despite strident calls for "geographies of sexualities," and claims to a "steadily growing body of work concerned primarily with gay (and to a lesser extent lesbian) geographies,"[44] it is difficult to find any concrete and specific examples upon which David Bell bases his own accusatory writings.[45] In the same way that the mantra of place does not automatically produce geographic insight, so chanting "Space, Space, Space" does not create geographies of the "contracepted female body [or] the visibly pregnant female body."[46] And like Brown's mantra to place, Bell's apparent insistence that any writing is "geographic," simply because everything exists in space, leads to nothing more than a postmodern version of Herman Krickism.[47] Even reviewing the work of others becomes one more excuse for posturing display and cathartic release,[48] and discloses Bell thinking like a seventeenth-century Cartesian running around with a "body" and a "mind" (?) somehow glued together, "uncoupling [his] rigidly held binary chains in the play of utopian postmodern polysexuality." Illuminating geographic insight requires more than mantras.

Hegemonic Radical Feminism

The third strand in the cathartic genre is hegemonic radical feminism, a subgenre in which circular footnoting is so incestuously prevalent that a reader can almost predict the next paragraph. A major mark of most of this writing is its wholesale, uninformed, and uncritical adoption of both traditional and contemporary forms of Freudianism. For any informed person, retaining even the slightest hold on moderate reason, this is an extraordinary choice of a psychoanalytic framework within which critical thinking has to be constrained. As Stephen Gould has noted, "Freudians . . . elevated a limited guide into a rigid creed that became more of an untestable and unchangeable religion than a science."[49] Yet much radical feminist writing claiming geographic status positively wallows in its patriarchal sexuality, nominally fingering its vocabulary for "penis," "phallus," "genitalia," "castration," and so on, while adopting "phallocentric" adjectivally to the point of self-parody. Gillian Rose, for example, a leading Lacano-Freudian acolyte, uses "phallocentric" to qualify the following: *accounts, arguments, challenge, claim, constitution, critic, definitions, desire, discourses* (2), *economy, fetishism, gaze* (3), *knowledge* (2), *look* (2), *masculinities, mirror* (2), *mirroring, modes, modes of intersubjectivity, organization* (3), *perspective, pleasures, posing, power* (2), *principle, relation, reiteration, resistance, scenography, self, space* (11), *space of self, space of self-knowledge* (16), *spatiality* (3), *structure, subject* (31), *subjectivity* (4), *terms, visions, visual depth, visual spaces* (3), *visuality* (2), *visualized space*, and *voyeurism* (2), all in a single essay of eighteen pages.[50] No parody spoofing the subgenre is required where such self-parody is present.[51]

Pointing to the uncritical adoption of a Lacano-Freudian framework by hegemonic feminists should not be interpreted as "Freud-bashing." There is such a huge literature today questioning the positing of imaginary things by Freud out of his nineteenth-century mechanical framework that another voice can only be interpreted as a clambering-on-the-bandwagon.[52] But critical theorists, by their own strict claims to self-reflective thinking, should be able to grasp that it is perfectly possible and consistent to honor a pioneer opening up new areas of human thought and understanding, while at the same time being highly critical, even dismissive, of early claims, positings, and writings. In science (and Freud insisted on his claim to scientific status), ideas are examined critically, only to be replaced by other, more effective frameworks. No one denigrates Galileo or Newton because they failed to be Einstein. The radical feminist adoption of a rigid, unyielding, and above all authoritarian and patriarchal psychoanalytic framework becomes so strange and seemingly contradictory that the question of such an adoption itself becomes one of great interest. This is particularly so in light of the formulation of other, far more open perspectives intensely respectful of the patient, so that interpretation (for example, of dream content) becomes a genuine and joint hermeneutic endeavor, and not an insistence that every troubled person conform to a nineteenth-century template formed out of neurotic, middle-class women in *fin de siècle* Vienna.[53]

But the influence of hegemonic feminism goes far beyond the printed page, and appears quite capable of employing fear to bring both men and women under its discipline. I know of two cases where the academic advancement of women into high administrative positions has been derailed because they would not join the radical feminist and women's studies groups on their campuses, and so were punished accordingly. And with e-mail today, what starts as supportive networking can be used as defamatory innuendo with little chance for an "accused" to face the accusers. In the course of a convocation address at Clark University, I noted that many women in the academic world wish only to be judged as scholars, and not as women.[54] At the reception at the president's house afterwards, hardly a woman faculty member dared to approach me under the watchful eyes of some members of a tightly knit group in the audience who had stirred audibly when I made such an outrageous and politically incorrect remark.

Power can always be misused: one male friend, applying for a tenure track position at a "western university," was taken quietly aside by two women vice-provosts and told that if he declared himself homosexual the position was his. They had reserved the position, they explained, for either a homosexual male or a lesbian woman. He refused to lie about his sexual preference, and the offer, which was never officially made in public, was canceled on the spot. Numbers, too, can be distorted and misused in such an any-means-justifies-the-ends atmosphere of propaganda. At a recent AAG meeting, lists were posted showing the proportion of women faculty in geography departments, but those making the counts made sure that all the male professors emeriti, who had long since retired, were included in the denominators. Since most departments are quite small, such inclusions could shift the rank of a department radically up or down. The simple truth is that many more women are continuing to the doctoral degree today than some decades ago, so only now is the pool of potential applicants much enlarged from former days.

Posturing Postcolonialism[55]

The final subgenre of cathartic geography is a most curious one, for analysis of the colonial era is employed not to understand other worlds in conjunction, but to display one's own attachment to political correctness, and on the way achieve some level of catharsis by being on the side of the angels. What is particularly curious is that those writing in the postcolonialist vein have never experienced colonialism either as colonizer or colonized. Which raises, of course, the question of the motivation for such inquiry, since simple curiosity can hardly account for the obsessive critical concern. It is now half a century since India and Pakistan became independent, and forty years since most of former British and French Africa achieved a similar status of self-responsibility, yet almost every ill that befalls one of these nations is laid by the postcolonialists at the door of the colonial past. Few come to the spatiotemporal "geohistories" with anything approaching a sense of openness to the evidence, trying to achieve a balance in attempting to understand an era in which geographies were radically restructured. There are plenty of dismal things to find in many colonial records, but colonialism appears more and more diverse and complex the more one examines the particular experience, and there is also much to be found that was motivated by fine and high ideals, decency, and courage. The trusteeships of the United Nations were not shams.

A major tributary feeding a geographic form of postcolonialism is a genre of literary criticism that seems determined to put its own spin on any text, turning Jane Austen into a rabid imperialist sucking sustenance from the slaves of the West Indies, or interpreting Kipling's finest novel, *Kim*, in ways that any knowledgeable reader must find utterly bizarre. One theorist, mercifully identified only as "an American scholar," described Kim's first encounter with the old lama by noting that Kim "with the insight and intuition of a born colonialist, recognizes that the lama is a new territory worth annexing."[56] One wonders whether the book, one of the finest novels ever written in the English language, had actually been read. A different view comes from Rukon Adrani, a senior publishing executive in Delhi, and a novelist himself, who noted that "Plenty of people all over India . . . still read *Kim*. It is frequently a prescribed text at the BA Hons level in many of our 140 or so universities."[57] One wonders what a postcolonialist would do with accounts of Pathan militias on the North-west frontier.[58]

But the postcolonial perspective can intrude itself into the most unlikely forms of writing. Alastair Bonnett, in an otherwise fascinating article on some quite extraordinary male bonding and initiation rituals that have emerged recently in the western United States, insists that the "bricolage" of themes used are reworkings of "colonialist myths" drawing upon "primitivist discourses rooted in the colonial era."[59] But the elements of the original rituals and *rites de passage* had absolutely *nothing* to do with colonialism, and were certainly not "colonial fantasies," as though imperialists had been rushing around Africa making up stories. The original rites of initiation, many of them for girls as well as boys passing through puberty, have been honored and respected parts of indigenous cultures for hundreds of years, and are still important thresholds in many young people's lives to this day. They are no more myths to be ascribed to nefarious colonialism than is the strange rite of circumcision with a prescribed male audience for Jews, or the equally strange practice of splashing water on some poor screaming baby in a Christian baptismal service.

Beyond Catharsis and Forward to Geography?

The requirement for emotional self-release in public is one rarely encountered in geographic writing before the last ten years, and it clearly presents some problems, not the least for anyone willing to swim against the tide. Critical theorists, claiming "theoretical" positions that are untestable, and therefore irrefutable, are long on criticism provided the direction is away from them and toward others. Swivel the critical eye around, and murmur a dissent to the currently politically correct wave, and you are branded immediately as premodern, homophobic, antifeminist, and imperialistic. It is apparently useless pointing out that within each of these strands there are many genuinely illuminating, but noncathartic writings. Turning critical discourse back upon itself touches a raw nerve, and perhaps transcendental medication is all that one can suggest to ease the pain.[60]

But cathartic writing does not get published without cathartic editors and cathartic reviewers, and it is no use hiding behind holier-than-thou editorials deploring trashing, while at the same time aiding and abetting precisely the sort of trashing polemic that is deplored.[61] Some might even say that you were two-faced. Geraldine Pratt declares that the journal *Society and Space* "is not a club. We relish differences in opinion,"[62] but she holds Matthew Sparke up as a poor and abused "trashed" example, when it is transparently obvious that his cathartic release against Gunnar Olsson could not possibly have been written without access to Olsson's essay long before publication. Such access is clearly the responsibility of Editor Pratt, who, along with others, is thanked by Sparke "for their comments on drafts of this paper."[63] It is clear that the thirteen-page polemic, with thirty-three references, was composed over a substantial period of time, with inputs of commentary and suggestions from the seven listed in the acknowledgments, one of whom, Gillian Rose, was in the United Kingdom, rather than Western Canada like most of the others. But as Olsson notes, "Exactly twenty-nine days after receiving the subscriber's copy . . . I get a letter from the editor asking me to respond to the critique she encloses. . . . [T]he . . . commentary on a lifetime's work had taken less than three weeks to write."[64] But, of course, aided and abetted by an antitrashing editor, the "collective effort" had been carefully planned and had taken much longer.

Is this, in turn, fair? Do I qualify under the Prattian guidelines as a Spivak-approved internal critic? I certainly accept "at least some of the . . . priorities and frame of reference" as a geographer,[65] for a lifetime commitment, a calling to geography, situates me within a debate and critique initiated and carried on by someone claiming to be a fellow geographer. And *this* claim should in no way be construed as a veiled attempted to draw a boundary line. Growing up at midcentury in the Age of Adoration, whose Maîtres adjudicated what was and was not "Geography," I rejoice that a new and much more intellectually secure and confident generation is reaching out to other scholars in all directions. One of the most rewarding professional experiences I have had was reading, a few years ago on sabbatical, scores of books and articles opening up and refurbishing a new social and cultural geography that asked pertinent and socially biting questions. But now is the time for all good editors to put their judgment where their publicly declared principles lie, rather than skulking behind editorials while egging on stalking horses for their own agendas.

It is time to move on: let us have genuine critique, not trashing; let us embrace, at least intellectually, all others who are trying to illuminate this strange and wonderful human condition of ours. However, *if* we claim to be geographers (and this is not a condition for immediate nirvana), then let us think what we ourselves can bring out of the geographic tradition to the still rather bare Table of Knowledge. Chanting mantras to Space and Place will not do it: it is time to move beyond Cathartic Geography to matters of geographic substance and illumination.

The Structure of Spaces

THE VERY IDEA of a "structure of a space" has intrigued me for most of my professional life.[1] It has many informing strands, some of them now past recall, but it started with Ned Taaffe, my graduate advisor at Northwestern. He was fascinated by the way air passenger movements measured connections between America's cities,[2] and even then he was speculating about the geographic effects of the then-unbuilt Interstate Highway System. The basic idea of spatial structure was certainly nurtured by my own dissertation research in Ghana, as I saw how building, graveling, and tarring roads during the brief "colonial presence" of less than sixty years differentially warped and completely reshaped the traditional map of geographic space into seasonally pulsating cost and socioeconomic spaces.

A little later, reflections on the "cosmic wastepaper basket" space of Newton, and the disquiet of Leibniz, who insisted that space was defined by relations between *things*, were reinforced by the arrival of multidimensional scaling. This was an approach that allowed you to work quite concretely with sets of things and measures of similarity (relations) between them, and to show these graphically in map-like form as distances in all sorts of spaces. Some of us had fun rereading the plays of Shakespeare, coding very crudely how the players addressed each other, and producing maps of Romeo and Juliet, Lear, Hamlet, Macbeth, etc., in "Shakespeare space," easily divided into Montague and Capulet regions for the first, or Alive and Dead regions in the case of Hamlet.

A very different perspective on spatial structure arrived in the 1970s when I worked with two mathematicians, Ron Atkin and Jeff Johnson, both algebraic topologists trying to make well-defined and operational some of those intuitive ideas we all have when we use, perhaps too casually, the word "structure." In formal ways, they tried to define sets of things and the relations between them very carefully, so that these connections formed what they called a "backcloth." Other things had to have a backcloth structure to exist at all, and sufficient "connectivity" so they could move, or be transmitted, from one part of the structure to another. Although quite abstract mathematically, the intellectual exercise of thinking these concepts through at an empirical level was often even more difficult: for example, thinking of a viewer's eye as traffic being transmitted over the architectural backcloth of an old village square; condoms as breakers of connections between sets of human beings structured by sexual relations over which the HIV could be transmitted; a set of subjects connected together to form the backcloth of a television program carrying a traffic of cultural values,[3] and so on. I have not actually carried out any formal analyses for years, but the *thinking* of backcloth, traffic, and traffic transmission is with me all the time as a geographer. Even in writing and *structuring* this book, I have to think about you, the reader, as traffic being transmitted smoothly over this backcloth, without too many structure-breaking literary condoms on the way!

In the geography department at Penn State, we often used to play Graham Chapman's wonderful teaching game *The Green Revolution*, setting aside most of a Saturday morning and afternoon for a single game. For many students it became quite an emotional experience, and the debriefings at the end produced many different perspectives on the same game that everyone had just gone through, playing an Indian villager struggling to exist in a marginal environment. How the player-villagers perceived the structure of their interactions with each other could be mapped to disclose "winds of poverty" blowing some into the rapacious hands of two moneylenders, or disconnecting others from the game in secret ways.

As for postal spaces, one day I realized that my own, had-it-up-to-here chagrin at the slowness with which letters and manuscripts arrived from around the world was shared with a number of colleagues across the campus. It transpired that budget-cutting directives from the central administration had produced a flurry of attempts to cut postage costs by employing private carriers and other bulk carriage devices. No one in the administration had thought about what happens to the transmission of traffic when you change the backcloth connections, but a modern university is a node of communication in a worldwide backcloth, and its fundamental functions of teaching and research depend on rapid communication. So with the help of the university's postal service, and a number of academic colleagues around the world, I conducted a simple experiment comparing the various carriers. Even the central administration understood the implications, and Penn State soon restored its more rapid connections to the outside world.

As for restructuring spaces, for many years now my wife, a ski instructor, and I have gone to visit our ski instructor son who teaches at Beaver Creek (Vail,

Colorado)—just to make sure he is all right. Ever since living in Sweden for a year, I have been hooked on cross-country skiing, not for competition, but simply to enjoy the quietness, the beauty, and the peace in a slow, rambling sort of way. At Beaver Creek there is a marvelous track between 9,000 and 11,000 feet, where you see eagles soaring across the valley looking toward the peaks of the White River National Forest. One of the most charming moments of my life came while skiing there alone. Suddenly a little white ermine launched himself in a white streaking parabola out of the soft snow on one side of the track, disappearing like a diver into water on the other. I stood still, absolutely transfixed, until up he came, standing on his hind legs, looking at me quizzically while he brushed the snow out of his ears. For almost a minute we stood looking at each other while he finished his toilet, and then, totally nonchalant and unconcerned, he scurried into the deep hollow around a tree half-buried in snow.

So where is the structure in this personal account? Nowhere: life sometimes is more than structures, and pleasures should be shared. Those who cross-country ski know the pleasures of being able to take off and ski for miles without backtracking, with always a fresh scene coming into view. Beaver Creek's lovely track had a topological structure that did not allow you to do this: a skier, me, as traffic being transmitted on a ski track backcloth, had to go over certain sections at least twice. What to do? Well, invite the grandfather of all topologists to ski along with me and make some suggestions for restructuring the course. Leonhard Euler, who really invented *analysis situ*, or what we now call topology, knew all about spatial structures from his earlier work in Königsberg, so together we suggested some changes so that all skiers could make an Euler circuit!

Microgeographic and Behavioral Space
The Game of the "Green Revolution"

> Search processes may be viewed . . . as processes for gathering information about problem structure.
>
> —Herbert Simon
> *Sciences of the Artificial*

The Structure of Spaces

THE WAY IN WHICH the structure of a space shapes the behavior within it is often a question of major interest and concern. Examples range across such things as the movement of spermatozoa in a uterine canal structured by an acidity gradient, the spread of fire in a building constructed from materials of varying inflammability, daily work patterns shaped by historical changes in land ownership,[1] the spread of rabies across a landscape structured by fox territories,[2] the diffusion of measles at the national level,[3] and the trajectories of subatomic particles warped by gravitational and magnetic fields. No matter whether the scale is microscopic or cosmological, an assumption is often made that the structure or "geometry" of the space influences the movement of things within it. The relationship between structure and movement can also be reversed: nonrandom movement may reveal underlying spatial structure.

Since, by tautological definition, "space" is an inherent part of all spatiotemporal study, it is not surprising that cartographic representation often forms an important and valuable analytic approach. While it is always possible to express geometry algebraically, a visual, graphic expression has the capacity to generate understanding with an immediacy unmatched by other sym-

---

Originally published as "Espace de comportement et espace microgéographique: Le jeu de la 'Révolution Verte,'" *MappeMonde* 1 (1987): 4–9.

bolic forms—as great mathematicians, like Jules Poincaré, well knew. He was renowned for his ability to combine powerful geometric and algebraic thinking, and apparently became quite cross at the thought of doing the dull algebraic proof when he had already "seen" that it was obviously true.

Now despite the very general context of these introductory paragraphs, this essay has a more modest, and essentially pedagogic, aim. It applies methods already well known in the literature to a specific example of the "geometry-behavior" problem in a microgeographic setting. In the process, it calls attention to a game of great pedagogic value created by a geographer, and points to some references that may be useful and suggestive as catalysts to creative geographic thinking.

The Green Revolution

The Green Revolution game, and its more complex, expanded versions, was devised by Dr. Graham Chapman to sensitize people to the difficulties of marginal farmers in many Third World countries.[4] Based on detailed and meticulous fieldwork in Bihar, India,[5] it is now used regularly in training programs in the World Bank in Washington and New Delhi, in agricultural and economic research institutes in India, and in academic programs of geography students in the United States, Canada, and the United Kingdom.[6] Its impact on the players is often quite extraordinary: both government officials and university students tend to approach the game in a condescending, even blasé, fashion, but all of them take part in the often heated discussions afterwards, and they admit that participation in this game has radically changed their understanding and sensitivity. In some games, I have had students who have become quite distraught and wept when their children died and their farm went under. Their particular circumstances had not been fair, they said, although the moment of their distress was hardly the time to note that this was precisely the point.

The game is played for six to eight hours, with 8–20 players, a banker, and a game manager. The latter plays the role of Nature, assigning by the chance drawing from decks of specially designed cards the initial farms and farmers at the start of the game as small plastic figures, fields, tokens, etc. (fig. 10). She also produces at random the rainfall and pest attacks during the early, middle, and late rice seasons; assigns births at the end of each round of the game; and determines the arrival of the monsoon. This occurs every thirty or forty min-

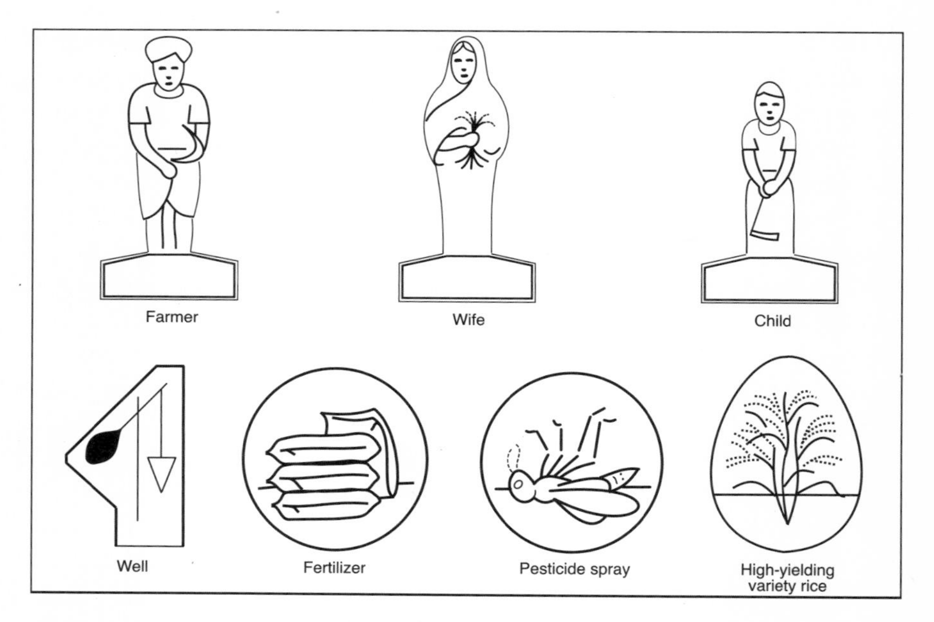

Fig. 10. Some of the figures used in the Green Revolution game: a farmer and his wife and child, and the tokens used for wells, high-yielding rice, fertilizer, and pesticides.

utes, and the game manager may announce five minutes ahead of time that she sees the black storm clouds looming on the horizon. This often produces a burst of frenzied activity. Players may protect themselves against drought and pests by sinking wells and buying pesticides if they have enough money, and they can increase their yields by planting the new hybrid rice. This, however, makes greater demands on water, pesticides, and fertilizer, and so it tends to be chosen by the richer farmers.

There are virtually no formal rules to the game, except that physical violence among the players is forbidden. There have been a few occasions when they had to be reminded of this by the game manager, since emotional involvement can become quite high. Otherwise, players are allowed to work alone or together; pool their resources; form cooperatives or unions; set up their own bank; lend money, seed, fertilizer, and pesticide—in fact, do anything that is not contrary to the laws of Nature. In other words, they may not claim to grow rice without land, to keep children alive without food, to survive severe drought without a well, and so on. The game is enormously rich in the virtually infinite variations it may generate, and it forms an excellent example of a system that is maximally complex. Chapman has been approached on a number of occasions by computer modelers, who thought that the game could easily be simulated by a computer drawing random numbers.[7] However, after taking part in one game, which always unfolds as a result of totally unpredictable human decisions, all the computer wizards decided to devote their intellectual abilities to simpler topics—like quantum physics.

In some respects, the most important part of the game comes at the end, when the players, banker, and game manager take part in a long, detailed, and very frank debriefing session. It is here that the overall economic performance of the village is summarized, and contrasted with the results of individual farms or cooperatives. Pointing to these often stark contrasts raises heated, even bitter, discussions about moral and ethical issues, and students may argue about them for days after the game. In addition, these debriefing sessions raise penetrating questions about a wide variety of other issues—minimum farm and family size, types of rice, investment decisions, access to information, and so on—issues that lead to a deeper understanding of the difficulties faced by marginal farmers. It was during such a debriefing session that the question of the microgeography of the "village" (seminar room) was raised. Some of the players began to suspect that *where* they had been located relative to others in the game had had some effect on the intensity with which they had interacted. Such interactions took place throughout the game, but particularly when the players were negotiating for loans, forming cooperative units, exchanging excess labor for rice, and so on. In brief, the idea arose that the spatial behavior of the players was not unrelated to the "structure" of the space, and that the structure was partially defined by the relative locations of the players themselves. The question was: how could the relationship between the "geometry" of the microgeographic space and the configuration of the behavioral space be examined?

Microgeography and Human Behavior

The students who played this particular game chose Indian pseudonyms, and these are used here to preserve anonymity. Specifying the structure of the microgeographic space of the room in which the game had been played was easy (fig. 11), for a simple map and a coordinate system was sufficient. But what about the behavioral, or interaction, space? To approach this more difficult

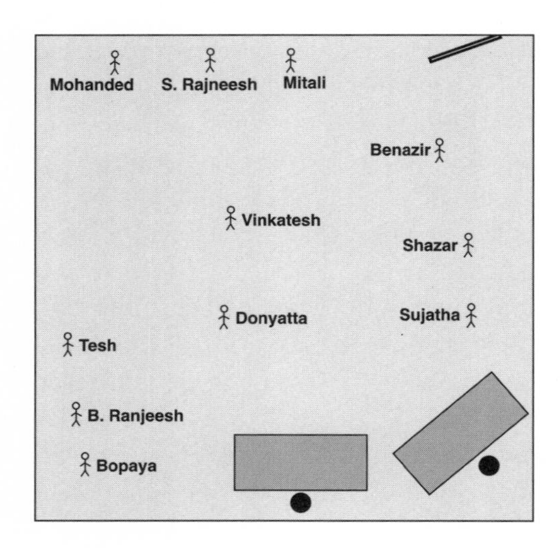

Fig. 11. The actual configuration of the microgeographic space, a seminar room, used by the players of the Green Revolution game. The chairs, and most of the small tables, have been omitted for the sake of clarity and simplicity.

	Bopaya	B. Rajneesh	Tesh	Donyatta	Vinkatesh	Mohanded	S. Rajneesh	Mitali	Benazir	Shazah	Sujatha
Bopaya		11	8	10	5	3	9	2	4	6.5	6.5
B. Rajneesh	11		4	8	6	2	7	3	5	9.5	9.5
Tesh	4	10		11	4	4	8	4	4	8	8
Donyatta	9	8	7		3	3	5	3	6	10.5	10.5
Vinkatesh	5	10	2	6		4	9	8	3	11	7
Mohanded	8	7	6	2	9		11	10	5	4	3
S. Rajneesh	5	7	2	4	6	10		11	3	9	8
Mitali	2	4	3	6	9	10	11		5	8	7
Benazir	4	9	3	6	7	2	10	5		11	8
Shazah	3	7	3	9	7	3	7	5	10		11
Sujatha	4	7	2	10	6	3	8	5	9	11	

Fig. 12. The matrix of perceived intensities of interaction by the players during the Green Revolution game recorded as reversed rank orders. Notice that the matrix is *not* symmetric, except in the case of a few elements on either side of the diagonal.

idea, I asked each player to make a list, placing themselves at the top, and then ranking all the other ten players below them in order of the intensity with which they had interacted with them during the game. This is a very crude measure of *perceived* interaction, and the asymmetries actually produce a non-metric space, not an uncommon situation in many areas of human geography.[8] The mathematically sophisticated reader will have to judge whether we can take some strictly unjustified liberties with these ordinal values. In principle, other measures of interaction could be devised, perhaps by monitoring the entire game on television, and then carrying out a very careful content analysis of the interactions, counting either the number of exchanges, the time they took, or even some weighted measure of their emotional intensity.

The perceived interactions form a "behavioral" matrix (fig. 12), in which the ranks have been reversed simply to make highly perceived interactions correspond to large numerical values. Obviously, there is no reason why the matrix should be symmetric: for example, Shazah perceives quite a low level of interaction [3] with Tesh, while Tesh has a much higher perception of his interaction [8] with Shazah. We shall examine and interpret such asymmetry later, but for the moment we need to create a behavioral space from this data. This is easily generated by multidimensional scaling (MDS), using the now standard hybrid program KYST, named after *K*ruskal, *Y*oung, *S*heppard, and *T*orgerson, who contributed various pieces to it.[9] The interaction matrix was folded along the diagonal, and the two values added together to provide a measure of the total interaction between the pairs of players.

The result is a map of the players in "behavioral space" (fig. 13), where players interacting strongly with each other are close together, while those having little to do with each other are far apart. The stress (a measure of how well the distances in the configuration match the interaction values) is reasonably low, but its value of 0.123 raises three important questions. First, the question of the dimensionality of the space: if the behavioral data are scaled in three dimensions, the stress will obviously be reduced, in this case to 0.051. However,

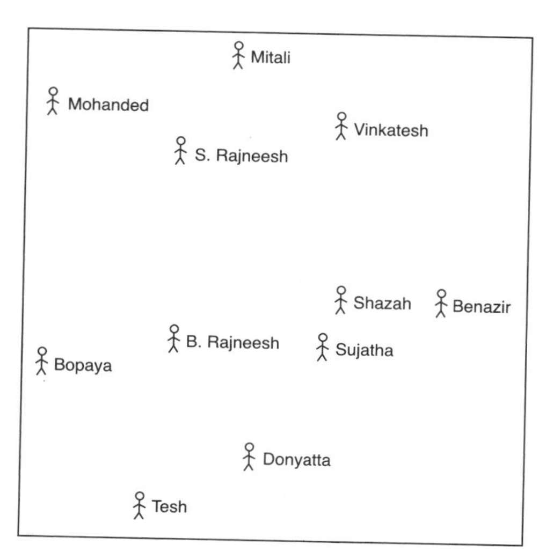

Fig. 13. The locations of the farmers in the behavioral space, created by multidimensional scaling.

with only eleven players we might expect such a reduction, since we have the additional freedom provided by a third axis, but we shall not explore this question further here.

Secondly, we must ask whether the stress is a global value, or whether the configuration has become stuck on a local minimum. While no formal mathematical proof of a global minimum is available, many experiments, involving different starting positions, were tried, and I am reasonably confident that the map presented here constitutes the global configuration. As a mathematician, Kruskal warns readers and users explicitly about this problem, but few human scientists appear to pay the slightest attention to him. A number of local minima were encountered during the analysis, and from considerable experience over the past couple of decades, I would guess that perhaps more than a quarter of the results published in the human sciences are *not* the global configurations they claim to be.

Thirdly, we must ask whether the metric of the behavioral space is Euclidean, implying $p = 2$ in the general expression

$$d_{ab} = [\,|x_a - x_b|^p + |y_a - y_b|^p]^{1/p}.$$

This is nothing more than the theorem of Pythagoras (the square of the hypotenuse is equal to the sum of the squares on the other two sides), rewritten in a very general algebraic way by the mathematician Hermann Minkowski. When $p=2$, we have the familiar Euclidean space of Pythagoras, and when $p=1$ we have a Manhattan, or taxicab space, since distance is now measured over a square grid of streets typical of movements in the highly structured space of an American city. When $p=3$, we usually refer to an Amorous space, since the geometry now draws two points closer to each other, like the eyes of lovers meeting across a room. The barbarians on the Outer Islands of Europe tend to $p=1$, and they make excellent and very practical taxis. Those who live in *la belle France* tend to $p=3$, and they make excellent and very impractical . . . but we do not have space to pursue this topic further.[10] A series of experiments, with p ranging between 1.3 and 3.0, indicated that the stress was extraordinarily insensitive to the p value employed, possibly because the entire space of interaction was so small in this case that little effect could be detected.

The question now arises: what exactly *is* the relationship between the configuration of players in behavioral spaces (fig. 13), and their locations in the "village" (fig. 11). You will recall that this question was raised intuitively by the players themselves, a question examined by bidimensional, or Euclidean, regression.[11] This is simply a two-dimensional version of the ordinary regression $Y=a+bX$, which relates one set of locations (x,y) to another (u,v) by:

$$\begin{bmatrix} \hat{u} \\ \hat{v} \end{bmatrix} = \begin{bmatrix} a_{11} \\ a_{12} \end{bmatrix} + \begin{bmatrix} b_{11} & b_{12} \\ b_{21} & b_{22} \end{bmatrix} \cdot \begin{bmatrix} x \\ y \end{bmatrix}.$$

The linear operation of translation is produced by the $\boldsymbol{a}$ vector, while the rotation and scaling operations are contained in the **b** matrix. It is important to note that the mirror operation, that of turning one map over, is *not* accomplished by these three linear operations. The bidimensional regression parameters are usually of little empirical value or interpretive interest, since they depend solely upon the coordinate systems employed for the two maps. Of much greater interest are the degrees of correlation ($r=0.65$) and the residuals. After linear translation, rotation, and scaling by the usual least squares criterion, the fit is reasonably close, but the residuals appear as vectors (the difference between the $[\hat{u},\hat{v}]$ locations predicted in the behavioral space from the $[x,y]$ locations in microgeographic space, and the actual $[u,v]$ locations) (fig. 14). The residual vectors may be interpreted as the remaining nonlinear, or "rubber sheet" stretchings and squeezings that must be undertaken to bring the microgeographic and behavioral spaces into total conformity. If the residuals do not cross, it implies that a smooth transformation could make the two con-

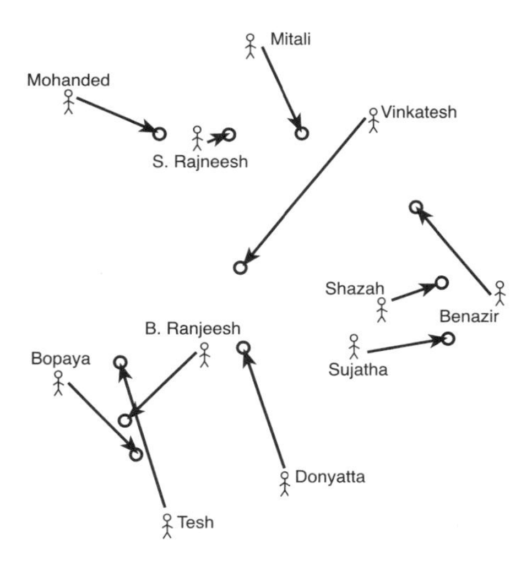

Fig. 14. The residual vectors from bidimensional, or Euclidean, regression after the locations of the farmers in the microgeographic space of the village have been used in an attempt to predict the locations in the behavioral space.

figurations conform, so that one was just a continuous deformation of the other. But in this game, the two vectors of B. Rajneesh and Tesh cross, implying that a rather drastic tearing of the microgeographic space would be required in that particular local area.

It remains only to examine and interpret the asymmetries of the interaction matrix (fig. 12). One way to highlight such asymmetry cartographically is to examine the net effect for each player as a "wind of influence" blowing across the space.[12] While a computer program is available for large matrices, the principle is quite straightforward, and for small problems computing the "winds" by hand has considerable value. Simply imagine a piece of transparent graph paper centered over each player, say Donyatta in this particular example. From the interaction matrix we see that Donyatta interacts with Bopaya [9], while Bopaya interacts with Donyatta [10], so there is a net flow of 1 from Bopaya to Donyatta. This is shown as a vector of unit length from Bopaya to Donyatta (fig. 15). On the other hand, Donyatta and Benazir both interact with each other with the same intensity [6], so the net "wind" is zero. All the vectors are drawn, their coordinates found, and then summed to give the resolution vector, the net force or "wind of interaction" blowing Donyatta roughly toward the right-hand side of the village (fig. 16). The other winds are computed in a similar way for all the other players.

To interpret these strange winds and crosscurrents blowing through the microgeographic space requires a detailed knowledge of this particular game. Bopaya, Tesh, Mohanded, and Mitali were very poor farmers, constantly seeking loans of rice for seed and food, as well as money to buy pesticides, fertilizers, and wells, sometimes from the richer farmers Vinkatesh and Donyatta in

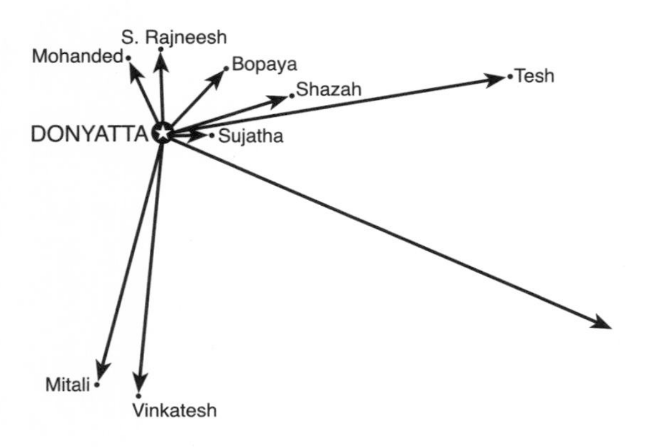

Fig. 15. Computing the net "wind of influence" arising from the asymmetries in the perceived interaction matrix. The example of Donyatta is used here.

the middle of the village, or, even more frequently, from the very rich Shazah-Sujatha combination. The latter was an enormously successful moneylending partnership that became the major source of loans in the village. Of course, "successful" here is meant in a purely material sense, with no moral or ethical overtones. The combined wealth of these moneylending farmers generated great resentment, and they received some very heated criticism for the selfish, grasping, and uncooperative way they played. They, in their turn, attributed their wealth to extraordinarily astute management and their own high and individualistic moral fiber. All players tend to obliterate any memory that chance plays a very important role in the game.

Given difficult lives and even despair to the "west," and moneylending wealth to the "east," it is easy to see why the winds of poverty are sweeping across the space, blowing the poor farmers toward Shazah and Sujatha as sources of material survival. However, two farmers, B. Rajneesh and S. Rajneesh, have winds blowing them away from the center of the space. This implies that some of the other players perceive themselves as interacting with these two farmers much more than they themselves evaluate their own interactions. In fact, these two "brothers" had made a secret agreement before the game started to steal as much as possible from the other villagers. They were highly successful in this antisocial activity, surreptitiously taking money, rice, fertilizer, and pesticide while the others were distracted, perhaps because these farmers were arguing among themselves, or negotiating for a bank loan. In this

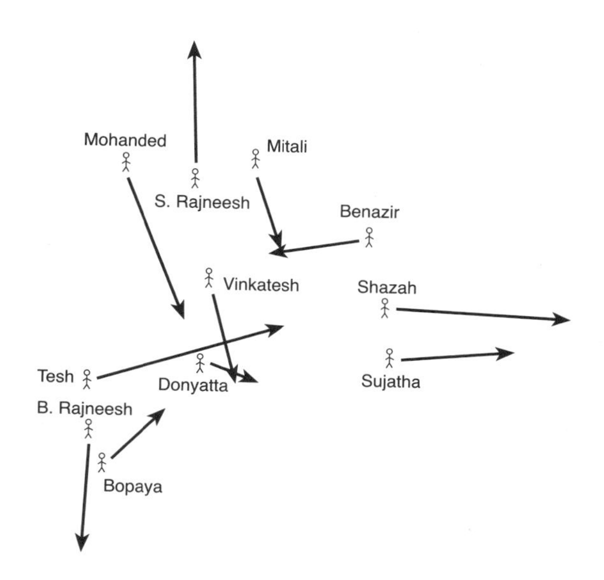

Fig. 16. The "winds of poverty" blowing through the microgeographic space of the village partly controlled by the movements of poor farmers toward the sources of loans in the villages.

sense, the Rajneesh brothers were not really in the same game, and the winds of interaction tend to push them away from the rest of the farmers in the village.

As noted initially, the intent of this essay was to be catalytic, indicating possibilities for cartographic expression and analysis to suggest provocative ways of thinking about spatial relationships. We all live in structured spaces at a variety of scales, and human geographers are acutely aware that the way these backcloths are structured may well effect the players on the stage. But what do we mean by "structure"? And how, quite concretely, does a structured space shape and relate to the behavior within it? What metric is appropriate? What representations do we have available? How can we examine and interpret asymmetries and their effects, whether these are differential perceptions and interactions, or even one-way street systems in towns that sometimes generate such frustration and stress for car drivers? On our planetary home, what are the global winds of perception? How do people push and pull other places away or toward them? What are the winds of interaction, the breezes of asymmetric perception, blowing in Europe today? Or between the departments of France? Whatever the mode of analysis, the final summarizing expression must be in cartographic form. Only then can someone say "I see! . . . and I never thought about it in that way before!"

Penn State in Postal Space(s)

> Here again the communities of scholarship beyond the borders of individual campuses or even of individual nations are increasingly essential to the university, as the only practical means of assembling and interpreting the pertinent data and, invoking the distinction drawn earlier, of turning this information into knowledge (if not also into wisdom).
>
> —Jaroslav Pelikan
> *The Idea of the University: A Reexamination*

Sharing Knowledge

As a university, in both its teaching and research functions, Penn State is a major hub of communication for the state, nation, and world, employing telephone, fax, e-mail, postal services, radio, television, publication, and oral presentation to share its knowledge with others. We are reminded that *communication* has embedded within it *con munus*—"with present," a sharing of gifts. Such a sharing of knowledge, of all types and at all levels, is the most vital task of a university, its raison d'être. A university disconnected from the world, unable to communicate, is a contradiction. A scholarly life may on occasion be cloistered; it cannot be disconnected.

It follows that all forms of communication must be as reliable and as efficient as possible, or the fundamental function of a university is placed in jeopardy. Few funds expended in the university can be so well justified as those allocated to the basic tasks of communication. If budgets must constrain, this is the last place they should bind, for such binding hobbles the teaching and research functions, both of which are defined by a dissemination of knowledge.

Written to clarify the efficiency of private postal services, which claimed equal service for less money compared to the United States Postal Service as a member of the Universal Postal Union.

Communication from a university is not easy, and it may be expensive. Increasingly, in a major university like Penn State, it is international. It cannot be purely electronic, even with an unlimited budget. It is quite usual for telephone, fax and e-mail messages to be followed up by postal communication, and the first two—telephone and fax—may be an order of magnitude more expensive compared to the half-ounce letter conveyed for $0.50 or less. Many professional communications must be in a permanent or semipermanent form, requiring hard copy, often with a signature of responsibility, if not legal commitment. One only has to think of scholarly invitations, letters of recommendation, submissions of scholarly articles, editorial decisions, letters of negotiation and appointment, and hundreds of other forms for which the telephone call, fax machine and e-mail are inadequate by themselves. Postal services are at the core of the communication task, and in general they are the least expensive.

Control of the System

In the last analysis, no university controls the forms and institutions it requires to fulfill its communicative responsibilities. We all need the complex hardware, physical structures, commercial services, and international agreements put in place by others. An international telephone call, for example, requires immense resources in national telephony, satellite placement, and international treaty. These are taken for granted; one hundred years ago they were not. As for postal services, a simple airmail letter from Penn State to the University of Darrusalam, Brunei, moves over an immense network and through numerous nodes, requiring bar coding, conveyance to local post offices, trucking to regional centers (Altoona), domestic air carriers to national points of assemblage, international carriers (often several), local sorting at intermediate airports, truck transportation at the destination, and final hand delivery, all made possible ultimately by the signatures of American and Brunei representatives to the Universal Postal Union (UPU).

Emphasizing such complexity is neither fanciful nor irrelevant. It points squarely to the fact that once a letter or package leaves University Park we do not have much control over the "system." Our own mail services are highly efficient, processing, billing, and moving nearly a quarter of a million pieces of *international* mail alone each year; but along the way, and at the "other end," there is little, if anything, that we can do to speed the mail to its final destination. Even from this necessarily limited postal experiment, we can infer delay as a result of ethically-inspired foot-dragging (France), quasi-chaos (Russia), simple lassitude (Turkey), sporadic delivery due to geographic remoteness (Costa Rica), translation from roman script (China, Thailand), religious observance (Ramadan in Brunei), and sometimes sheer, if not totally irresponsible, inefficiency, all of these acting singly or in concert to thwart the rapid and efficient conveyance of mail. Apart from making representations to the UPU, there is nothing we can do to make international postal communications more efficient.

Except one thing: the initial choice of a carrier, a choice now available in the United States as commercial firms have entered the mail conveyance market, offering cheaper rates than the United States Postal Service, but promising as efficient (perhaps even more efficient?) service.

Choice Points

In order to lower international postal costs running around $350,000 a year, Penn State has tried a number of options, including regular postal service, international priority airmail (IPA), and a commercial carrier conveying much of the international mail in bulk to Kastrup Airport, Denmark, where it is stamped at a special contract rate and sent on its way. Regular postal service is self-explanatory, and is simply the equivalent of paying the standard airmail fee at any U.S. post office. International priority airmail requires accumulating mail to a minimum of 10 pounds (4.55 kilograms), at which point it is taken by the U.S. mail service to Jamaica, New York, for stamping at a lower fee (15 cents for the standard half-ounce airmail letter), before being flown in bulk to its destination. Total costs of IPA run approximately 75 percent of regular postal service. The private commercial carrier charges a set fee, assembles pieces of mail from many clients, and flies them to Denmark, where the pieces are stamped and sent on to their destinations. Costs of commercial carriers are approximately 58 percent of the regular postal service.

Domestic mail from Penn State uses the U.S. Postal Service: except for Express-type mails there is no cost-effective alternative. International airmail to Latin America, Australia and New Zealand, and, significantly, France, uses IPA. All other destinations in Europe, Africa, and Asia are processed in Denmark. International postage costs the university more than the simple half-ounce charge of $0.50 times the 225,000 pieces, or $112,500: many communications are scholarly manuscripts, reprints, documents, etc., exceeding the half-ounce weight many times.

It must be emphasized, however, that the fundamental question of cost is not answered in mere dollars. The basic question is: what is the *true* price, who really pays it, and in what coin? If the private commercial carrier can offer services equal to, or better than, the U.S. Postal Service, clearly it becomes the rational cost-effective choice. Why pay so much more to the U.S. postal authority for the same service? But is the private carrier via Denmark providing equal service? This is something very difficult for a client to judge independently from statements of good intentions and assurances provided by the private carrier. Stated baldly: do letters and pieces of mail from Penn State get to their destinations in the same time, or even sooner, as those sent via the U.S. airmail and IPA systems? And can this be tested, given the vagaries, not to say idiosyncrasies, of other national systems subscribing to, and making up, the Universal Postal Union? It was precisely to test this question that this small postal experiment was designed.

A word of caution, which may not please certain economists, and others who prefer to reduce everything to either a concrete cash nexus, or to a theoretical, but practically useless, concept labeled "utility." What price a day's delay, or two, or three, or four . . . ? How do we reduce to cash a delay in scholarly publication that may obviate the priority of a scientific claim? What price a delayed letter of recommendation, which may alter a young scholar's life? What is the true cost of a communication from one scholar and researcher to another? How much for page proofs of an article or book? What price a communication with serious moral, as well as merely legal, consequences? How is delay at the postal level amplified into delays in communication with all sorts of scholarly and other human ramifications? What price for the loss of prestige claimed by an "international" university that appears to be unable to communicate with a speed and efficiency demonstrated by others in the same league? These consequences of inefficiency and delay are not to be measured in dollars, but judged by standards and values informing the fundamental task of the university, so allowing it to justify its claims as an international center of scholarship, research, and teaching.[1]

Testing Time: A "Feynmann" Experiment[2]

Serving on the committee of inquiry after the disastrous *Challenger* explosion, the distinguished physicist Richard Feynmann dunked some O-rings in a glass of ice water, and proceeded to demonstrate how they lost their elasticity and ability to seal at temperatures close to those experienced on that terrible day at the launch site. Experiments need not be complex, but must simply have the ability to point toward a truth that is convincing. Our own postal experiment could not have been simpler. On March 1, 1994, 95 duplicate letters were sent to potential respondents in 54 countries. The first (Letter A) was posted at 9 A.M. at the State College post office; the second (Letter B) was placed in the university mail system at the same time. Respondents were told to expect the two letters A and B, and were asked to record the date on which they were received. Shipping labels used on the envelopes also noted that the day of receipt should be recorded in case the recipient was not immediately available.

Of the 95 duplicate letters sent, responses were received from 76; of the 19 for which there was no response, five were in Africa, seven in Latin America, four in Europe, two in Asia, and one in Ukraine. Since all respondents were known personally, it may well be that some of these letters never reached their destinations.

Matters of Efficiency

From the 76 responses in 42 countries, we can begin to get a fairly reasonable estimate of the efficiency with which the international postal system works as a whole, as well as the potential delays and accelerations in delivery times due to IPA and commercial carrier.

Given that there are few airports in the world with regularly scheduled services more than 36 hours away from State College–University Park,[3] it seems reasonable to expect a considerable degree of efficiency from the international postal system working under the agreements of the Universal Postal Union. Unfortunately, no complex global system or chain is stronger than its weakest link, and once again we are reminded that the individual sender has no control whatsoever after the initial carrier is selected. In these terms the global network might be usefully regarded as a huge delay amplification system. For example, dilatory sorting at one node in the network may mean mail sits over a weekend, which means a domestic flight is missed, which means an international flight is missed, and so on; the initial delay is amplified at several other points along the way. Many combinatorial variations of this theme could be written, including delays by weather and destruction by warfare. At the time of writing (May 1994), an airmail letter to Kigali, Rwanda, is unlikely to be delivered with dispatch—if at all.

Unfortunately, international mail from State College–University Park does not appear to move with anything like the dispatch of an individual traveler.

Posted on a Tuesday morning (March 1, 1994) at the State College post office, the earliest Type A letters were delivered in Europe seven days later, with an average for the continent of 8.72 days. Russia is excluded here; a regular airmail letter to Moscow took 31 days to be delivered to the Russian Academy of Sciences, but the correspondent noted his despair of the current condition of local services, and asked to be informed of the time for his reply. Eastern Europe appears to be well integrated: letters to Estonia, Poland, and the Czech Republic took nine, eight, and eight days, respectively. Thessalonika (Greece) experienced considerable delay (14 days), but two letters to Athens took nine and ten days, so it seems safe to conclude that further delay was within the Greek postal system. Istanbul, Turkey, also seems to be a delay amplification node: two letters were received 20 and 21 days after posting.

In marked contrast, Type B letters, carried by private carrier to Kastrup, Denmark, were received on the average 12.7 days later, implying a mean delay of four days. But such an average is deceptive, since it includes six letters to various destinations in France, all of which were mailed via IPA from Jamaica, New York. When these are removed from the sample, the mean delivery time via Denmark rises to 13.3 days. The greatest delay appears to have been caused by the commercial carrier and the Danish postal services at Kastrup: from the postmarks of envelopes returned, it is clear that all the letters to Europe were only processed and mailed on March 9, or eight days after being sent on their way by Penn State's mail service. By this time, 67 percent of the Type A letters via the U.S. Postal Service and IPA had already reached their destinations. Smug complacency about the efficiency of the U.S. Postal Service may be misplaced, however: IPA letters to France were not stamped at New York until March 6, a "domestic delay" of five days. In contrast, they were handled with great efficiency and dispatched by the French postal system to reach nearly all the scattered destinations in that country only two days later, including time for handling at the airport terminals and the transatlantic flight. Not all European services can match this: a Type B letter to Orbaek, Denmark, just a day's hike from Kastrup, took five days to be delivered by the Danish postal system.

How the Kastrup-processed mail moves in Europe is not clear: presumably, rail, truck, and air are all used, depending upon the destination. What is clear is that the intra-European system as a whole needs to tune up its operations. Type B letters from Denmark took a further six days or more to reach such cities as Athens, Barcelona, Basel, Bergen, Brussels, Istanbul, Thessalonika, the Vatican, and Puerto de la Cruz in the Canary Islands. European mailings to the U.S.A. are somewhat better: based on 25 returns for which the postmark was legible, letters from Europe took an average of 7.36 days, with record times to University Park of three and four days from Amsterdam and Brussels, respectively. The former should be contrasted with an exasperated letter from a fellow of the Royal Netherlands Academy of Arts and Sciences to a Penn State journal editor, noting intolerable delays of two or three *weeks* in the reverse direction via commercial carrier and Kastrup. As it turns out, these are nothing: delays of journal articles of up to *six* weeks have been recorded.

Mailings to Latin America via commercial carrier are sorted separately and sent via IPA. Excluding a remote biological research station in Costa Rica without daily service, and a greatly delayed letter to Coyoacan, Mexico, caused by misspelling, there was no significant difference between the U.S. Postal Service and the IPA. Indeed, it appears likely that both employed the same bulk service. Unfortunately, deliveries by both systems took an average of 15 days, at least five days of which were in the United States before being postmarked at New York. The remaining ten days were used by the national systems. Such delays do not point to a high degree of efficiency anywhere in the Western Hemisphere. Australia and New Zealand do no better: both regular mailing and IPA took 12 to 15 days (allowing for the international dateline), with five of these accounted for by delays in the United States.

Letters from Penn State to Asia and Africa, however, use the commercial system via Kastrup, although it should be noted that *all* postal services make a mockery of the idea that we now live in a "global village." Letters to Asia via the regular U.S. postal system took an average of 14.9 days, while the Kastrup service took 17.1, eight of which can be laid squarely at the door of the commercial carrier, since the letters were only processed in Denmark on March 9. These values exclude the exceptional delivery after 27 days at Bombay via both systems, due to the egregiously dilatory nature of the Indian postal system at this particular delay amplification node. Africa is even worse off: regular airmail services average 16.6 days, while delivery via Kastrup takes 24.1 days, the delay amplification of over a week being due to processing in Denmark eight days after mailing in the United States. With reasonably favorable winds, an oceangoing catamaran could make it to Cape Town in about the same time (14–21 days).

Apart from the eight days lost due to processing in Denmark, it is difficult to know exactly where the remaining delays to Asia and Africa lie. Ramadan, one correspondent noted, probably caused some delay in Moslem Brunei, and may have been a factor in Jakarta, Indonesia, where it took 17 and 18 days to deliver Type A and B letters to a correspondent in the American embassy. China,

Thailand, and Japan appear remarkably efficient in delivering the roman-letter addresses, which require translation into local scripts and characters, and were among the shortest delivery times in Asia, with the exception of Hong Kong (9 days). As noted above, delay times between the two systems were small.

In Africa, however, the delay times were significant between the two systems, and for five countries no response at all was generated by either the Type A or the Type B letters. Direct flights from the United States to South Africa probably account for the 10–14 day delivery times compared to the 15–21 via Denmark, but delays greater than 8 days must presumably be due to inefficiencies in the national postal systems of Africa. Regular airmail took 17–21 days to Nigeria, Niger, and Malawi, but 23–36 days via Kastrup. The record delay of 36 days to Niamey, Niger, is worth commenting upon: given the lack of direct international flights from Kastrup to most of Africa, it seems reasonable to infer that letters are sent to other nodes in the international system with the best and most numerous direct flights to African cities. In the case of Niamey, Niger, this would be Paris. Here, it is suggested, the "ethically inspired foot-dragging" comes into play. The French do not appear to take kindly to American commercial carriers essentially bypassing, and thereby subverting, the agreements of the Universal Postal Union. They do not refuse to handle mail to France or Africa stamped under special contract rates in Denmark, but since internationally agreed upon airmail rates have not been paid, the French appear to be in no particular hurry to move things along. In a similar vein, letters from the United States bearing Danish contract rates will not be returned via airmail if they are not deliverable. Indeed, the French appear to have a knack for choosing the slowest boat available, so that several months may elapse before the original sender knows the letter never reached its intended destination.

In contrast to these delays, the correspondent in Niamey noted: "In fact the mail in Niger works very well with Europe. It's not the same with the Americas. . . . A letter from Niger to France takes three-four days only." The French ability to subvert those who they feel are subverting them seems to come into play, and it seems to be the major reason the commercial carrier makes France the singular exception in all of Europe by using IPA rather than Kastrup. Since the French are abiding honorably by international agreements, they cannot be faulted. Such agreements are important, since at the end of every year financial imbalances between all pairs of countries, due to small asymmetries in postal flows, are paid up. The French clearly feel the commercial carriers are "cheating the system," since their flows never show up in the financial statements indicating debts and receipts due at the end of each year.

Penn State in Delay Space

Exactly what the penalty is from using the commercial carrier can be highlighted by mapping the differences in delivery times between "ordinary postal space" and "commercial postal space" (plate 3), thereby showing Penn State as the center of a "delay space." Destinations with one or two days acceleration (Singapore, Hiroshima, and Santa Cruz, Argentina), and the Moscow and Istanbul aberrations are shown in gray. The most dramatic delays are to Europe, the very destination accounting for the bulk of Penn State's international communications. The exceptions are the French destinations using IPA, the same service that produces little or no delay in Australia, New Zealand, and Latin America. The apparently marked delays in Mexico and Costa Rica are aberrations due to misaddressing and an exceptionally remote location. Africa suffers acute delay via Kastrup, but 42 percent of the correspondent destinations never responded to either letters. All were in Francophone Africa (Cameroons, Guinée, Mali, Senegal, and Tunisia). This could, of course, be a coincidence.

Some Implications

Although small and simple, the postal experiment points to certain general and specific conclusions. First, in these days of rapid air travel, the international postal system appears to fall far short of its potential. Delays can occur at any point, and for a variety of reasons, but as a whole there appears to be much room for improvement. Second, the commercial carrier via Denmark may produce lower costs, but at the expense of unpardonable delay that jeopardizes grievously Penn State's ability to meet its international communication responsibilities. The university cannot legitimately lay claim to international status if its scholarly, research, and general communications with the rest of the world are so chronically delayed. This would appear to point to IPA as the mode of choice, since it provides 25 percent savings in cost with little or no delay over the regular postal service. A commercial carrier can cut this by a further 17 percent, but the savings in cash are paid for in other, and generally much more valuable, coin—namely, the ability of scholars to share their understanding with others.[4]

Skiing with Euler at Beaver Creek

> Die geographie ist mir fatal.
>
> —Letter from Euler to Goldbach
> August 21, 1740[1]

It is the delight of cross-country skiers to be able to complete a ski trip without backtracking at any point over a portion of the course. There is nothing rational about such delight; the experience is purely aesthetic. Whether such a redundant-free circuit is possible depends upon the topological structure of the cross-country ski course and the point at which the skier starts.

All ski courses over a three-dimensional topography may be projected onto the plane, and represented as a simple map or *planar graph* by a series of vertices (V), joined by edges (E), cutting the plane into separate faces (F). Since Leonhard Euler's remarkable pioneering work in 1736,[2] we know that for *any* planar graph $V-E+F=1$.[3] Simply as an example (fig. 17a), if we take a small set of nine vertices, we may join them by nine edges. Obviously, a skier may start at any point and complete a circuit without passing over any edge twice. This result seems, and actually is, utterly trivial, but things can get more complicated quite quickly. For example (fig. 17b), if we add two edges by joining vertex 5 to vertices 2 and 9, our skier can still complete an aesthetic circuit, but only if she starts at vertices 2 or 9. All other starting points force her to backtrack. Notice that by adding two more edges, we have also created two more faces, so that Euler's rule that $V-E+F=1$ still holds.

Suppose now (fig. 17c), we join vertices 1 to 8, 2 to 7, 6 to 9, and 3 to 9, at which point we realize that we have actually created seven more vertices where the lines cross (small open circles), and 18 more edges, and 11 more faces. Euler is still safe at $16-29+14=1$, but who would dare to hazard a guess that our skier can make an aesthetically appealing circuit now? And from what starting points?

Originally published as "Du ski de fond avec Leonhard Euler à Beaver Creek (Colorado, Etats Unis)," *Revue de Géographie Alpine* 85, no. 1 (1997): 101–114.

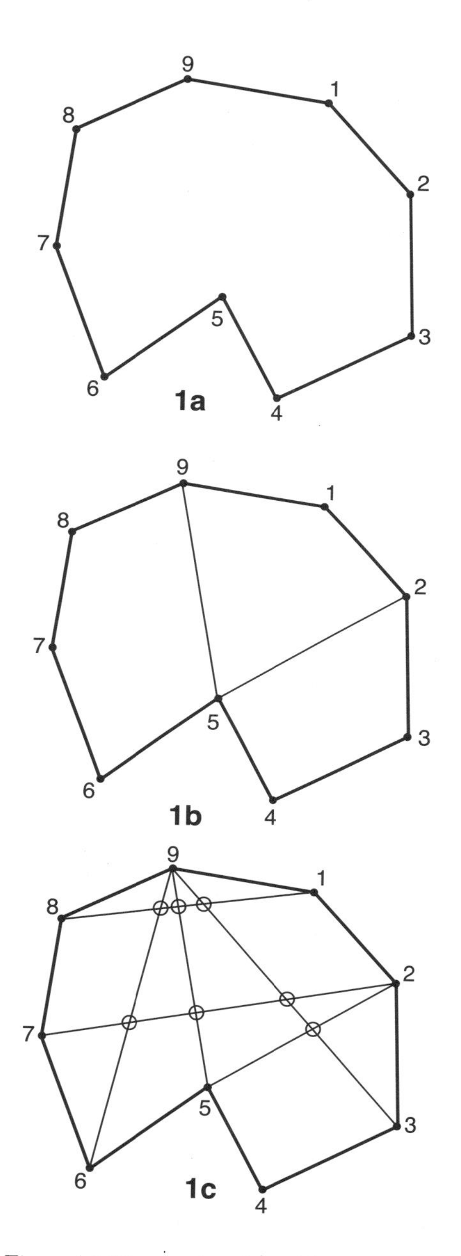

Fig. 17 (a) Nine vertices joined by nine edges to make a circuit. A skier can start anywhere. (b) Two edges added, confining a skier to vertices 2 and 9 to make an "aesthetic circuit." (c) A more complex graph, with new edges and vertices, raising the question of whether an "aesthetic circuit" is possible.

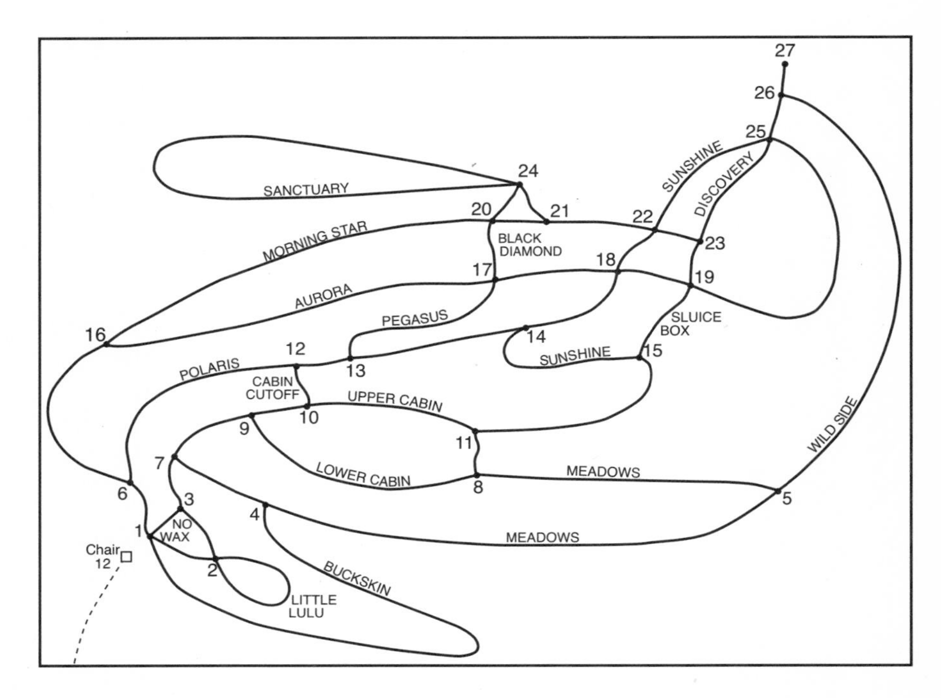

Fig. 18. The cross-country skiing course at Beaver Creek, Colorado, as a planar graph.

Now let us take a big jump into the complexity of the real world, and consider the actual cross-country skiing course at Beaver Creek, Colorado (fig. 18). The rather stark black-and-white graph of 27 vertices, 44 edges, and 18 faces (you can check that Euler is still safe!) is not as pretty as the oblique picture-map (plate 4), with the tracks shown in color against the white snow of the mountain with its green trees, but it has been meticulously field checked by a geographer getting snow on his boots, and it is more accurate topologically. And the topology, meaning how the vertices are connected by edges, is all that interests us here, although distances are more or less correct. In the case of vertices 8, 14 and 18, small delta (Δ) junctions with sides of a few meters have been reduced to simple wyes (Y), since the deltas were only formed by the snow tractor's forward and backing maneuvers. As for a starting point, vertex 1 is the only practical possibility, because chairlift 12 arrives there. One could try to start at, say, vertex 27 by coming over the more than 12,000-foot peaks in the White River National Forest, in four or five meters of soft snow, with a high probability

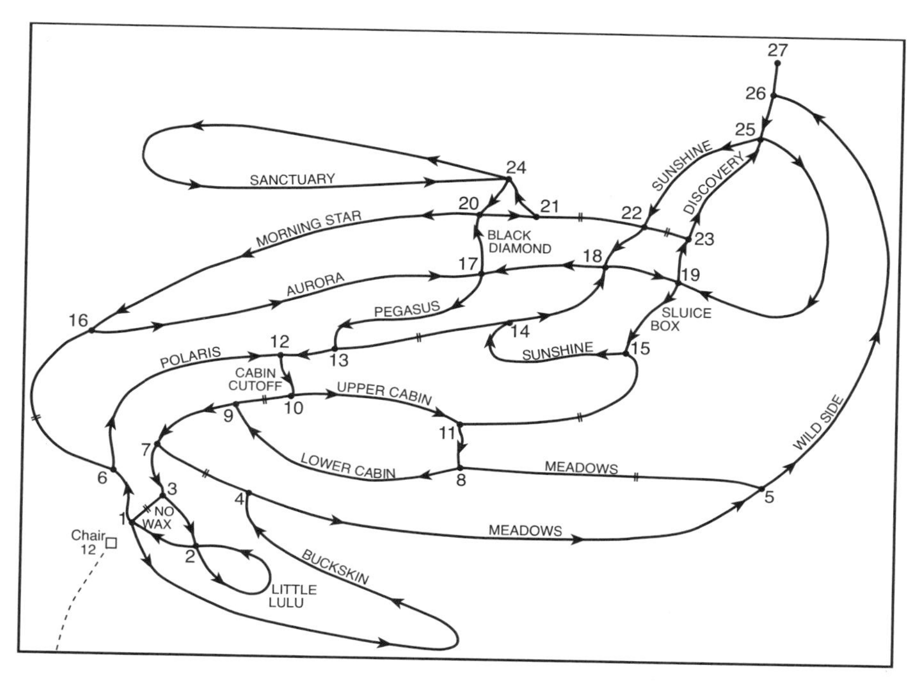

Fig. 19. A first attempt to find an aesthetic circuit on the Beaver Creek course, one that touches all vertices but leaves nine edges untouched.

of avalanche, but the attempt would probably result in death, at which point aesthetic circuits have little appeal unless they spiral straight upwards. In the absence of any empirical evidence, such analyses are left to the faith of the reader.

Let us try a few examples (fig. 19), starting at vertex 1, and heading down and up Buckskin to vertex 4. Already we feel a tension: if we now head toward vertex 7, we sense immediately that the lower Meadows edge 4-5 is going to be a problem in the future, unless somehow we can wriggle our way all around the course and end up on vertex 5, to make the last edge of our circuit 5-4. On the other hand, if we head initially down lower Meadows to vertex 5, we somehow have got to get back to vertex 7 at the end of our ski tour to incorporate that edge 7-4. Notice that to complete an aesthetically pleasing circuit, there is no particular need to end up on vertex 1, our starting point, although it would be very convenient.

Already we have developed an uncomfortable feeling that the more we ski around the course as it is presently constructed, the more we are constraining our future choices and possibilities. In this first example,[4] we have touched all the vertices, but we have missed 9 edges, and have not ended at vertex 1—not a very encouraging start! But let us try again, this time taking off from vertex 1 towards vertex 6 (fig. 20). At vertex 6 we again sense some discomfort, rather like the unease we felt at vertex 4 in our previous example. Because if we head along Polaris to vertex 12, we realize immediately that somehow we will have to get back to vertex 16 at some point to complete the edge 6-16. On the other hand, if we head uphill to vertex 16, in order to ski the first edges of Aurora or Morning Star, we will be forced at the very end of our skiing to vertex 12 to complete the edge 12-6. Vertex 12 is the only one giving us access now to that piece of Polaris, because we cannot get back to vertex 6 without passing over edge 1-6 a second time.

So right at the beginning of this second example we have the sense of branching and looking ahead, rather like Karparov playing chess against IBM's Deep Blue computer program, except that aesthetic skiing is much more serious and far more difficult. Notice two things:[5] first, we have used all the vertices again, but we have left 12 edges untouched (things are going from bad to worse!); and, secondly, we sense that it is those wye junctions, where three edges meet, that cause us the most decisional stress. The vertices with an *even* number of edges, say 4 of them, cause fewer problems for us. We feel we can zip or branch right through these, more or less confident that if we come back to them at any point we can always escape using an edge we have not used before.

It turns out that our intuitive feelings about the vertices and edges of Beaver Creek are perfectly sound, but fortunately our Swiss companion Leonhard Euler is still a superb skier for an old guy, as well as a fine mathematician who pioneered in what used to be called *analysis situ*, which we now call topology.

"Come on, Pete," said Leo (his friends call him Leo), "remember the bridges of Königsberg? And all the nice results that followed? You know perfectly well that a graph can only have a Euler line [here he blushed modestly], a path using all the edges just once, if all the vertices have an even number of edges springing from them.[6] And just look at the course at Beaver Creek. It has 18 'odd' vertices!"

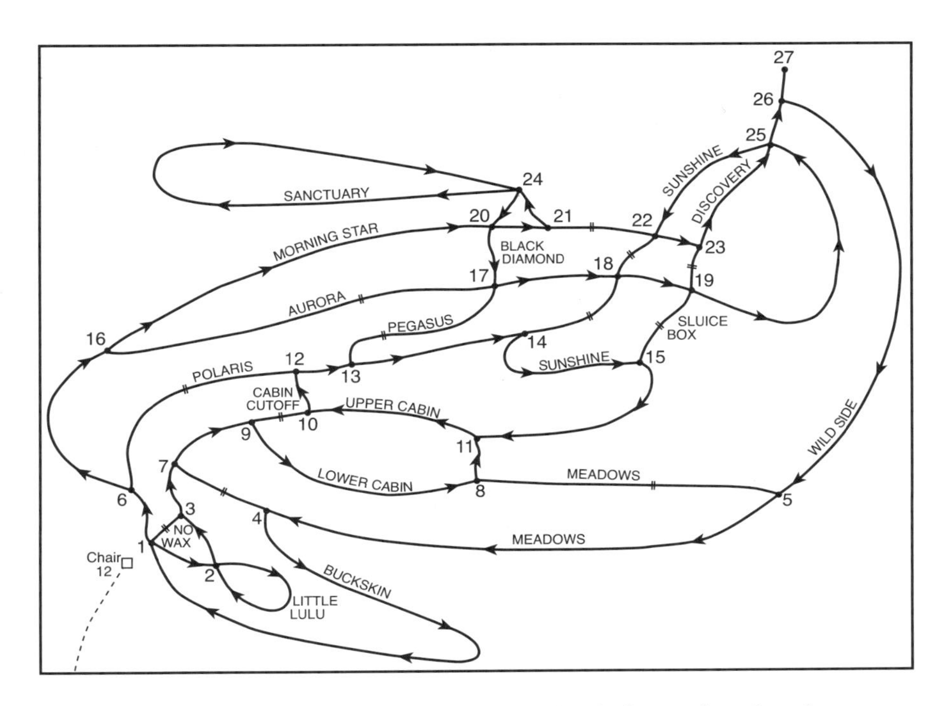

Fig. 20. A second attempt, which leaves us worse off than before, with twelve edges untouched.

"But what about that little example we constructed as figure 17b?" I said to Leo. "Remember how a skier starting at vertices 2 or 9, both of them with odd edges, *could* complete a nice circuit named after you?"

"Yes," said Leo, "but that's the only exception.[7] One or more of my circuits does exist on a graph with *two* odd vertices, provided the skier starts on either one. And actually," he added, "we really ought to throw out vertex 27, because there is no way you can ski down that short edge 26-27 to the little picnic table, from which you have that fabulous view over the mountains of the White River Forest, and then get back without crossing over the edge that got you there. Of course," he continued, "you could keep vertex 27 with its single (odd) edge, and somehow make vertex 1 odd too, and all the others even, but this really wouldn't be very pleasing, would it? You might complete your circuit, but at the end of an already tiring afternoon you'd have to ski all the way back from vertex 27 to 1 in order to take the chairlift down to the village. So let's agree that vertex 27 'doesn't count'—OK?"

I readily agreed to this sensible suggestion, and then began to wonder how I could help the people who laid out the course each year to modify the topology they had created. I could suggest that they give each vertex an even number of edges, and that would be rather nice, because a skier could start at vertex 1 at the top of the lift, and then finish there after tracing her "Euler line." On the other hand, making vertex 1 odd, and keeping the other odd vertex not too far away, might also give a pleasing solution. But how to modify with the least disruption? Even with the elimination of vertex 27, we still had 16 odd vertices to get rid of! Obviously, the easiest way was to link up those odd vertices with new edges, so turning their 3s into 4s, but a practical solution was not easy to find, since any skier who knew the topography of the course realized at once that there were terrain constraints. This was real-world stuff, not just playing with pencil and paper. On a sheet of paper you can draw lines all over the place, but on the actual terrain of the Beaver Creek course some of these would lead skiers out of bounds, over cliffs, or into dangerous avalanche conditions.

As a reader, you might like to make some copies of figure 18 and see what you can come up with, but you will have to draw your lines pretty much within the existing bounds unless you want to kill someone. After lots of trial and error, Leo and I came up with what seemed the simplest and most sensible solution (fig. 21); notice that it creates three new vertices, A, B, and C. Unless you want to redesign the course completely,[8] we can keep the existing layout and add eight "cuts," turning our 16 odd vertices into nice even ones. Making them all even has several advantages. First, you can start and end at vertex 1, which is both aesthetically and physiologically satisfying. But best of all, skiers can design their own courses, so that each day of skiing becomes a different aesthetic (but a similar physiologic) experience. Depending on our mood, Leo and I like to start "down," by taking Buckskin to vertex 4, and then gradually work our way up and around to the wonderful views from Sanctuary and the upper part of Sunshine.[9] On other days, we head up immediately to vertices 6 and 16 to Aurora or Morningstar.[10] The sky is almost the limit, because all you have to do is mark your track with little arrows and make sure you do not go back on an edge you have already used.

"But you know what?" I said to Leo, "there is one little constraining rule."

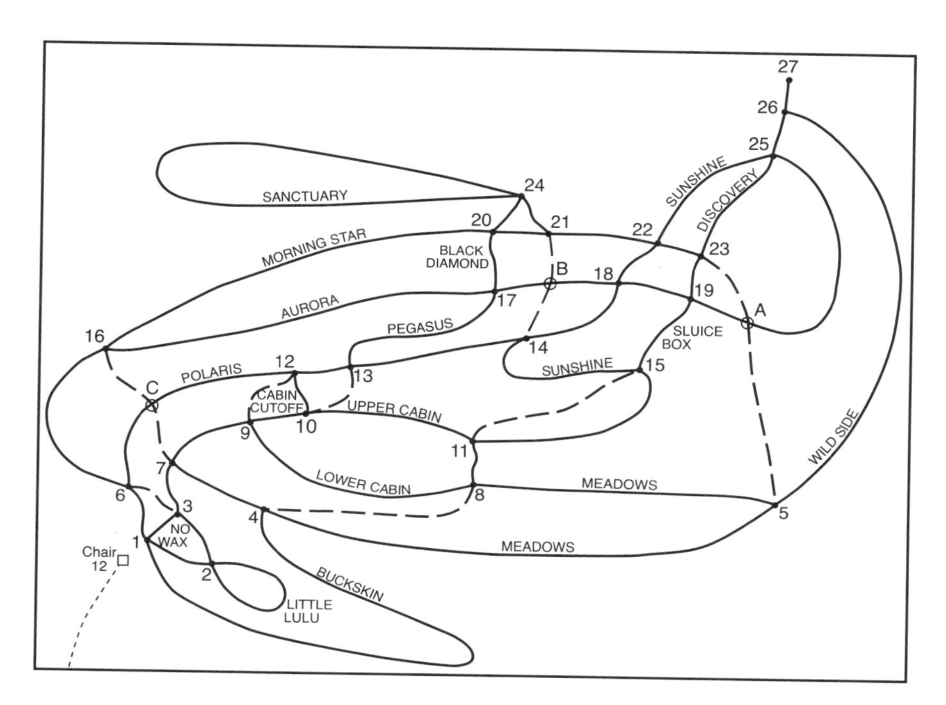

Fig. 21. One way to modify the Beaver Creek course to allow skiers to make aesthetic circuits.

"What's that?" he said.

"Well, you always have three choices of edges the first time you arrive at a vertex, except in the case of a loop, like those starting at vertices 2 and 24. When you hit those, you *must* take the loop, otherwise you'll never get back."

"Know what?" said Leo, "I never noticed that before. Hmm . . . let's call it Pete's Rule, OK?"[11]

"But given that single rule, how many possible but different paths could a skier design?" I asked him. "In other words, how many days could she ski without repeating herself?"

"Well, that's easy," he said, "but let's not give Fermat and Moebius all the fun."[12]

"No, of course not," I said confidently, hoping that he would not notice I was bluffing.

"And just think," said Leo, "how much Vail Associates are going to save. Every afternoon one of the ski patrol has to 'sweep' the entire course on a snowmobile to make sure no skier is still there, perhaps lying injured with a broken ankle or worse. Now they can do it in one of my continuous lines without wasting gasoline, oil, depreciation, and time going over an edge they've already swept."

"Funny, isn't it?" I replied. "That means about 71 cents a day, times the 119 days they're open between December 22 and April 18. . . . why, that's almost the same as the $84 I spent on seven half-day lift tickets to do the field work for this research in applied mathematics!"

"Hmm," said Leo, "I wonder if they'll think it's a snow job?"

Pro Bono Publico

HOLDING an Evan Pugh professorship at my university is a tremendous privilege, giving a person a mixture of new freedoms and new responsibilities that can be difficult to handle sometimes. I received mine in 1986, a few months after the terrible nuclear accident at Chernobyl in Ukraine, and the award made it easy to complete an already half-formed decision. Those holding an Evan Pugh receive a small amount of money each year which may be used for "scholarly purposes," and I decided to use mine to do things that I thought were important, but that were unlikely to receive research funding from the usual Washington and other funding agencies.

The effects of Chernobyl diffusing through physical and biological systems, and metaphorically through human bureaucratic systems, were the first topics to be tackled, with fieldwork in Sweden and Norway.[1] Research on the AIDS epidemic came next. All the epidemiologists seemed to be busily playing with their differential equations in their computerized sandboxes, producing useless streams of numbers down the time line. Funding for this sort of stuff was easy to get, but the problem was that no one could actually do anything with the results. People had forgotten that every epidemic always has a geography in space, as well as a history over time.

So at Penn State, with Joe Kabel and a number of other students, we put together, at unbelievable costs of time and frustration, a series of maps at various scales showing the spread and intensification of the epidemic. No funding was ever made available, but the map sequence appeared in many publications overseas, as well as in *Time*, *Forbes*, *Atlantic*, and even *Playboy* in the United States. ("Aha!" said some of the graduate students in my department when *that* issue appeared, and they refused to believe that I was not a subscriber! All I could do was point out to them that *Playboy* had eight million readers, so if you wanted to get an educational message across, what better place? Given the primitive state of sexual education in the United States, particularly when compared to Scandinavia, which seems to have virtually shut down HIV transmission, *any* outlet reaching young and sexually active people is welcome.)

The question of error in *any* map, let alone a series over time, is something many people are not used to thinking about very carefully, mainly because they have been brought up to assume that a map is always "true." Given the general antagonism at the time to any geographic portrayal of the AIDS epidemic, I decided to use the U.S. series as a concrete device to bring the whole question out into the open, raising deliberately some of the issues about interpretation, hermeneutic perspective, and trust. The AIDS/HIV epidemic has raised these and other questions to a degree that no other epidemic, including other sexually transmitted diseases, has done before. Issues of confidentiality that no caring person would dispute have been taken over by incompetent medical people, who have the authority, but no knowledge, about a matter that turns out to be almost purely geographic.

Sometimes it seems that good geographic research is almost impossible to achieve. For example, trust in testing, especially in "small town" states, has been low, so that sometimes the AIDS or HIV cases of one state, for example, Wyoming, are assigned to Denver, Colorado, where big-city hospitals provide greater anonymity. With federal money allocated on a number-of-cases basis, Colorado retains the numbers, but Wyoming is grossly underestimated.

What we have learned from the AIDS epidemic is something we have always known, but are reluctant to think about too much. Many viruses and bacteria need human beings to survive, and they travel wherever the people carrying them go. Recall traffic being transmitted over a backcloth. As a world pandemic, the HIV has raised enormous concern about transmission on a global basis, particularly as few places seem to be more than thirty-six hours of flying time apart. Given resources for traveling, the world can be an awfully small place these days, which is why the viruses go where the people go.

Sources of Error in a Map Series, or Science as a Socially Negotiated Enterprise

O hateful error, melancholy's child,
Why dost thou show to the apt thoughts of men
The things that are not?

—William Shakespeare
Julius Caesar, v, iii, 67–69

ALTHOUGH terminating in 1990, the five maps (plates 5, 6, 7, 8, 9) illustrating this article still constitute the most detailed graphic expression ever presented of the geographic diffusion of the AIDS epidemic in the continental United States.[1] Based on approximately 2,500 spatially varying units,[2] most of them counties, they have focused attention on the oblivious ignorance of the traditional medical and epidemiological communities to the importance of series in the spatial domain, and the way the spatial perspective was totally ignored by public health authorities during the first decade and a half of the epidemic. They were shown at the International AIDS Conference, Amsterdam, July 1992, by the physician and AIDS researcher Dr. Mindy Fullilove, who pointed out the almost exact correspondence between the geographical development of the epidemic and the assumed sources of original infection specified by several hundred people interviewed by her in North Carolina. After the cartographic presentation, one epidemiologist in the audience got up and said, "I believe we have a new paradigm here!"—only 7,000 years after maps of Babylon were incised on clay tablets.

These, and similar map sequences at larger scales, have also shocked audiences here and abroad in such public forums as *Time*, *Forbes*, *Atlantic*, and *Playboy*.[3] They are also available in animated form for educational television directed at young people, who now form the cohort of the population most at risk. It is worth emphasizing immediately that the contour-color interval is geo-

This essay originally appeared as "Sources d'erreur dans une série de cartes, ou: la démarche scientifique, objet de négociations," *MappeMonde* 2 (1993): 22–27.

metric, each change up the natural spectrum from blue to red multiplies by 7.5 the previous value, with the red areas simply "over 2,000." However, if you made a three-dimensional map in 1995, and used 0.25 millimeters to represent each person dead or dying from AIDS in the five boroughs of New York City, you would have a spike fifteen meters high, and only slightly smaller ones at Los Angeles, San Francisco, Miami, etc.

Taken as a whole, the sequence constitutes a powerful rhetorical statement, using the word "rhetorical" in its old and honorable sense as the "art of persuasion."[4] Upon reviewing the AIDS explosion in the cartogeographic domain, many people are persuaded for the first time that AIDS is not something "out there," remote and far removed from them, but may well be all around them. To a geographer, the sequence is a classic case of spatial diffusion, with strong evidence of both hierarchical diffusion, controlled by relations of interaction in the urban system or hierarchy, and spatially contagious diffusion from regional epicenters—the "wine stain on the tablecloth" effect.

The cartogeographic perspective challenges the totally aspatial view of traditional epidemiological modeling confined exclusively to the time domain. Since the construction of such a series illuminates aspects previously hidden by bureaucratic obtuseness parading in apparently impeccable ethical concern for confidentiality, and since actually seeing the explosive diffusion in the first decade may be politically delicate, especially in an election year in the United States,[5] we may expect attempts to throw doubt on the series, attempts emphasizing that errors are present. It is important to deal with such efforts to denigrate in a direct, firm, scientific, and philosophically aware manner.

Types of Error

In any scientific statement there is always error, since science is a mortal, rather than a divine, enterprise. However, after examining a large literature, it becomes apparent that little has been added to any theory of error since Karl Friedrich Gauss made maps for the duke of Hanover during the eighteenth century. Those who boldly pronounce judgment about the amount of error always retreat behind a cloud of assumptions after realizing that one cannot specify any quantity or degree of error without actually knowing the truth. Even such a correspondence theory of truth (and, therefore, of error) has been is disarray since the days of Kant, and no one living in these hermeneutic days could take such an approach seriously. What we can do is make some quite open judgments about the types and sources of error, and then argue that these would not alter any major conclusions about the geographic processes at work and the cartogeographic representations these produce. Whether a reader finds such scientific rhetoric persuasive or not depends on the hermeneutic or interpretative stance he or she is prepared to take. In the end, scientific truth is always socially negotiated, including the construction and interpretation of maps, as Brian Harley was able to teach us before his tragically early death in 1991.[6]

What are the sources and types of error that take this research from the error-free summit and realm of the immortal gods down to the foothills of Mount Olympus where ordinary geographers dwell? There are essentially five, none of which can be cleanly separated except for purposes of exposition. First, there is the problem of underreporting, particularly in the early years of the epidemic. Less was known about the various ways an infected person could convert to symptoms diagnostic of AIDS; tests were less reliable, and some doctors (perhaps up to 50 percent during the early years in Germany) were prepared to sign death certificates for "pneumonia," "cancer," and so on to spare the feelings of shame that some families expressed. Early maps are likely to reflect such errors of omission rather than commission.

Second, there are temporal errors, usually delays in reporting that make it extremely difficult to monitor the course of the epidemic properly and so use forecasting techniques that rely on recent information to make appropriate parametric adjustments. No matter how sophisticated the methodology and technology used, it is no use monitoring junk. We may even find ourselves in the curious situation that model predictions, far from overestimating the course of the epidemic, actually turn out in the future to be closer to some unknown truth than the official figures reported by medical bureaucrats. This constitutes a nice philosophical, not to say political, problem in its own right. Even when AIDS is a legally reportable disease, errors of temporal specification may be gross: in June 1991, for example, 75 percent of the AIDS cases reported in Washington, D.C. were not days, weeks, or even months late but had been diagnosed in previous years.[7]

Errors over time are clearly related to the third type of errors—definitional errors. After the first decade-and-a-half of the pandemic, as more has been learned about the HIV and its effects on the human immune system, we are able to recognize, in lowered T4 cell counts and other diagnostic approaches, the earlier signs of conversion to opportunistic diseases. New definitions in 1993 interrupted the previous time series, inflating cumulative totals to the point today (1995)[8] where the 400,000 mark has long been exceeded and the totals are still growing. In the previous year (1992), scientific advances in diagnostic tests were found to be politically unacceptable, and so were socially negotiated away by centralized power structures within the Republican Party. For one more year, a nation breathed a sigh of relief that things were not so bad after all. How many people are infected today we do not know with any reasonable degree of assurance.

The fourth kind of error is spatial error, a form lying in a domain of thinking familiar to the geographer, but an intellectual arena where doctors of medicine and epidemiologists have little if any experience. Unfortunately, such ignorance does not prevent them from making judgments, some of them catastrophic for our deeper understanding of the epidemic and for our ability to intervene with education and health care planning. Spatial error is simply misplacing in space values reported, and it may be thought of as the geographical equivalent of delayed reporting in the temporal domain (i.e., "misplacing" people by a month or a year). Like any other error, it is unavoidable to some degree. Even if we had a dot map of each person, the individual dots would only stand as a spatial mean for a probabilistic smear or "field of movement" created by individual human lives.

Spatial error is particularly likely to arise in connection with the fifth kind of error—errors of estimation. Many of these arise because of the quite proper and understandable ethical concern to protect the identity of people with AIDS. I wish to make it quite clear that I am in total agreement with this ethical ideal, while noting at the same time that it has been carried to quite absurd extremes. Spatial errors of estimation arise when we want to move to finer levels of spatial resolution with data (numbers of people with AIDS) that have been deliberately aggregated spatially, ostensibly to preserve confidentiality. Notice, however, that even the ability to observe scientifically now becomes socially negotiated, with the negotiations directed and informed by the ethos of a society and the power it is willing to allocate to certain groups of professional experts. In the United States, most states now report regularly by county: some, like Virginia, by zip or postal code; others, such as Kentucky, by aggregation of counties into regions; one (Nebraska) by three, totally irrelevant economic areas; another (South Dakota) by two regions east and west of the Missouri River, although this physical feature is not known to be an effective barrier to the diffusion of HIV. A few states (Wyoming, Wisconsin) still report only state totals. Generally, there is a tendency to report by smaller and smaller spatial units as the epidemic, measured by rates of infection, intensifies.

Errors in Texas, Florida, and Iowa

I want to illustrate the problem of spatial errors of estimation, and the way these problems are produced and convoluted by a mélange of other problems, by focusing upon three states in the map sequence—namely, Texas, Florida, and Iowa. Texas is prepared today to report by county, and from 1986 onwards the map sequence uses the officially reported, updated, and corrected figures. But before 1985 there was no consistent database, and even today the state medical authorities know the cumulative county totals only from 1985 onwards. As a result, we have to estimate the 1982 and 1984 values in this part of the country. This actually requires no sophisticated mathematics or computational model: the annual totals for 1985–1990 plot with classical smoothness and regularity and they can be extrapolated back with a plastic curve to be anchored at 1981, when what we now call AIDS was first recognized (although not the HIV, which was only discovered in early 1983). Thus, we can estimate, with what must be only the very slightest error, the cumulative totals for 1982 and 1984.

The question then is: how do we assign these totals spatially? Since we have the map distributions (i.e., the spatial series) for 1985 onwards, we can simply deflate county values, say for 1984, by the ratio of the cumulative totals for 1984 and 1985. This is a simple linear extrapolation backwards, but the difference between the linear and nonlinear approximation will be minute over this time span. Some informed and educated guesswork is involved: a county may appear on the map in the lowest category (blue) with one or two people with AIDS a year before or after it really did, but recall that we have no idea what the reality was in those days and no better way of capturing it. The reader of this text and map must simply judge whether this is plausible, whether it is reasonable, whether it is persuasive. And note: this judgment will be informed by what the reader brings to these written and graphical texts; in other words, it will be related to the hermeneutical stance one is prepared to adopt. I suggest that an experienced geographer will find such spatial estimations acceptable. I further suggest that the ordinary layperson, viewing Texas in the entire sequence, will accept the 1982 and 1984 maps without comment since they reflect what I can only call a "spatial logic" that is to be found everywhere else on the map (California, Washington, New York, New England, etc.). Each map seems to develop quite logically out of the previous one, like a photographic plate developing in the darkroom. On the other hand, bureaucratic epidemiologists, trained to think exclusively in the temporal domain, and slowly realizing that they may have been sitting on scientifically valuable spatial series without knowing what to do with them, may try to exaggerate the minute errors injected by such a procedure. Science becomes, once again, a socially negotiated endeavor.

Florida presents other human, not to say politically charged, problems. Many polite inquiries to the Florida State Health Authorities for cumulative county values produced replies to the effect that they were quite capable of handling the geographical analysis of the epidemic themselves, that only state totals were available to outsiders, and that they needed no help whatsoever—thank you very much! Fortunately, one state health worker, who clearly must remain anonymous, thought this attitude was unreasonable, defensive, and even unethical since, in a state where the epidemic was rapidly becoming catastrophic, it made a major database the private preserve of a few researchers who had no appropriate geographical and methodological ways of using the data. At that time, no geographic modeling had been undertaken, let alone published. We received a photocopied packet of county values as they had appeared at the end of each year, and we inflated these by a factor based on corrected state totals after the database had been revised by incorporating late reports. In constructing the Florida sequence from these revised figures, we were also fortunate in having perhaps the only geographical analysis of the diffusion of HIV at the time, an analysis that used Florida as one of four case studies.[9] The result is the only published sequence of AIDS diffusion, a sequence with very small, though still unspecifiable, errors.

Iowa presents another problem, one shared by states like Montana, Wyoming, and South Dakota. These states are characterized by rural populations of very low density, interspersed by a few urban centers connected by ribbons of interstate and major state highways. Only state totals are available for states like these. Even states such as Wisconsin, which has a much higher population density, and a reasonably developed central place structure focusing upon Milwaukee-Chicago, exhibit this problem. In cases like these, the cumulative state totals must be assigned in proportion to the county populations. Such an estimation procedure is based on the perfectly reasonable assumption that AIDS is, in large part, density dependent. Detailed analyses of somewhat similar states, like Ohio and Pennsylvania,[10] confirm this procedure as reasonable in the absence of any other information. In actual fact, other information for Montana, Wyoming, and North and South Dakota was made available to me under the standard and strict ethical conditions governing the scientific reviewing procedure. It is clear that in the early stages of the AIDS epidemic—the so-called "seeding" stages—the appearance of AIDS cases in areas of extremely low rural population densities consisted almost entirely of young homosexual men coming home to die from major urban epicenters on the East and West Coasts. I cannot and will not use such information to correct the earliest maps, so here spatial error must be knowingly left literally in place.

In the early years, the cumulative numbers for these sparsely settled states are very small, counted in tens or less, and in later years, as the epidemic takes hold, the maps become more and more reliable (i.e., less prone to error). In a year when Montana had ten cases, the national total was already in the tens of thousands. The overall relative error is minute: the local spatial error may be initially quite large but reduces quickly. Notice it is not the totals in a state that are in dispute (except for the other sources of error discussed above), but the exactitude of the spatial allocations. With the exception of Waldo Tobler's error ellipses in the very different area of multidimensional scaling and cartography, I do not think we know much about measuring such errors. And, once again, how can you measure error without knowing beforehand what the truth is?

On Matters of Belief and Trust

Turning back to the five-map sequence, what effects might such errors have on our belief that the sequence is a reasonably accurate representation of the diffusion of the AIDS epidemic at this scale? I emphasize "at this scale" since we recognize degrees of generalization in any cartographic representation. No one would attempt to use these maps for analytical purposes appropriate to much finer levels of resolution. The thickness of a contour line may well exceed the size of some of the townships reporting in a state like Virginia. Some degree of generalization is inevitable in any scientific statement. Indeed, and perhaps almost by definition, a scientific statement in words, graphics, or algebras is a generalization where we can see the forest rather than the individual trees. Notice that in regions where the county database is reasonably fine, and the official values reported considered reasonably reliable (over much of the eastern part of the country, for example), the unfolding sequence generates a high degree of trust and therefore belief. The spatial logic appears reasonable and truthful mainly because the information content in such spatial logic arises precisely out of the plausible spatial autocorrelative properties in such relatively local areas. But why should we believe that Iowa, Texas, and Florida, and other states where estimations have been made, are any different? Yet notice further how words like "trust," "belief," "reason," and "truth" have entered the discussion. Whether you trust the map sequence, whether you believe it to be reasonable and close to some unknown truth, depends upon you and what you bring to the hermeneutic task, a task that faces us as human beings as a condition of possibility. Thus, and as thoughtful scientists, it is necessary to negotiate in a communicative discourse.

As People Go, So Do the Viruses

> There are certain circumstances, however, connected with the progress of cholera. . . . Its exact course from town to town cannot always be traced; but it has never appeared except where there has been ample opportunity for it to be conveyed by human intercourse.
>
> —John Snow
> *On the Mode of Communication of Cholera*

ALTHOUGH conventionally trained epidemiologists assiduously ignore their nineteenth-century heritage, a heritage acknowledging the importance of the spatial domain,[1] I think we have to bypass their obdurate lapse and start with a simple fact: all things, even epidemics, exist in space and time, so everything has not simply a history of how it came to be, but always a geography of where it is. Better still, everything has a spatiotemporal geohistory, for ideally these two fundamental domains of human concern should be kept together. Split them apart, and the result is too often the impoverished static analysis of the conventional map, or the aspatial decompositions and banal computer games played with time series.

But we have to push this matter of spatiotemporal thinking even further: it is not just a matter of marrying process with traditional views of geographic space, but realizing in concrete and analytical terms that the conventional space of the geographic map is socially constructed. Unfortunately, simply proclaiming this vast postmodern insight does little good, except to generate a glow

---

An amalgam and partial summary of several presentations, including "Social Physics Is Alive and Well . . . and the HIV Knows It," the 1995 Ralph Brown Lecture at the University of Minnesota; "Spreading HIV Across America with an Air Passenger Operator," *Proceedings of the International Symposium on Computer Mapping in Epidemiology and Environmental Health*, 1997, 159–162; "Geo-Demos-Logos: The Outlook for Operational Models," Conference on Global Molecular Epidemiology, 1995; "The Dynamics of HIV and AIDS," lectures given at the University of South Carolina and the Open University, England; and two papers, "Il virus nel villaggio globale," *Sistema Terra* 4, no. 2 (1995): 19–22, and "La géographie du SIDA: l'étude des flux de population permet de prédire," *La Recherche* 280 (1995): 36–37

of profound intellectual warmth in the proclaimer. We shall have to do better than this, go beyond such simplistic declarations, and specify in highly practical and concrete terms how geographic space is constantly structured and restructured by the human presence. Historically, we know this has happened as massive technological innovations in transportation have warped the travel time paths of people, paths that were already complex enough from the effects of physical features such as mountains, rivers, and deserts. The homogeneous spaces posited by theory are tough to find. The physically complex space of the United States was structured and restructured by turnpikes, canals, and railways in the nineteenth century, with clear effects on the movement of diseases,[2] and the process continued into the twentieth century with airline and interstate road systems. Many other parts of the world can tell a similar story. Indeed, the world as a whole has been radically restructured in the same way by sailing, steam, diesel, and atomic ships, as well as the now ubiquitous airplane. "Global village" may be an exaggeration, but the space-time shrinkage has been profound over the past 150 years.[3] In "human-technological" space, New York is now only one unit away from San Francisco, compared to 500 units in 1860.

Practical Descriptions

It is people who structure spaces: by their technology, by their movements, and by their very presence. And so if various pathogens—bacteria and viruses—need people as suitable ecological niches to exist and reproduce, we are really talking about "disease spaces," enormously complex spaces that control the movement of viruses in much the same way that the hills and valleys of the familiar topographic map control the flow of water. Which means we need to map these "human-disease topographies" and use our knowledge to stop the viruses in their tracks. The question is how.

People by their very presence structure a geographic space by the simple variation in the densities they create. At a certain scale, a scale of observation that may be critical, many epidemics are clearly density dependent. In the United States, the HIV pandemic certainly is, for the more people there are, the more chances there are for sexual and needle-sharing interactions, and so the greater the chances of HIV transmission. In general, and leaving aside specific and well-known concentrations of marginalized people, the cities are great soaring exponential peaks of HIV infection, while the rural areas have much lower rates. Take a three-dimensional map of population in America and submerge it in a swimming pool. Now raise it slowly and make an overhead movie of the population surface as it comes out of the water. You will have an almost perfect animated map of the spread of HIV and AIDS in the United States.[4] Millions of local interactions, like a shimmering Brownian movement, ensure that the HIV is transmitted, and the higher the density the higher the rate of transmission.

And if this description, which results in the "winestain on the tablecloth" movement that geographers call spatially contagious diffusion, seems too rarified, theoretical, and abstract, we can always focus down on almost any city in America and describe concretely and practically how the daily movements of commuting transfer the HIV from a raging epidemic in the marginalized inner cities to people living in the commuting field of the suburbs. In fact, it does not matter where the HIV enters the "system"; it is quickly sucked into the high-density, poor, and impacted inner city, where it proliferates, only to be spewed out again over the region. In New York's metropolitan region, for example, a simple 24×24, to-and-from matrix of daily commuting generates indices of "commuting accessibility" that predict AIDS rates over six years about 92–94 percent correctly.[5] In brief, if you can find the space of transmission, and map its topography, you have a rather insightful spatiotemporal handle on how the viruses move at that scale. Put more simply, if you can say something sensible about how people move, you should be able to say something sensible about how the viruses they carry around with them move. It is really not a very difficult idea to grasp.

Except, of course, for those who have been intellectually "potty trained" in major universities to think about epidemics exclusively in the time domain. When an epidemic comes along, these people reach up on the shelf of conventional thinking, pick out some differential equations whose nonlinearity gives them paroxysms of joyous profundity, and proceed to play with them in computerized sandboxes. For all the nonlinearity, the results are utterly banal: if you hand out free condoms and clean needles at t_0, guess what happens to transmission and infection rates at t_5! I mean, that was the whole point. And note that nothing *useful* can be done with streams of single numbers generated by "guesstimates" down the time horizon. No kid from the inner city will be the least bit impressed by a single number, a value equally useless in planning any extension for health care delivery. In contrast, an animated map showing the diffusion and increasing intensity of the epidemic generates the "Wow,

Man! I never realized it was so close!" *All* planning decisions require not simply *when* but *where.* To paraphrase the old Scots engineer in Rudyard Kipling's wonderful poem "McAndrew's Hymn":[6]

> A raging epidemic will never bide a wee,
> To stop the virus in its tracks, yer need geography!

Aye, so yer do.

Predicting and Postdicting

One thing that has proved very difficult to get across is that there is an enormous amount of valuable scientific information in a spatial series; in other words, a simple map showing the way some variable, say AIDS cases, varies over a geographic space. Why so many should have trouble grasping such a simple but profound idea is either a mystery, or a sad acknowledgment of the power of graduate schools to mold thinking into narrow channels. Because even if you have been taught how to massage a time series with all the harmonic, Fast-Fourier, and spectral software around, the great intellectual leap from the temporal to the spatial domain is not really that demanding. In a time series, it is obvious that a succession of random numbers contains no information, that it is simply the "white noise" of the electrical engineer. Only when such a series has some *structure* to it—a trend, some periodicity, some pattern—do we acknowledge that there is some information. And that, in turn, means we have *some* chance better than just the average of the series of predicting what the next few values will be. In the single dimension of the time domain, values next to each other, or at certain "distances" (periodicities) away, may not be random and independent, but form clusters or patterns in which values are related (autocorrelated) with each other.

So why, if this is so simple and obvious, is it difficult to *think* exactly the same thing in the two dimensions of the spatial domain? We could imagine a map covered with grid cells into which we place numbers drawn from a random number table. This would be simply the geographic equivalent of "white noise," by definition and construction a random spatial series without any information. But let there be a regional trend, or periodicity, or clustering, in other words some *pattern,* and the numbers are no longer independent of each other but related (autocorrelated) to each other. So now, if you know the value in one grid cell, you ought to be able to predict, better than just the average of the series, the values in the cells nearby. We have a geographic signal in the noise, and we have to do our best to find it.

Of course, the old-fashioned geographic space of the map may not be the most appropriate space in which to search. If spaces are socially constructed, then perhaps a transformation of geographic space to a space structured by human interactions may simplify things. But transformations are *mappings*, in both a geographic and mathematical sense. For example, we might transform France under a simple gravity model rule into an "AIDS space" (fig. 22), in which the big cities like Paris, Lyon, Marseilles, Bordeaux, and so on are close to the center, while smaller towns are on the periphery of a multidimensional cloud of points. While such a mapping might horrify the French, it may help us to understand how the HIV actually moves at this scale. We could blow a multidimensional bubble from the center and observe the time at which various cities and towns were captured, interpreting this as the time of the first arrival of the HIV. In the inverse mapping, from AIDS space back to geographic space, we would see the virus hopping around the map, controlled by hierarchical diffusion, and then slowly spreading to smaller places from the major regional epicenters by spatially contagious diffusion. This is, of course, precisely what we observe in the United States, and anywhere else where we can accumulate fairly reliable empirical data.

But the very best thing we can do is to combine the patterns in time and the patterns in space, and by using *all* the information try our hand at spatiotemporal prediction, predicting not just numbers but where they are. In other words, predicting the next maps. We can do this today with spatial adaptive

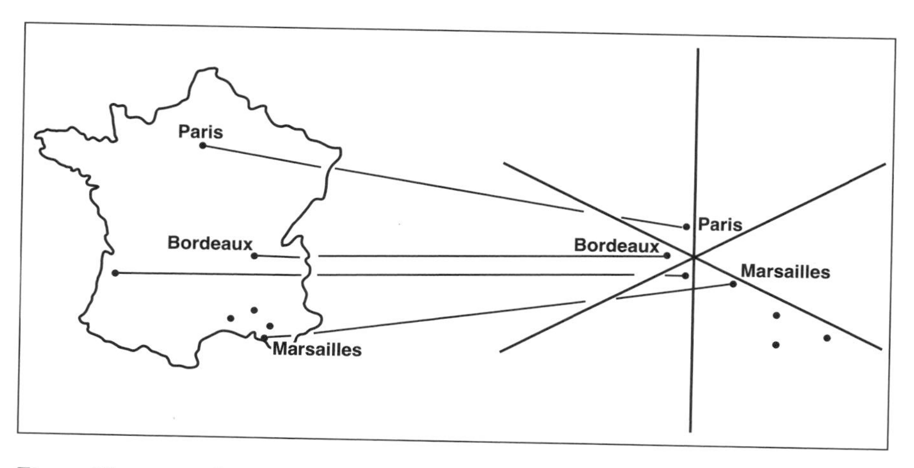

Fig. 22. The geographic space of France transformed into "AIDS space" under a simple gravity model mapping.

filtering and other approaches that come under the general heading of "expansion methods."[7] There are many technical wrinkles, and for large problems, like predicting the next maps of AIDS in the United States with over 2,500 geographic units, it is helpful to have a large and fast computer. Nevertheless, *conceptually* it is fairly straightforward.[8] In essence, we want to extract as much information as we can from a three-dimensional spatiotemporal "cube" of data, which is really nothing more than a pile of maps in sequence, with the numbers, say AIDS cases, specified by their (x,y,t) coordinates. We then take a little function relating AIDS cases to time (a cubic does nicely because it can imitate both the upsurge in the initial stages of an epidemic, and the slowing down later), and unleash it like a small mathematical bloodhound over the first slice (map) in our spatiotemporal cube of data. Moving from cell to cell on the map, our bloodhound-function adapts its initial parameters, which were first set globally over the entire map, so that it can give a better *local* estimate. When it has adapted as best it can, we store its "locally best" parameters and let it move onto the next cell (or township, or county, . . . or whatever units we are using). Here it goes through the same adaptive procedure again, and so on, and so on. After many sweeps of the map it can do no better, so we take the next slice in our cube and repeat the whole thing again. And so on, until all the map-slices have been analyzed.

The next part is simple: we look at the parameters stored for each cell and see if they track in any regular way over the times we have chosen for our map-slices. If the AIDS cases had been a random phenomenon, a variable changing across the map like numbers pulled out of a hat, we would expect the parameters to jump up and down without any rhyme or reason. And you cannot predict much with a jumping-up-and-down "white noise" parameter. But in all empirical cases examined so far, this is nothing like the case. The parameters of our little bloodhound moving over the map, sniffing out spatial structure, track with quite extraordinary smoothness. This means we can extrapolate them from "now" into the future, and use the predicted parameters to estimate AIDS cases not just *when* but *where*. We can predict the next maps. In Ohio, for example, we predicted the maps up to 1995 from the information between 1981 and 1989. "How ridiculous," some said. "Ohio will never reach values that high in so short a time." When 1993 came along, it turned out we were about two percent too low.

Today America . . .

Predicting the next maps with spatial adaptive filtering relies heavily on local clustering or spatial autocorrelation, although not exclusively so because our adapting function can also use nonlocal information if we allow it to. For example, in Ohio the values of big cities like Cincinnati and Cleveland may allow Columbus to adapt its parameters more precisely. Even so, we know we are in a country characterized by enormous mobility, much of it by air, so that diseases carried by people to people may spread by a form of "city-jumping" diffusion controlled strongly by relations between cities in the urban hierarchy.

One way to examine this is to take the largest 102 metropolitan areas for which AIDS rates can be estimated fairly reliably, and form the 102×102 matrix of air passengers. The matrix can be scaled to give us the sort of probabilities that would let us say, "Pick any air passenger in America at random, and now what is the probability that they will fly from A to B?" Of course, the probability of someone flying from New York to Chicago is very large compared to someone flying from Little Rock, Arkansas, to Burlington, Vermont. Intuitively, one can see that the probabilities of interaction are strongly related to the sizes of the cities interacting, and how far they are apart. Which is precisely why air passenger traffic can be estimated about 92 percent reliably from such simple "gravity model" ideas. But recall that viruses like the HIV also move around the same way because they travel along as baggage inside the people.

We can think of our air passenger probability matrix as an operator that can multiply a list of cities called a vector. Suppose we start with a list-vector of zeros, implying that there is no HIV around. Then suppose we put a one (1) in the vector at Chicago, interpreting this as the first appearance of HIV. If we multiply our vector by our air passenger operator, it will probabilistically smear the virus all over the urban hierarchy in accordance with the probabilities Chicago has of interacting by air with all the other 101 cities. If we multiply again, then we shall probabilistically smear all the previous probabilistic smears . . . and so on. Eventually the values in our vector settle down, and it happens that they predict AIDS *rates* years later up to about 80 percent accurately.

For example, if we inject the HIV at New York, then 85 iterations or probabilistic smearings later we can predict AIDS rates in over half the population of the United States in 1986 about 75 percent accurately. On the other hand, if we inject it close to the bottom of the hierarchy, at Peoria, Illinois, it takes 500 multiplications or smearings before we get the same predictive level. From Peoria, the HIV has to work its way up to the big cities, at which point their huge interactions with each other and with smaller places ensure that the HIV is eventually everywhere. It is just like the old spiritual, "There's no hiding place down here," although I suggest we can leave out the "Hallelujah!" If we were to inject HIV into the urban system as $1/\sqrt{5}$ at New York, Los Angeles, San Francisco, Miami, and Houston, a subsystem exchanging in 1992 nearly 13 million people by air, then after only four multiplications we can predict AIDS rates in 1986 at just over 80 percent (plate 10).

The residuals, those cities whose rates are somewhat higher or lower than we would expect from our smearings, form a provocative pattern (plate 11). Taking plus or minus one standard error as our category for centers whose rates are reasonably well-predicted (yellow), a number of southern cities (Austin, New Orleans, Atlanta, Charleston, Savannah, Fort Myers, and West Palm Beach) have considerably higher rates than we would expect. They are joined by Albany and a number of smaller centers in the Northeast embedded in the mass of Megalopolis, where spatially contagious effects may be adding to their rates achieved by hierarchical diffusion from the air. Conversely, a number of old rust belt towns—Detroit, Buffalo, Grand Rapids, Cincinnati, Lansing, and Toledo, as well as a number of others in the "well-predicted" category with yellow, downward-pointing triangles—all of these have rates less than we would expect. Many are characterized by blue-collar, industrialized work forces, many with Catholic and late immigrant backgrounds. They are joined by a group in the Southwest: Salt Lake City, where Mormon influences may be reducing the rates; Las Vegas, where the condoms seem to be working in the legalized brothels, which have declared that no sex worker would be financially penalized for refusing unprotected sex; and Albuquerque and El Paso, where rates are also considerably below what we would expect. Is there some underreporting going on?

So closely does the human geography of America predict the diffusion of HIV and its deadly consequences that we can use our independent variable "smearings" to predict the AIDS rates in such defined urban areas as Omaha, Nebraska, and Madison, Wisconsin. These are cities in states that refuse to report at the county level (from which we can aggregate up to metropolitan areas), but our model "postdicts" that their rates roughly tripled between 1986 and 1988, and then doubled again by 1990. Perhaps forty or fifty years from now, when these things are of historical interest only, we may get confirmation of these predictions.

. . . *Tomorrow the World?*

There is great and growing concern about the appearance of "new" viruses and pathogens, and the whole question of global transmission is rapidly coming to the forefront. If there is global warming, it is likely to extend the geographic range of vectors like the anopheles mosquito, and with it such diseases as yellow fever and malaria. As human populations continue to explode and press ever harder on delicate and previously isolated ecological systems, pathogens like Bolivian hemorrhagic fever will spread from an endemic state in animal populations to cause new epidemics in humans.[9] At the global scale, how can we even begin to think about and describe the potential pathways of disease, to limn the "global topography" caused by human and other carriers?

As we move to the global level, many new things come into view, including the linking of previously separated ecosystems (ballast water from China pumped out at Peru); technological innovations, like the doubling of aircraft size quadrupling the probability of inflight infections; unpredictable refugee movements; shifting ranges of disease vectors, and so on. But not the least is the fact that the 4,028 airports with regularly scheduled services link cities exploding with people, many of whom do not have even a minimum of basic sanitary facilities. In brief, we are talking about global networks connecting thousands of places that are almost perfectly designed incubation chambers for the rapid growth and local transmission of once-rare pathogens. And although my search has been far from exhaustive, I have been unable to find any pair of airports more than 36 hours of flying time apart. If you go from Ushuaia, at Terra del Fuego, to Zamboanga, a town on an outer island of the Philippines, it will take you exactly 36 hours; from Invercargill, the southernmost airport of New Zealand, to Vagar in the Faroe Islands, will take you 35 hours and 10 minutes; and from Anadyr in eastern Siberia, to Nome, Alaska, only 800 kilometers away, it will take you 33 hours and 10 minutes total time to travel the 25,000 kilometers via Madagan, Moscow, Helsinki, New York, and Anchorage. A healthy Eskimo in a kayak with the wind behind him could make it in about the same time. This reminds us that political matters have a hand in structuring our global transmission space, as eastern Europe is rapidly finding out. Although eastern Europe was relatively well-protected against HIV during the 1980s, transmissions are soaring when miles of trucks at Romanian and Bulgarian borders are served by sex workers in a region where Western and Japanese condoms go for well over ten dollars each on the "informal market."

Let me suggest, as a simple and straightforward starting point, that we need to examine the structure of global air passenger space, not as it appears conventionally as a globe, but as a multidimensional and strongly shaped topological space defined by air passenger interactions. We have seen, at what we must now consider to be the "regional" level of the United States, that these flows are highly predictable from simple gravity model considerations. I do not think we can hope for such geographical predictability at the global level: too many economic, cultural, and political constraints would distort such regularity. But this means we shall have to collect data on air passenger flows, and put into place international agreements to monitor and upgrade these on an annual basis. There is no technical constraint on making a 4,028 × 4,028 global operator, with its more than 16 million potential flows, and moving between various iterative experiments and direct visualization of the consequences with modern computer cartography to point up the major pathways, the superhighways, not to say the "chokepoints," of disease transmission. These are going to be crucial to identify for early warning systems, because even if a new and deadly virus starts in the global equivalent of Peoria, Illinois (where, not incidentally, it is most unlikely that it will be diagnosed), we know it will work its way up the air equivalent of a capillary feeder road, until it comes to an interchange giving it access to one of the global arteries or superhighways.

Notice that each iteration or probabilistic smear can be mapped, producing initially faint pathways on our globe, but getting stronger and stronger with each multiplication, intensifying as the process proceeds. We could link, say, 100 or 1,000 of such images together with animation, and so begin to visualize the very complex spatiotemporal diffusion process at the global level. The Centers for Disease Control at Atlanta could have such general, and readily visualized, simulation models grounded in real, well-monitored data. We could specify a starting point with a latitude and longitude, and see immediately, in a two- or three-minute film, the most likely outcome. Such a presentation might even remind people at CDC and NIH what all geographers know: that every epidemic always has a geography as well as a history.

It is clear that we need air passenger data just to start to understand what global transmission pathways are now, and what they will be in the future. Such data are not available, although it would not be difficult to collect them under present international agreements, particularly as almost every booking is already computerized. The International Air Transport Association and the World Health Organization, both agencies of the United Nations, live side by side in Geneva, Switzerland. Like the two Englishmen on the desert island who never spoke to each other because they had never been properly introduced, I suggest we say "Messieurs, Mesdames, je vous présente."

Trying to Be Honest

ONE OF THE nice things about being asked to contribute to a book on a fairly broad topic, one that can be approached from many different angles, is that you have a freedom to write in open and honest ways more or less forbidden by the stuffy professional journals, whose editors too often want you to write in *their* way rather than your own. So when John Eyles approached me about his book,[1] he emphasized the question of how I *really* went about a piece of research, his emphasis implying, of course, that it might be rather different from the strictures laid down by the philosophers of science, who never actually do any research themselves, or the inculcating *rites de passage* through which so many poor graduate students are forced these days. Somewhere in our exchange of letters the word "exposure" appeared, and I immediately thought of an amusing poster of a man dressed only in a raincoat standing in front of an art shop, "exposing himself to art." Such merriment would be immediately condemned today as yet one more manifestation of the hegemonic phallocentric gaze, but in my insensitive, Neanderthal days, when I was still politically incorrect, I confess I chuckled. And so "Expose Yourself to Geographic Research" was written.

It is, I think, important to place it, and John Eyles's larger concern, in a broader context, one that I find particularly disturbing for a life of genuinely free inquiry in the university. There are enormous pressures at work today to force research into channels of funding, and more and more people are being measured by the funding dollars they raise, rather than by the research they do. The pressures on young faculty approach the obscene as "mentoring committees," too often a euphemism for a looking-over-your-shoulder monitoring committee, recommend lines of (always fundable) research. Sometimes they even have the effrontery to give points to various journals where it might be advisable to publish.

What with teaching, committees, meetings, and all the other aspects of a busy academic life, it is difficult to salvage from a week those precious *blocks* of time for research and writing, rather than the useless fragments left over from all the other things that always appear to have priority. Until, of course, promotion and tenure time comes along, when all the priorities are radically reversed, and decisions are made on the *necessary,* but not necessarily sufficient, conditions of research and publication.

And so proposal writing itself has become an "art": there are actually short courses taught on the subject, courses that have little to do with scholarly substance, and everything to do with dissimulation and obsequious flattery. I have even been told by respected peers in my field that you are a fool if you write a proposal before you have actually done the research—a marvelous example of intellectual honesty. Moreover, you should always find out who is on the reviewing committee, and who the likely reviewers might be, so you can salt the proposal with kowtowing references to the giants upon whose shoulders you are standing in trembling awe. The whole, intellectually dishonest business soon makes its way down to graduate students. Those intending to do fieldwork, or even archival research, overseas are taught to state what they are going to do and how they are going to do it, before they have even tried it, in a country and culture they have never seen, ten thousand miles away.

Of course, for yourself, and for any graduate student you are trying to help, you do the best you can in thinking through what you want to investigate and how you are going to go about it. And, of course, you do what you can with background sources in libraries and so on. By definition, if the research is really original, these sources are likely to be meager. But what has to be recognized is that you cannot foretell, or rather you should not be able to foretell, the results of your research before you do it. Ultimately, it comes down to a matter of *trust:* you choose the best people you can, people you have worked with and with a good prior record, and then you let them loose *trusting* that they have the intellectual flexibility, the fortitude, and the insight to make all necessary changes once they get immersed in the topic, or find themselves in an always-unanticipated field situation. And if you have chosen and judged well, such an approach works. Some of the best students I have ever had were trusted—in real places such as Philadelphia, Uganda, Ghana, Malaya, Sierra Leone, as well as in other intellectual "places" ripe for exploration. But that was in the old days, when people were judged by their ideas, not the dollars they raised by sucking the Establishment's "flavor of the week" lollipop in Washington.

Another opportunity to be open and less chafed by constraints is to give an address to a group of people whose invitation implies that they are interested in what you have to say on a topic not necessarily appropriate for the usual research journal. When the invitation from Lorraine Clarkson and Anne Lowe arrived from the Ontario Association for Geographic and Environmental Education, I took it as an opportunity to think together with a large group of highly professional teachers of geography in Ontario's schools. Unlike the United States, where the country and the National Research Council appear to be once again in the process of "Rediscovering Geography,"[2] geography in Canada is a prominent component in the initial grades, and becomes an optional, specialized subject in high school through the thirteenth, preuniversity year. Teachers of geography hold degrees in geography, often to the master's level. I was addressing a highly knowledgeable, sophisticated group of fellow professionals.

So I sat quietly one day, looking out, as I so often do, at the flower beds in our backyard, the woodpeckers and nuthatches doing their ecological duty up and down the trunks of the old trees that had never been cut over,[3] and I decided I would try to touch, honestly, upon things that I felt were important, but that never seemed appropriate for more professional outlets. Quietly jotting things down, I felt a sense of calling, communion, disquiet, relevance, justice, the spatial, need, caring, and right. And these were the headings that ended up in the address.

Expose Yourself to Geographic Research

Is the scientific paper a fraud?

—Peter Medawar
The Listener

[Answer: You better believe it!]

By Way of a Somewhat Lengthy Preamble

If you take a course on Research Methods, it is unlikely that you will be told that introductory paragraphs of papers, or first chapters of books and theses, are often written last. As we shall see, you will normally be obliged to set out the entire form and structure of what you want to say long before you know what it is that you are going to say. That this ritualistic requirement is all back-to-front and topsy-turvy is . . . well, precisely the point of this essay. And in writing this essay I have been forced to come back to the beginning just as I was ending it. This is because the Truth, whether derived from experience, thoughtful reflection, or research (often in various combinations), sometimes makes its own demands on us. The person who approached me about this essay asked me to set out how I *actually* went about a piece of research, what were the ideas, what were the real difficulties, what were the joys and sorrows of a concrete piece of geographic inquiry.[1] But as I was setting down some of the things that happen during a research project, I started to think. This, of course, is often fatal. In thinking about research, what I came to see was the enormous amount of pretentious ritual that students are made to wallow through before they can undertake imaginative inquiry about something they find intriguing. So to be Truthful, I really felt I had to undertake a sort of intellectual slash-and-burn operation, and make a clearing in which people might start to think afresh about inquiry—*all* inquiry, including the geographic.

The Joys and Sorrows of a Geographer

I am going to assume that you have been "called to geography," although I will be the first to admit that this expression has a rather funny ring to it these days. The idea of being *called* to a particular way of looking at the world may be appropriate for people in some form of religious or artistic life, but it seems a bit excessive for others. Most of us can accept that poets are called, at least the really good ones who disclose and sing with words; and so are the artists who paint, sculpt, etch, and draw simply because they must. Dancers, too, those who are driven by a disciplining rhythmic demon inside of them; and composers and musicians, the ones who seem to have no choice but to set down what they hear, to bring forth for others what has been granted to them. Religious and artistic ways of life have a certain feeling about them that makes the notion of a "calling" appropriate. But can it really be applied to geographers?

Of course it can, although only you can tell whether you feel that sense of calling, that compulsive drive that tells you that you do not want to do anything else. Only you know whether you will become a professional geographer, despite the obstacles in your path. Which means you have already seen a clear distinction between commendable amateur enthusiasm and truly professional dedication, and have decided that you will be responsible for pointing your own life in the direction of the latter. After all, life is not a rehearsal—one performance only.

Your decision to follow this particular calling gives you great freedom. It also, quite paradoxically, imposes some considerable constraints. The freedom is given to you because the geographic way of looking places few bounds upon the actual subject matter you may inquire about. There are few things of interest at the human scale that cannot be illuminated by thinking about them as a geographer. However, that same decision also places some constraints upon you. The first is sheer professional responsibility. The very eclecticism of geographic inquiry, and the feeling that many different things are connected together, can easily lead to superficiality. Geography, as a way of looking at the world, touches upon many other, much more limited and specialized, ways of considering people and things. Along the geographic way you are going to meet many others with a particular passion for inquiry in much more limited domains. If you choose to look at these things from your own geographic perspective, you owe it to yourself, and to the way of looking to which you have been called, to gain a reasonable grasp of what is already known in these more limited areas.

Unfortunately, by taking on the responsibility to inform your own inquiries with the more specialized knowledge of others, you immediately expose yourself to another danger. That same specialized knowledge you acquire from others can also constrain your own inquiries from a geographic perspective, and severely limit your thinking by channeling it down a certain way of already established procedures that address already established questions. So at this point things look pretty bleak; you may feel that there is no way you can win, that you are damned if you do, and damned if you do not. If you do *not* expose yourself to more specialized and established ways of thinking and inquiring, you may miss something important, and so be labeled superficial; but if you *do* thoroughly investigate these more specialized ways of inquiry, then your thinking may be so conditioned that you end up as a psychologist, hydrologist, or even an economist. At this point, genuine and open *geographic* inquiry becomes practically impossible, because your thinking now has become trapped in those particular paradigmatic ways of accepted procedures that metamorphose genuine inquiry into the rituals and forms of professionally accepted research. These rituals and forms have the capacity to shape severely the questions, procedures, methods, and presentations, and we should take a moment to look at them more carefully.

Rituals and Forms

It must be difficult being a student today. There seems to be so little freedom to go your own way, to choose your own questions, and to make your own mistakes. It always seems that there is a right way of doing things, an approved-by-your-adviser-or-tutor-or-committee way of doing things. Follow *their* way and all will be well, but stray from this path of righteousness, and you will be crunched. Of course, this do-as-I-say way is fine for the person who likes being told what to do. In fact, if you do it well, you can even plod all the way to the Ph.D. Unfortunately, it is a way almost guaranteed to frustrate the really imaginative and curious person, and nowhere is conformity more apparent than in the outlining and writing of research proposals, whether for an undergraduate paper or thesis, or for a postdoctoral research program funded by a national agency.

The problem is that no one is allowed to say openly and honestly, "Well, what I really want to do is poke around in this area and see what I can find."[2] Before you can do what you want to do, you will have to posture, pose, and distort as you formalize and disguise your geographic curiosity in the great ritual of the Research Proposal. This document may have little to do with what you really want to accomplish, or the way you will actually go about it when your adviser or committee have been lulled into thinking that it is all right to unleash you. Perhaps the great Research Proposal is really designed to protect the adviser, committee, or the dispenser of funds. The "system" holds them responsible for judging whether you can be unleashed or not, and the best way for them to cover themselves is to make you tell them what you think you are going to do, how you are going to do it, and what you are going to find . . . *before* you have done it. Yes, I know, it all seems a bit backwards, but I am trying to be honest about what actually takes place. You will be told that all this is designed to help and protect you, but as you struggle to squeeze your enthusiastic curiosity into thoroughly irksome ritualistic forms, the feeling will grow upon you that this cannot possibly be right. And you will certainly be partly correct. What you are going through is a *rite de passage* to ensure that you conform to the accepted (ahem!) *paradigm*. Invariably, this paradigm is the one into which your adviser was inculcated when he or she was a graduate student. This is why most academic research is so sluggish and slow to change. The priests of one generation train the acolytes of the next.

What are the pieces of such a proposal? First of all, you will have to outline it, and the more formally you do it the better. A detailed hierarchical structure, with all the Roman numerals and subsections of subsections, will do much to convince Them that you are on the right track. Of course, you will have to fill in all those hierarchical subsections with things that you invent out of thin air, for the simple reason that you have not done the research yet. How can you possibly know whether you will find X when you have not looked, or that method Y will work when you have not tried it? If it all seems senseless and back-to-front, just remember that there are people in the academic world who spend their lives doing this. They outline, collect bibliographic references, and do literature searches because they are much more intrigued with the forms of the ritual dances than actual inquiry. All too often, those who do little research themselves are the ones who teach others how to do it.

Now, quite obviously, the time to outline and structure your thinking and findings is *after* you have done the research, when you really have something to say, and so have the responsibility to say it clearly, fluently, and well to your reader. But I doubt that anyone will tell you this. It would undermine one of the great rituals of the day, a ritual that can only make you feel fraudulent, uncomfortable, and perhaps humiliated. I use the last word advisedly. When I first started to teach, one of the "Great Figures" of the day said, with characteristically unquestioned authority, that "the purpose of writing research proposals, and holding doctoral examinations, is to give the student a good dose of humility." I cannot think of a more inappropriate or humanly disgusting attitude for a teacher to have, and I despised the man from that day on.

Secondly, you may be forced to pose hypotheses, no matter how banal these seem, although what you *really* want to do is observe carefully, closely, openly, and with as little prejudice as you can muster, and then describe, with insight, skill, and thoughtful imagination, the topic you have chosen. In the process, you hope that along the way all sorts of interesting things turn up that you could not possibly have anticipated. But such a natural hope and expectation that things might come to light that were not seen before is something that the formalists do not talk about. Hypotheses it must be, even if these have to be winkled out on the end of a blunt paradigmatic pin.

One problem with such hypothetical mouthings is that they tend to be dissimulating. After all, what on earth is the use of posing a hypothesis if you are pretty sure it is going to fail? If you think it is going to fail, why pose it in the first place? Still, you can always get around this by posing the alternative hypothesis as your original hypothesis, and with this intellectual pirouette demonstrate your positive approach to scientific research. Unfortunately, there is another problem: once you set the hypothesis down in the accepted form, you find yourself trapped into having to test the wretched thing in some way that is equally acceptable. This, almost invariably, means Statistical Methodology these days, because we increasingly confuse a highly constrained and limited way of marshaling certain types of evidence with *the* Scientific Method. This confusion, and how it came to be, is a deep and important topic for the intellectual historian of the twentieth century, but we cannot pursue it any further here. Still, it is worth pointing out that science made splendid progress for three hundred years before the statisticians came along, and that most of the really great names in the history of science never posed a statistically testable hypothesis in their lives. Of course, they certainly had hunches, and they pursued these with great skill, imagination, and diligence; but few of them ever set down a formal hypothesis and then went on to test it at the five-percent level of significance. Most really good scientists have the intellectual self-confidence to put this limited domain of methodology firmly in its place, and bring it out only when it is needed. Only people who are intellectually insecure hang onto to the coattails of the statistician in order to make their often banal questions and findings acceptable to Them. I have the uncomfortable feeling that the more statistical tests of significance are reported, the more banal, plodding, and pedestrian the research is likely to be.

What is continually ignored by those who cling to the forms and rituals is that most "eureka" experiences come from *descriptions* of the physical, biological, and human worlds that allow us to see something we did not see before—like $e=mc^2$, models of DNA molecules, huge water ripples in Washington scablands, gaps in chemical tables, jumping genes, and subducting plates. The eureka experience does not have to be exactly Archimedean in its explosive power—there is no need to expose yourself to that extent[3]—but it should mean that something not seen before emerges into the open clearing of our thinking. Something *dawns* on us, and we say "Oh, I *see!*" I think it is worth pausing for a moment and thinking about the phrases we use when something like this happens.

Notice how we constantly use images of light whenever we try to express our sense of coming-to-understand-something. We direct our thinking, which means we direct ourselves, like an illuminating beam to light up something that was there all along, only *we did not see it before*. In every science, and in all successful research, there are moments of sudden seeing when something that was concealed from us becomes unconcealed. Which is exactly why *aletheia* for the Greeks was the word for truth, the *a* negating the *Lethe*, the dark underworld of concealment, to make the truth un-concealment.

Statistical Significance and Meaning

Now contrary to what you may be thinking at this point, this is not a tirade against statistical methods and statisticians. Sometimes these ways of marshaling evidence are perfectly appropriate, and may be the only feasible possibility of bringing evidence to bear on a particular question. But this *possibility* should not force all research into it. We have to be very careful to distinguish between what is simply significant in a statistical sense, and what is meaningful. Often, in a statistical approach, something, sometime, somewhere will seem worth reporting as "significant," simply because "significance" is tautologically defined within the totally closed system of statistical thinking. Something is significant because the N (the sample size) and the p (the probability level) say it is so. And N and p say it is so because that is what has been defined as significant. The trouble is that *significant* does not mean *meaningful*, and "meaningful" means capable of being given persuasive interpretation—at least for the historically contingent moment.

So what do we *really* do when we undertake a piece of research? First, there is no point in undertaking research at all unless the topic, the general area of inquiry, fascinates you. To do research simply because some requirement for an academic degree says you must is one of the most anti-intellectual and soul-destroying things I can think of. And to undertake a piece of research simply because it happens to be the Establishment's "flavor of the week" down in Washington desecrates the human freedom to inquire that should be the University's most priceless possession. Unless, of course, you like being manipulated and want to join the Marionette's Chorus Line. Secondly, interesting research, the sort that discloses unexpected things and perspectives, is nearly always a difficult-to-talk-about and strange mixture of successes and failures, hopes and disappointments, times when you wish a day had a hundred hours, and other times when you groan because you know it does. In any research, there are times when you have to grit your teeth to get through some miserable task at hand (gathering, coding, plotting, drawing, programming, computing, etc.). But perhaps the best way to demonstrate this is to tell you the story of one piece of research. Not because it was particularly meritorious (no DNA or $e=mc^2$), but simply because it was my own and I know it best.

The Story of Mental Maps

It really started one afternoon while I was doodling on a blackboard, and thinking in what I can only call a "gravity model mood." Like many geographers, I nearly always have to draw pictures when I think, or before I can convince myself that I really understand something. I was wondering how so many forms of human interaction were predicted quite well by very simple gravity model ideas. In those days, and long before Alan Wilson was to derive the entropy maximizing version, these were straightforward Newtonian, $I=P_1 \cdot P_2/D$ analogies, used to describe journey-to-work, shopping, and air traffic movements. These were rooted in the migration studies of the nineteenth century, which were themselves standing in an even earlier tradition of social physics.

The early 1960s were the days when geographers were starting to drive the questions back to the human beings who were actually doing the deciding about locating, moving, growing, making, investing, buying, and so on, and notions like decision making, optimization, normative models, information, satisficing behavior, and things like that were in the air.[4] It was out of this intellectual flux that something called Behavioral Geography was to emerge, and in this embryonic context it seemed quite reasonable to ask individuals where they might like to live if they had an absolutely free choice. To what extent were broad patterns of movement simply a result of many people doing what they wanted to do? Could you get them to rank places in order of preference? Could you actually *measure* people's preferences? And what would you do with the responses if you got them?[5]

The last question was interesting to someone who was intrigued with methodology for its own sake, although you never actually dared to admit this. Remember, you started in the great Research Proposal tradition, posed a problem in a pretentious series of null hypotheses, and only then decided "what was the most suitable methodology"—a phrase intoned in a solemn voice of quite impeccable transcendental seriousness that nearly always means the established way of doing things is driving you rapidly towards thoughtless hypocrisy. In fact, it often worked the other way around; you came across a then-new and intriguing method, cast around for a problem (actually with a body of readily available numerical data that would let you use it), and then proceeded to work things through, hoping that something interesting and interpretable would emerge at the other end. Not surprisingly, it often did, and some of the most imaginative and illuminating research of those days came from precisely this much scorned and disparaged "methodology first" way of going about things. Actually, if you are really prepared to think about it, rather than simply pouring scorn on it, it is really not so surprising that some "illumination" of a problem should appear. When we look at something in a different *way*, from a different perspective, we often see things that we did not see before. And the reason I italicized *way* in that last sentence is precisely because methodology means a way *(hodos)* towards knowledge *(logos)*. If you explore a new *way*, if you take a different path, it seems quite likely that you will see something from a different angle. And if a definition of scientific research does not encompass seeing things differently along new ways, then I submit that there is no hope for any of us. It means that today's Establishment, bred in the graduate schools of the late 1960s and early 1970s, have got such a grip on our thinking that we can no longer call it our own.

Now those were factor analytic days, when the world was a great data cube of things, places, and times,[6] and depending upon how you sliced the cube, you could run the gamut from *O* to *T* modes.[7] So people and their preferences for places immediately suggested an *m*-places versus *n*-people data matrix, and although places were unique (and, therefore, according to the Traditional Authorities, were not capable of being studied *scientifically*),[8] and although everyone knew that people were also unique (and, therefore, according to the TAs, not capable of being studied *scientifically*), it was rather fun thinking about what it might *mean* if you factor analyzed such a places v. people matrix. At least it seemed worth a try, and since I already had a Ph.D., and in those days did not have to waste half my time writing Research Proposals, but could just say to myself "I wonder what would happen if . . . ," I went ahead and did what I wondered. At least a Ph.D. was good for something then.

So I asked students in a beginning class to put themselves in a mood of totally free, money-doesn't-matter choice, and then to rank order their preferences for the states of the U.S.A. I then factor analyzed the preference matrix, doing all sorts of totally forbidden and undefined mathematical operations, computed the scores for the states, and plotted them on a map. It was that map of the "preference surface" that was the interesting thing to drop out at the other end. When this experiment was repeated at other places across the country, some really very interesting and unexpected things started to appear. Although taken from groups of young adults thousands of miles apart, the mental maps from Pennsylvania, Minnesota, and California were almost identical except for a dome of residential desirability right around the local area. A little later, I found the same thing in Britain, thanks to Rod White, who did most of the dirty work, and Peter Haggett, who arranged for the schoolteachers attending the Madingley Lectures at Cambridge to take part. In Nigeria, Dan Ola found the same sort of thing,[9] and also uncovered the possibility that young people were learning about places in rather regular and predictable ways.

As a result of this rather haphazard poking around, factor analyzing, and "wondering if . . . ," I reached the point where I wanted to undertake a full-fledged study, a study that would allow me to ask all sorts of questions whose answers could not even be contemplated without lots of data. For this I needed a reasonably small country, with a good schooling system, and a tradition of geographers being involved in practical and useful research. To cut a very long story short, I found such a country in Sweden, together with a group of geographers at Lund who could open some very necessary doors to get people high in the educational system to grant me permission to approach principals and teachers for their help. But for this I needed money, and so I found myself back in the grip of the great Research Proposal. Although it was eventually dressed up in the necessarily pretentious phrases, implying that if the research were not conducted the intellectual life of all humankind would be greatly impoverished, what it really said between the lines was "I want to go to Sweden, gather very large quantities of data relating to people's preferences and geographic information, and then see what I can find." But, of course, I could not just say that openly and honestly. One reviewer objected to the proposal with the question "But where is the *model?*" meaning, in essence, where were the equations that should have been hypothesized before any research had been done to explain why Person X at Place Y liked Place Z—at least to a certain degree of probability. As usual, all these equations had to be written down beforehand, and it did not matter particularly whether they made any sense ultimately or not. The important thing was that you had something in the accepted *form* that could be tested with equally accepted statistical procedures. No one really gave a damn whether you accepted or rejected your essentially meaningless null hypotheses and intellectual pirouettes at the end of the research. Form was everything, and trying to uncover the Truth was not only a poor and distant second, but was considered by the cognoscenti to be really quite distressingly naïve as well. After all, *everyone* knew that Truth was either relative, or really did not exist anyway. The important thing was to put forward a model, and then accept or reject it at the five-percent level. Fortunately, there were enough reviewers who thought that trying to find out something new was still acceptable science, so eventually some money came through.

What came out of this research, research based on about 11,000 returns of residential preferences and geographic information from Swedes living in 58 of the 70 regions, ranging from seven-and-a-half-year-olds to adults? First of all, things that were expected, expected in the sense that what was found confirmed some of the things we already suspected from the U.S.A., U.K., and Nigeria. But not all by any means, and this large data set began to reveal some problems we had never suspected before.[10] Ideas that appeared rather neat and tidy when people were sampled at only a few locations now started to unravel, and peoples' mix of information seemed to vary widely, but in quite regular geographic ways. None of these things could possibly have been hypothesized in Research Proposal fashion before the huge data sets were collected, and before hundreds of patient research hours had been expended.

But one of the rewarding things about carrying on empirical research is that sometimes the unexpected things that turn up do connect with ideas that have emerged from more theoretical perspectives. Some people would assert that this is back to front; it is the theory that should guide the empirical research. Happily, this is not how scientific research always and *actually* takes place, but only how logicians, who never do any research themselves, say it should take place. In fact, it is a mixture of the two, the theoretical and the empirical; and a discipline, and even a person, may swing from one viewpoint to the other. Any time a theorist gets too conceited, just point out that there is nothing in all those deductions that was not contained in the original definitions. It is "just" a matter of grinding out the consequences—a merely mechanical task. Which says, of course, that theory is really grounded on fruitful definitions, not the relatively trivial binary operations of logic that follow. Even a computer can do these, if it has been properly programmed.

On the other hand, any time an empiricist gets too uppity, and sneers at all that abstract theoretical stuff, just point out that without systematic, coherent, and connected ideas and concepts, all our "facts" will remain unstructured heaps, like untidy hayricks scattered around an intellectual landscape. This means that to see all the possibilities of your research, to bring these to light, you must try very hard to keep your own thinking open for the unexpected things that might emerge from the empirical realm, and also try to see how these might fit in, or perhaps even reshape, the theoretical frameworks that are currently available. To do research that genuinely illuminates requires a lust for both the theoretical and the empirical. One is useless without the other, in the sense that one unconnected to the other has no *meaning*. There is simply no point in carrying out empirical work if its ultimate purpose is to throw up another factual hayrick. The problem is that the theorist may become so enchanted with the formal logical operations that the *geographic* purpose is forgotten. Equally, the empiricist may become so enchanted with the factual details—another hayrick for the collection—that, once again, the geographic purpose passes beyond recall. As a geographer, you must learn to ride two horses, and if they decide to diverge, and split you up the middle . . . well, surgeons can do wonders these days.

However, if you can become highly aware and sensitive to theoretical problems confronting geographers before you enter an area of empirical research, you obviously enhance your capacity for seeing connections between the concrete things emerging from the substantive topic you have chosen, and the more abstract concepts that may deepen your understanding and enhance your appreciation for the difficulties involved. On the other hand, in becoming aware of such theoretical perspectives, you have to be constantly aware that they may trap your thinking along certain, and by definition pre-scribed, channels, some of which become incised very quickly into intellectual canyons. It is difficult to see over the rims of deep canyons to look for other possibilities, and remember that not much new light reaches their floors. Which is just another way of saying that most inquiry is dangerous, and constitutes another instance of being damned if you do and damned if you don't.

Geographic Structures

For nearly three decades now, various pieces of theoretical and empirical research have been pointing towards the fundamental implications of the structure of geographic space,[11] and these are going to be found at a variety of scales and times. At the macro scale, both information and preference are conditioned by the distribution of population that constitutes the source of human interaction. At the meso scale, the actual structure of the transport network comes into play, shaping the way in which the information surfaces of children emerge in highly regular and predictable ways. At this scale, there are often clear corridors of travel, or perhaps warped probability fields, shaped by the way geographic space is structured for human purposes. Some of these structural channels may be extraordinarily stable, influencing the spread of innovations and diseases in very consistent ways over long historical periods.[12] Confirmation comes from studies of historical epidemiology,[13] but we still know very little about the way the geographic structure of one time influences and shapes another. Indeed, we really do not know very much about geographic structure at all: how it maintains its form and influence; how structures at one scale aggregate, or disaggregate, at other scales; how societies maintain their geographic structures, and are shaped by them in turn. In fact, we are not very good yet at making our strongly intuitive notions of *structure* actually and concretely operational. Yet it is only by seeing these different things in the context of common structural concern that apparently disparate and unconnected pieces of geographic research begin to connect up, and we can begin to limn the still fragile, but gradually strengthening, fabric that *is* human geography. As one who has been called, you must attempt, in every waking (and even dreaming) moment of your professional life, to see connections that make up the structure of geography. Otherwise, things fall apart . . . because the center cannot hold. As the poet Yeats well knew.[14] But then, poets have that capacity to expose themselves for Truth, rather than Sooth, saying.

Thinking like a Geographer

> The Old English *thencan*, to think, and *thencian*, to thank, are closely related. . . . Is thinking a giving of thanks? . . . Or do thanks consist in thinking?
>
> —Martin Heidegger
> *What Is Called Thinking*[1]

COMPARED to the geographic wasteland south of your border, geography appears to have fared well in Canada, both as a formal discipline extending our understanding by illuminating the human and physical worlds, and as a subject taught at all levels to create informed and aware citizens. I know you think much remains to be done, and perhaps things always look a bit greener on the other side of the hill, but I hope you will not mind my somewhat envious gaze. In fact, when I received the five volumes of the "curriculum guideline" for geography, issued by the Ontario Ministry of Education for the intermediate and senior divisions of your high schools,[2] my feelings were not so much envy as panic, a panic that rapidly induced something close to intellectual paralysis. Because in a moment I can only describe now as utterly weak and foolish, I started to read them, only to learn that your senior students "analyze, interpret . . . explain . . . design and develop networks, systems and simulations that involve six or more variables."[3]

In retrospect, I still think my panic was justified. After thirty years of hard postdoctoral work, I have reached the point where I can think about the interactions of maybe three variables, but hardly the combinatorial possibilities of six. Such ineptness would mean that I might just squeeze into your tenth grade, and from there slowly work my way over the next two years towards those Olympian heights of analytic thought where dwell Ontario's high school graduates.

Given initially as the Plenary Address at the meeting of the Ontario Association for Geographic and Environmental Education, Toronto, 1989, and subsequently published in the *Canadian Geographer* 35 (1991): 324–33

A Sense of Calling

Well, of course, I am teasing you—at least just a little—but I can assure you that my panic was real. "What," I said to myself, "what on the face of the earth can I possibly say to these people?" And then I calmed down a bit, and in one of those conversations with oneself that we call thinking, I thought, in good dialectical fashion, exactly the opposite. What, as a teacher and as someone who loves teaching, what could I *not* say to my fellow teachers and geographers? After all, we had made the same choice, going into teaching for the vast material rewards that are so familiar, those huge paychecks piling up each month at the bank as we desperately try to spend them.

No, we know that was not the reason: our hearts are pure, even if our bellies are half full (which is probably better for our cholesterol count anyway). No one goes into teaching—no one *should* go into teaching—without having felt that tug of calling, that sense of commitment, that deep intuitive knowledge that teaching young people is a privilege—and the most important thing in the world. In a sense, they are us, but perhaps, in a deeper sense, we are them, because they are our hopes for a more decent future.

Oh yes, we get tired, of course we do. There is not one teacher here today who has not returned home exhausted, sucked dry, after giving, giving, giving all day long, with half the day's work unfinished and still facing you after the evening meal has been cleared away. It is tough being married to a teacher. The homework has to be corrected by tomorrow; those papers and essays, whose grammar, spelling, and structure do much to increase the sale of strong drink, have to be read and corrected without the sarcasm of fatigue showing through. And sometimes, in our marginal comments, we fail, leaving us remorseful for our misuse of power, or perhaps somewhat pugnaciously gleeful that the careless little nitwit now realizes that we are human too.

A Sense of Communion

But we have all felt at one time or another that empty feeling, when our personal storehouses of intellectual strength and variety seem empty, when our geographic granaries have only a few grains blown across the floor by a chill wind. Then we need renewal, and in our thoroughly human endeavor of teaching, that means contact, connection, and communication. There is a lot of talk today—or perhaps "jabber" would be a better word—about communication, but let us stand still for a moment and listen to the word. At its root is *munus*, a present, so *com-munus* is "with present," a sharing, a giving of gifts.[4] We share our professional and personal gifts with each other in many ways: in the quiet solitude of an intense encounter with an author who, by publishing, by the courage of a "making public," invites us to share in his or her own delight of discovery and thinking; in daily contact with colleagues and students, who act as intellectual antennae, who say excitedly, "Hey, did you see this . . . ?" or who ask the apparently naïve questions that stick like burrs and irritate and itch until we are forced to scratch them; and in special occasions of professional meetings, not just in the formal sessions, but perhaps in individual conversations, or in the small groups that form transiently like eddies out of the flow and flux along hotel corridors, dining places, and small corners out of the wind.

I would not dare to bring and share with you presents of geographic substance, of opinions about curricula, of advice in your daily tasks of teaching. To do so would be impertinent, if it were not so pathetically naïve. What I can perhaps share with you are what I would like to call the "wellsprings of the teacher," reminders of the larger, informing context to our daily, substantively pedagogic, lives down there in those classroom trenches. Some are presents I hope you have not opened before, created out of the delight of my own encounters with those who held presents out to me. Others have been passed down, with love and respect, from an Academy now over 2,500 years old. Like one of your lovely Eskimo statues, they have acquired their brilliant polish of relevance and shine of meaning for us today through the passing of many hands upon them.

A Sense of Disquiet

There seems to be some sense of disquiet about the relevance of the geographic perspective today. We have to hustle and display our wares in the marketplace; we hope our "show and tell" sessions will meet with the approval of . . . THEM. I do not think anxiety has ever been so misplaced: we stand today at the threshold of a time when the geographic wheel of fortune has come full circle. Or rather the hermeneutic circle has spiraled upward, for we are not where we started, and our interpretive stances are different today. In this day and age, we will not find our relevance in the presence of Humboldt's *Kosmos* or Reclus' *Géographie Universelle* in every educated home—and there are many more of these today than in the nineteenth century, thanks, precisely, to universal schooling and concernful teaching. No, our contributions, our Kosmoses and Géographies, lie today not on the coffee tables and bookshelves, but out there in a world increasingly aware that an understanding of the once-neglected spatial and geographic dimensions of human existence is crucial for working toward, for enlarging, the conditions of possibility for more decent, more humane lives.

We live in "interesting times," conditions, you will recall, of what once formed an ancient Chinese curse, in a culture that prized tranquility, that induced a don't-rock-the-emperor's-boat mentality above everything else. I would give that saying, "May you live in interesting times," as a blessing. I dare you to say the reverse. One life to live, no rehearsals: Would you really live it in "uninteresting times"? What child of our Western world, of the driving Greek inheritance, longs for ennui? Not me. More than just interesting, these are exciting times, especially for geographers, full of possibilities, all capable of writing geographies on the face of the earth, and they demand our best efforts in teaching and research. And, perhaps especially, in transforming research into teaching.

Let me explain what I mean in a sort of bricolage of geographic thinking that may capture some of the ferment of geographic concern today. I use the term *bricolage* advisedly, because we are going to roam all over the intellectual map—which is exactly what geographers ought to do. There are not many fields today that allow such intellectual peregrinations and explorations. We all know we live in an increasingly specialized world, where knowledge, and therefore understanding, becomes more and more fragmented. To cut deeper in our sciences, we have to break the disciplines into ever-smaller shards of specialization to keep the edges bright and sharp. I ask you: What other disciplines, other than philosophy and history, still have the capacity to combine the scientific with the humane, to inform rigorous inquiry with human care?

A Sense of Relevance

We live in democracies, and it is fashionable to sneer at their failings, until we see what happens to people in authoritarian states where "the party knows best." Not all the students in Tiananmen Square, not all the Poles, not all the young people moving from East to West Germany can be misguided. We take too lightly the world into which we have been thrown, a world of possibility wrung by many lives out of tyranny—as Edmund Burke knew only too well. Oh, it is not perfect. For all the tautological skill of a Voltaire, it is not Candide's "best of all possible worlds." But in the last resort, even as a country passes four ever more tightly controlling National Security Acts in one century,[5] even as civil servants sidetrack one Freedom of Information Act after another,[6] we can still throw the rascals out. Providing, of course, that geographic space is partitioned so that each voting member of the *demos* has equal weight—unlike the state of Georgia until quite recently, where one vote in the country counted for ten votes in the city.

It is no accident that geographers have been called upon to solve difficult reapportionment problems, essentially optimization problems of geometrical combinatorics under constraints.[7] Nor is it an accident (for example, in Britain) that a number of geographers are in the forefront of informing a wider public of the dangers to democracy of gerrymandering,[8] the twentieth-century equivalent of creating rotten boroughs, with the resulting disenfranchisement of more and more people. In Britain, the Social Democratic Party wins 26 percent of the vote and 3.5 percent of the seats in the House of Commons. In the European elections, two million votes are cast for the Greens, but not a single member is seated in the European Parliament. Democracy, my Michael Foot.

These, of course, are geometrical solutions undertaken for moral ends by employing the combinatoric searching power of large computers. Is it an accident that geographers provided the combinatoric evidence used in a federal court to break the spatial machinations of school boards to disenfranchise major ethnic groups in Detroit?[9] Is it an accident that geographers of great imagination have broken the logical backs of tame statisticians employed by national atomic power and weapons production consortiums,[10] statisticians who chorus "clusters of children with leukemia can occur just by chance," to which we might reply: "Around the Sellafield reprocessing plant, declared by international authorities to be the filthiest, most dangerous, most polluting source in Europe?" Is it an accident that geographers have nailed to the wall the tame statistician's excuse that "You can get any degree of clustering you wish just by changing the recording area?" How did the geographer do it? By examining the clusters of children with leukemia at *all* geographic scales, generating in northern England eight million rigorous Poisson-based tests, and plotting the quadrat circles whenever the stringent threshold of significance was reached. In an old and fine scientific tradition, the evidence is visual, cartographic; it lets you say, "I see!" as the light of understanding dawns on you when you see a great black blob, circle overlaid upon circle, significant clustering of dying children at all scales. No wonder it was published in the *Lancet;* no wonder the medical profession today is raising the question again, despite government inquiries and reports that tried to lay it in its quiet Establishment grave.[11]

But there is more. Out of this act of geographic imagination, using the computer to do something we could not do before, rather than just doing the same old stuff faster, came an act of original geographic discovery: the unexpected, the serendipitous, came out of concealment—like Columbus bumping into the Americas. On the East Coast was another huge black blob, significant quadrat piled on significant quadrat, totally unsuspected. In the middle of it was a huge urban incinerator, firing, it is suspected, at much lower than designed temperatures, spraying carcinogens over the area. Is it a cluster produced only by pure chance? Try GAM, the Geographical Analytical Machine, with Wilm's tumor, a cancer we know is genetically caused, and nothing happens. Try it with randomly generated artificial data—no black blobs. Put that in your pipe and smoke it, Hired Statistician. And remember, your smoking of that pipe does not *cause* cancer of the lip and jaw: correlation is not causation; may you and your consultant's check from the Tobacco Institute rot in hell.

A *Sense of Justice*

So who says geographers have nothing to say? The deep ontological roots of *dikē*, of justice, that we prize out of our Greek heritage, the *dikē* that transforms to a "meet and right" in a *Book of Common Prayer*, that is reshaped as fairness and inalienable right in an Enlightenment that still deeply informs and underpins our sense of democracy today, these possibilities for just solutions rest on a thoroughly geographical foundation. The sense of caring for children thrown into an atomic age is so lit up by geographic inquiry that even the most cynical Establishment is forced to take notice.

Let me give you two further examples out of my own recent experiences. There are honorable traditions of inquiry in our eclectic field and we choose between them, to the exclusion or denigration of some, only at the price of public, not to say intellectual, relevance. Theory-praxis, regional-systematic, quantitative-qualitative, human-physical, analysis-synthesis, space-time—these are choices to be made, stances to be taken, paths to knowledge to choose in light of a particular problematic that we see. Or does the problem sometimes reach out and grab us, like a melody granted to a Mozart?

When Chernobyl went off, and that radioactive plume deposited radionuclides over Europe, it was clear that literally scores of those specializations into which we have been forced to partition our knowledge—and let us admit we have no choice—scores of these were going to be involved. The fields include agriculture, anthropology, atomic physics and engineering, biology, botany, chemistry, economics, forestry, game management, law, limnology, medicine, meteorology, ornithology, pediatrics, politics, zoology—literally scores of those sharp disciplinary shards into which our world of knowledge is broken today, including, quite tragically, psychiatry, to help those whose "world" had been so shattered by that blast in Ukraine that they no longer wished to be. I know: I talked to such scores, from Academy of Science members to young Sami reindeer herders in Sweden and Norway. Because the problematic I saw, as I wrote *Fire in the Rain,*[12] was a level of intellectual fragmentation that only the old and honorable tradition of synthesis in geography could approach. And it was no accident that the subtitle was *The Democratic Consequences of Chernobyl.* I give you my word, the book did not start out as a muckraker, but the lying was so blatant and unashamed that no one could ignore the relation between national atomic dependence and Establishment dissimulation. As the director of France's electricity board said, "You don't tell the frogs when you're draining the marsh." So much for the *demos.* Even Aristophanes' chorus of frogs is stilled after two and a half millennia.

I had no choice about that subtitle: sometimes books take over and write themselves and you are simply the one holding the pencil. It is almost certain to be a most unpopular book, disliked by both the pro- and anti-nuclear groups. But presumably we are not called professors for nothing. Writing geography as synthesis, trying to pull the threads of the fragmented physical, living, and human worlds together again, is not a popularity contest. We try to tell the truth. And since "the truth" may leave us slightly embarrassed today and draw forth the knowing smile of condescension at such distressing naïveté—poor innocent fellow, surely he still doesn't believe in Truth?—then dare to think the reverse: let us hire teachers who are good at telling lies, whose training has been a fine honing in the art of dissimulation and the creative generation of false data. Let us hire Doctor Liar to further our cause of education for democratic empowerment.[13] In our postmodern sophistication we may be embarrassed at the naïveté of truth, until we recall what a world wholly based on lies would mean.

A *Sense of the Spatial*

The second example, omitting all the details for the moment, concerns AIDS. For a year now, with colleagues at Ohio State and Carnegie-Mellon, we have been modeling the spread of AIDS, trying, out of that rich tradition of spatial diffusion in geography, to predict the next maps by making mathematically well defined the structure contained in spatiotemporal series of counties and quarters.[14] To the best of my knowledge, we are the only people in the world doing this and it looks as though we can predict the next map about 96 percent accurately. The implications for planning, for meeting the crisis, are obvious. We passed 100,000 AIDS patients in the United States last July (1989) and a conservative estimate predicts 329,000 by 1992.[15] Hospitals in major cities are already overstressed, and we must think about new facilities, hospitals, hospices, and new forms of care for terminally ill patients.

But simple numbers, generated by traditional epidemiology down that time horizon, are not enough: to plan we have to answer that most geographic question: *Where?* And that is the problem: only geographers appear to see the human condition, to see an epidemic in space as well as time. One distinguished mathematician at the White House 1988 AIDS Conference in which I participated thought that "spatial modeling" meant running around from Alabama to Wyoming with a differential equation clutched in his hot little hand. And, of course, if the equation is nonlinear, so much the better, because it can lead to all sorts of exciting possibilities mathematically, none of which is of the slightest relevance for our understanding of AIDS in space and time. Not a single person has been made well, not a single transmission has been prevented, by this nonsense.

In contrast, geographic modeling helps directly with the task of effective intervention. First, it helps in the allocation of future, and always scarce, medical resources, by calling upon another rich tradition in our field, where spatial allocation and assignment algorithms are well known. But more important, it turns out that these map predictions, when incorporated into animated map sequences on a television screen, have a very strong impact, the very effects that people in health education call "cues to action," that slam in the psychic guts that makes someone, particularly a teenager with the immortality syndrome of the young, stop short, become self-reflective, and think, deeply and personally, about forms of protection and behavioral change. Here, I am convinced, we are on a cartographic frontier of animated maps showing the spatial dynamics of all sorts of geographic phenomena. It is an area, we are rapidly learning, where you can throw out much that was passed on as unquestioned cartographic wisdom from a time when maps were static and printed on a page. There is little to guide us here at the moment and we shall have to find our own way.

Let me give you two examples. Who needs all that written text on a map when you have a voice accompanying it to explain? As for color, this is no longer a costly rarity but something that is taken for granted. And the cognitive psychologists can play with these problems as much as they like: in the final analysis, their science has nothing to say in this essentially aesthetic domain—as Vincent van Gogh, writing to his brother from Arles, well knew.[16] Sitting at my color palette at a Macintosh personal computer, I discovered, in a naïve and simpleminded way, what he knew instinctively: colors look different, and mean different things, when placed next to each other.

Let us admit quite openly that we are concerned here with graphical rhetoric, and I use that term in all its old and honorable sense of the "art of persuasion." We want those animated, dynamic images to reach out to young people and persuade: there is no cure, no vaccine, only highly experimental drugs that prolong. To fight this dreadful disease all we have is persuasive education. Graphical rhetoric might persuade and so save some young lives. The secondary infections of AIDS are not pretty and the deaths are not quiet.

A Sense of Need

So who says geography and geographers have nothing to say? Where geographers acknowledge a genuine problematic—and our journals tell us that this is not always the case—where genuine problems reach out and grab us and are treated with all the imagination and illuminating power we can muster, then we have things of relevance to say that are the equal of anything in the physical, biological, and, perhaps especially, the human sciences. With pollution, hazard and risk assessment, epidemics, historical preservation, soft energy, pointing to the starkness of homeless lives, linking a global economic system to the specifics of place and restructured human lives, no wonder we hear a growing chorus of pungent and increasingly intellectually informed concern to reestablish the spatial in the social, to place human society back in its geographic dimensions from which it was torn by nineteenth-century sociologists and political economists.[17] Calling upon the best of our tradition—and why call on anything else?—we have nothing to be anxious about and certainly nothing to apologize for. With the world going to hell in a handbasket, we need geographers, we need people open, aware, and alert to the spatiality of the human world, as we have never needed them before.

And I wonder how many of you would disagree with what only appears to be a provocative statement—initially, before quiet reflection acknowledges the obvious—namely, that geographers are born, not made. Like mathematicians, poets, and musicians, you are either granted that fascination with the spatial, and called to the geographic, or are you not. Oh, of course, much more is eventually involved: one has to be thrown into a world where geographic inquiry as we know it stands as a condition of possibility, and, despite an initial calling, there is much to learn in formal ways, much to understand about the heritage in which one finds oneself. But many of you know the experience firsthand and can recognize it in a young person: that early fascination with maps, that wondering what lies beyond the horizon, that sense of relatedness in spatial juxtaposition, that intuitive awareness of geographic pattern and spatial structure, and how it allows and forbids. Some appear to have a sense of intuition, of intentionality, toward the geographic domain that translates into a caring and concern that marks a geographer's life path to the very point where the line in the space-time diagram is extinguished. Let me suggest that we would not be teachers of geography if we had not experienced such a granting to us ourselves. And out of our own wonder and excitement of the geographic realm, in all its eclectic ramifications, we know it has to be a component of education in a modern, democratic society with a genuinely empowered and informed citizenry. Just as poetry, music, mathematics, and a sense of historicity—of standing, alert and aware, in a particular historical heritage—just as these form crucial components of an educated life, so an informed awareness of the geographic domain must characterize responsible people inhabiting an increasingly interconnected world.

If we can back away from the immediate for a moment, what is it that grounds our concern for the geographic realm, either as teachers of others of all ages—kindergarten to adult education—or as those driven to inquire in the spatial domain? The answer should really come as no surprise. We are, after all, human, and when we ask a question of "ground" we are no longer in an ontic realm, the realm of things, but the ontological realm, where we are asking about the conditions of possibility for something to be. So there is nothing strange, nothing unusual, nothing esoteric in asking how geography, geographic teaching, geographic awareness, geographic inquiry "come to be" as thoroughly human endeavors. Questions of human ontology are as concrete as anything we are likely to think toward; we do not use words like "ground" and "fundamental" for nothing, or unthinkingly, or merely abstractly.

A Sense of Caring

These questions are very old: Martin Heidegger, drawing upon an ancient fable, tells how a personified Care went to a river's bank and shaped a piece of clay, and then asked Jupiter to give it spirit, which he did, providing that the newly formed creature would have his name.[18] Typical male chauvinist.

But Care wanted her own name to be used (typical female chauvinist), and then Earth arose and demanded *her* name be used because she had given the clay out of which the creature was made. Eventually, Saturn came along and became the arbitrator and ruled: Jupiter gave spirit and shall receive it back at death, and Earth shall receive the body back into herself. But because Care shaped the form, it shall be hers as long as it lives; and it shall be known as *homo,* for it is made of humus, or earth. Human-earth relationships begin very early in our Western world and are permeated with Care from the beginning, a care that has the capacity to be translated into responsibility, husbandry, and stewardship.

And, of course, but regrettably, the equal condition of possibility for their opposite: for courageous Brazilian ecologists to be murdered in the Amazon, for the rape of an ocean by nylon nets 50 km long, for toxic dumping, for the *Exxon Valdez,* for loading plutonium 239 into light water reactors, . . . I do not have to go on. Perhaps you can begin to feel why we use this little parable in the beginning of a graduate seminar at Penn State, coupled with a beautiful, almost allegorical short story by the French novelist Jean Giono, *The Man Who Planted Trees.*[19] Sandwiched between David Harvey's *Manifesto* the week before,[20] and John Imbrie's Fourier decompositions of 400,000 years of ice and seabed cores the next,[21] some of the poor students are in a state close to catatonic shock—that numb look that says "What am I doing here, Mother?" But by the end of the semester, they have read and discussed slowly, thoughtfully, and thoroughly some thirty to forty professional and intellectually relevant papers and essays, with faculty and second-year students often sitting in. The students were, in fact, the ones who begged us not to drop the weekly readings and discussions at the end of the fall semester but to continue them throughout the academic year. And why? Because they care, and it is that grounded, ontological sense of caring that ultimately informs much of their own reading, thinking, and inquiring.

And it is, of course, that ontological condition of possibility that informs the moral domain. We do not talk about this much in geography: we perhaps talk about it too little as teachers. We are much more sensitive today to that "unexamined discourse" we call ideology.[22] We are wary, and properly so, about indoctrination, and how easily power can be misused to slide education that way. Yet we care deeply to pass on a sense of heritage, and we believe, perhaps more profoundly than we care to acknowledge sometimes, in that Enlightenment ideal of a liberal education as a liberating education. Yet as teachers we are always "moral tutors," to use the old designation of Oxford, still in use today;

not the least because we are present as teachers and our own caring is watched; not the least because we try, within the always contingent historical circumstances in which we find ourselves, to tell, to profess, the truth—at the very least, a truth. So our discourse, our thinking, our inquiries, and our writing are not just touched by, but are embedded in, the moral domain. Geography *is* a subject saturated with moral and ethical dimensions. We do care; we do try to teach "meetly and rightly," even if we know that to some degree we always fail. After all, that is human too.

A Sense of Right

But take heart: we are not alone in our caring, in our understanding of the earth as the home of humankind. I think it is no accident that for the first time in what we might call the "moral history" of the Western world, philosophers are beginning to think toward an ethic of the future. Ethical questions always arise from, are embedded in, a given *ethos:* a humanly constructed set of moral obligations and sensibilities of what it is to do the right thing. They are, of course, culture- and place-bound: like any set of human constructs, they are not eternal, they are not immutable; they can change as a human culture itself unfolds into an unknown future. But all ethical "systems"—although I am uncomfortable calling them that—have dealt with present problems, problems of immediacy, problems of correct behavior, arising in the here and now. What's past is past; don't cry over spilt milk—even if we might feel guilty or ashamed. There is not much you can do about the past, and the future can take care of itself, because as finite mortals we are not gods and goddesses and cannot foretell what fate will bring us. This, by the way, is not *fatalism:* rather, it is the *amor fati*, the "love of Fate" of Nietzsche that lives joyously and resolutely in the times in which it finds itself.

But that blind stance toward the future is no longer good enough. Not because in our hubris we think we can now predict the future, that the world will unfold in predetermined ways if only we can find the right cogwheels in the human condition and compute their mechanical turnings with ever-larger computers. That stance simply cannot hold today for any morally touched person because we know what we are doing to our home, we know how we are fouling our own nest. Stream and lake sediments are laced with PCBs; aquifers are polluted with carcinogens; two kilograms of plutonium 239 if "properly distributed"—and here the quotation marks indicate that language itself breaks down under the obscene load it has to bear—could kill every person on earth, and we have made tens of thousands of kilograms. Species variety shrinks with every chain saw started in the tropical forest; and . . . so . . . on. The "crazy ape" is on the loose. It is no wonder that Karl Otto Apel starts his seminal essay on a future-oriented ethnic with the global pollution problem, with the responsibility we have to future generations.[23] Radioactive iodine 131 has a half-life of 8 days; caesium 137, 28 years; strontium 90, 30 years; plutonium 239, 25,000 years.

It means we have to steer toward a cleaner, more decent world—not for our own sake so much as for our children and grandchildren unto the thousandth generation or more. Steering, as you know, is cybernetics, but that is really not what I am talking about here. In the short run, the run of simple trend, interpolation and extrapolation expressed in simple mathematical forms, yes, we can make some good guesses—some times and, locally, in some spaces and places. But in the long runs and global spaces I am talking about here, even the most rabid social engineer and technical "fix-it" turns silent. But we do not have to predict the future: what we require is not computer simulations that chaos theory tells us are doomed to failure anyway, but rather moral reflection on the consequences of acts that we know, out of our own ethos, are immoral and unethical. If, for the long run, we may not know exactly, but know nevertheless in our bones that things will not be right, then we can act only with prudence, intelligence, and judgment. And here we are right back in Aristotle's *Nicomachaen Ethics*,[24] where he calls them *phronesis*, *sophia*, and *gnome*. These are not old texts to be dusted off in a philosophy course; they are more shiningly relevant to the human condition than anything we can think towards. Ignore them and there will not be a human condition much longer.

But prudence, intelligence, and judgment, thought and acted out by ethically touched and caring people, find themselves frequently in a world permeated by relativism and cynicism.[25] Both are deadly and are challenged by different geographers in different ways. Some announce the "only perspective" of a David Harvey,[26] sharing Marx's moral outrage at a human condition not fit for beasts, a condition ignored by virtually all the nineteenth-century churches, whose clergy passed by on the other side. After all, the poor will always be with us. No wonder Marx excoriated religion as "the opiate of the people." Other geographers hold to a postmodernity that categorically denies a single "way, truth, and light" and rejoices in a world of challenging plurality and different perspectives. This too can be morally informed, because plurality does not mean an anything-goes relativism. After the horrors that the tribes of Europe have brought us this century, we *know*—and how? from our ethos—

that not everything goes. It is a plurality of respect, of tolerance, of caring for ways not our own, *providing* within those ways there is seen to be a grounding respect and caring for the individual human person. Tolerance toward a pluralism is not an "anything goes": it does not stand aside and look the other way when drug barons bring a country to its knees; when children in Londonderry are blown apart by others, whose own children are slaughtered in their turn, both in the name of the same god; when women are treated as chattels and physically mutilated by a religion preaching brotherhood but denying sisterhood;[27] when 15,000 young people disappear in an Argentina; when students in a modernizing China are shot down. Plurality can be a morally informed respect, a tolerance of difference that grants the marvelous mystery of place, of culture, and the vibrant weave of human life in its kaleidoscope of geographic settings. Canadians know this better than most, and to the degree that they can gently and caringly support each other in their identity and difference, they have much to teach others.

But here we touch upon another, perhaps final, wellspring of the teacher and geographer. It is the ability to feel and stand in awe at the very mystery of being, the mystery of our own human presence, of our being-in-the-world, and our capacity to give meaning to a world that is always interpreted by the human presence. What could be more geographic than that? It contains the mystery of human intentionality that leads to purposeful scientific inquiry. It holds the mystery of beauty that carries the aesthetic capacity to flood us with awe and joy. It leads to what George Steiner has called "the naked wonder of life,"[28] and contains the potentiality of the Not-Yet of Ernst Bloch.[29] We do not talk about these things very much. They touch, of course, the religious, but not the religious in any formal sense, not the religious of dogma and strident evangelical faith, tendencies to authoritarian gestures that require any democratic society to separate church from state. Rather, it is a religious sense capable of being shared by all human beings, toward the mysteriousness of being that leads to thinking and thanking. Thinking like a geographer is thanking like a geographer, and in the English language, with a major root in the Germanic, it is not surprising that these two words are so close. It is a sense that raises the question posed to Dante, "Chi fuor li maggior tui?"[30] translated very freely in its context as "Out of what world did you arrive?" or "What is the nature of your journey across time?" And we know as teachers, even if we do not dwell upon this question each day ourselves, that it has that shock-of-recognition immediacy for many of the young people who surround us; young people trying to learn who they are, where they are going; trying, Socratically, to know themselves a little better. I have seen no mention of such informing ideas, of such wellsprings of inquiry and teaching, in any geographic publication. After all, what would a poor reviewer say to an editor when asked, "Are these things suitable for publication, for a 'making public'?" But why remain silent to that which called us to geography, that called us to teaching, to contribute to the human family in the way we were called? We do not have to sell geography. We only have to remain true to that calling and teach and learn in every way we can.

Thinking about Teaching

I HAVE ALWAYS loved teaching, especially those who want to learn, and I will go to hell and back to help students who are really trying, especially if they are building on a rather shaky foundation. I can recall *that* feeling very easily from my own schooldays. On the other hand, I become very impatient with the teach-me-I-dare-you's asleep in the back row, those who expect me, or some poor teaching assistant, to give them hours of private tuition the day before the exam, when lectures have been missed and carefully planned exercises left undone. And if they vent their chagrin on the teaching evaluations at the end of the course, let them. Most students are fair to the point of being generous, and if you lop off the top five percent who think you walk on water, and the bottom five who think you are a swine in swine's clothing, most teaching evaluations come out pretty fair.

The question of what to teach has always been difficult for me, and generally I tend to project my own delights and difficulties on the poor souls sitting expectantly in the lecture hall. I have always felt a bit of an idiot giving factual, narrative-type lectures, which, after all, are generally culled from someone else's work. Reading interesting stories and accounts has seldom presented much difficulty to me, and I assume others can read and summarize main points as well as I can. On the other hand, I remain an unreformed (although today perhaps a somewhat more reflective) methodologist, delighting in finding new ways of approaching problems. Then, when I am thoroughly familiar with them, I enjoy exposing their weaknesses, the traps into which real thinking can be led.

Two of the essays in this section were written nearly twenty years apart for the same journal, the *Journal of Geography in Higher Education*. It has proved to be a wonderful forum for the exchange of ideas about teaching, from the very first issue, in which I was able to reflect on what was worth teaching in geography, to the present day. But what a difference in the journal itself! Starting as a mimeographed and stapled go-for-it-on-a-shoestring effort riding on the convictions, not to say the sheer guts, of David Pepper and Alan Jenkins, it has become the impeccably produced professional journal it is today.

In 1977, I wrote out of my own experience at the time (what else?), and was concerned with translating the things I taught from words to pictures to algebras—the "languages" of our investigations, with a heavy emphasis today on those quotation marks around "languages." Nearly twenty years later, I have changed in many ways, while still remaining an unashamed methodologist. A big difference has been reading, almost every semester in formal and informal seminars, text after text of Martin Heidegger with a fine teacher. Heidegger writes of such things as "world," "horizon," and "Being" as all the conditions of possibility for human thinking. Teaching, it seems to me, ought to open up a person's "world," enlarge the "horizon," and try to explore new conditions of possibility for thinking about things in different ways.

It is dangerous work—all teaching is dangerous—and there is a damned-if-you-do, damned-if-you-don't tension in the fact that methodological teaching can open up possibilities and at the same time close them down. Not that choosing a particular methodology for this particular problem at this particular moment necessarily condemns you to thinking only along these particular ways. Robert Frost, sitting on his horse in a snowy wood, may have lamented a way not taken, but his horse knew perfectly well you could cut through the woods from one beaten track to another. So while some themes resonate from the earlier essay, particularly that of understanding by seeing, some of the concerns for teaching have changed, particularly in light of the incredible personal computer revolution that has taken place in the meantime.

Invoking Robert Frost was no accident: for years, in my two-semester sequence on spatial analysis, I have given the students a weekly poem or two, mainly just for fun, but partly to remind them that they were in a university rather than a for-profit research institute. Distinctions do get blurred. Sometimes the poems had a geographic feel to them—Tennyson on Ulysses sitting bored out of his skull in Ithaca after his years of adventures returning from Troy. Sometimes I used a seasonal theme—Keats on Autumn, Ezra Pound's "Winter is Icummen in/Lhude sing Goddam" for the start of the course in January, and a selection of mild but quite delicious erotica if the lecture happened to fall on St. Valentine's Day. Two bonus points were awarded on exams for recognizing "Who wrote ________." Occasionally poetry reinforced my own exhortations about the merits of the Protestant Ethic, advice that pointed to the necessity for hard work involving "wrapping a cold towel around your head and getting down to it."

One morning, on my birthday in November, I walked into the lecture hall and was preoccupied as usual with thinking about the lecture, making sure blackboards were clean, notes in order, and so on. I was a minute or two into the lecture before I turned from the blackboard to realize that most of the students had converted to the Sikh form of Hinduism. I faced a sea of white turbans, the famous "Cold Towel," around each head! Each turban towel was a token of good faith that they were earnest students, knowing that I had received an anonymous petition in verse to temper homework with mercy during the coming weekend. At the next lecture, I replied in the only way I could, with a leg up from John Keats to get me going. Those are the moments you remember, and to invoke a line from Kipling, "The rest can go to Hell!"

What Is Worth Teaching in Geography?

> Teaching, therefore, does not mean anything else than to let others learn, i.e., to bring one another to learning.
>
> —Martin Heidegger
> *What Is a Thing?*

I KNOW there are certain conventions of style to be observed when one writes for a professional journal. *I*'s indicative of real people must be suppressed, even at the cost of substituting royal *we*'s. Passive (*suffering, submissive*, says the dictionary) voices tend to smother any further hopes of the reader that a flesh-and-blood writer is trying to communicate. But never mind the conventions: this journal is a forum, a place of public discussion, and I would like to use it as a means of reaching out beyond the bare abstraction of this page. You may not be able to see me, or hear my voice, but at least I want you to be aware that another human being is alive and kicking in the scramble of symbols your eyes are recording. I am a teacher, and so are you, and whether we agree or not is of much less importance than the fact that we communicate about our common concern for the teaching task.

And what a bloody, miserable task it is sometimes! There are those who may romanticize it, there are others who *appear* to do it effortlessly, but you and I know that to teach well is sheer hard work. From this obvious but too seldom acknowledged fact, I think two things follow. First, we owe it to ourselves, and to those we teach, to teach *efficiently*—a word invoking all sorts of ugly connotations today, inspiring visions of production lines and people being processed like things. I do not mean that at all: I do not mean more people learning with fewer teachers, productivity measured in student credit hours, high graduation to entry ratios, low dropout rates, and all the other indices so casually bandied about today. Not at all. I mean we have a duty and responsibility to think hard and constantly about what is *worth* teaching.

This essay originally appeared in the first issue of *The Journal of Geography in Higher Education* 1 (1977): 20–36. It has been modified only very slightly. Reprinted with the permission of the editors and publisher

Secondly, I think it means we must confine our teaching where possible to those who genuinely want to learn. Most of us will go far out of our way to help people who are really trying, even though their abilities and previous preparation are not the best. Conversely, I see no reason why time should be wasted on the casual, the lazy, and those who want three or four years of mild, anecdotal, and not-too-stressful entertainment. The notion of intellectually efficient teaching is simply contradicted by such people, because teaching is a two-way affair. So we assume teachers of goodwill, people with a genuine desire to learn; but what is worth teaching?

Teaching How to Learn

Let us back into this rather uncomfortable, soul-searching area in a cowardly and negative way by saying what is *not* worth teaching. Although many of our lectures are crammed full of factual materials, facts are not worth teaching. Not for their own sake, certainly. They clutter up the memory, they get out of date, and they can be read about in books. You can collect them out of antiquarian interest, but there is no need for a literate person to compose factual lectures, culled from a set of fairly standard sources, and waste every other literate person's time standing in a lecture hall delivering them verbally. Such facts are only scribbled down by others, memorized, and dragged out during an exam, and then promptly, and quite rightly, forgotten. Of course they are forgotten: they were never worth learning in the first place, and they will never be referred to again for intellectual refreshment. Geography used to be like that. Perhaps in certain places it still is?

But now to be more positive—if categorical. In general, the only thing I think worth teaching is how to learn. Of course, unfurling a great banner of a statement like that does not really help. We can all give a faint and slightly cynical cheer, and then rush off and give the same lecture we made up when we were young and keen ten years ago. No, I have the duty to be more specific, for teaching people to learn is the hardest task of all.

First, I am convinced that the brain is a marvelous, although today an increasingly inadequate, pattern-seeker. Obviously there are good evolutionary (that is, survival-value) reasons for this, but much of the drive is based upon the pleasure, the aesthetic eureka, to find satisfying, if temporary and contingent, ordering structures. This is why I feel compelled to teach the spatial theories, methods, and models of our field right from the beginning. They are the only intellectually satisfying and meaningful things we have, the only things that help us to make sense out of the plethora of facts we have recorded about the world. But we must teach them honestly, placing our ordering constructs in a larger context of human pattern-seeking, in which all the theories and models we have created are truly presented as constructs of human thinking, rather than final truths to be accepted. In the beginning the ordering structures will tend to be static and geometric, for the static is generally easier to understand, and the geometric allows us to use the graphical languages of the map, illustration, and graph to supplement the inadequacies of the spoken word. But this, of course, raises another aspect of teaching people how to learn: the languages of our investigations.

There are many languages, but to investigate the world around us we tend to confine ourselves to three—the verbal, the graphic, and the algebraic. The verbal we know about, or at least we know enough to realize how little we know. Thanks to many philologists, philosophers, and linguists today, most of us realize the way in which verbal languages both liberate our creative imaginations and constrain our thought along certain, quite rigid, lines. As for graphic languages, they seem to me to be peculiarly geographic, and I want to take them up in a minute. For the moment, let us consider the algebraic, because these seem to cause the most trouble and generate the greatest antagonism and difficulty.

I am talking, of course, about mathematics, an area of many "languages" and "local dialects," where truly creative work that extends the known boundaries appears to be very much a matter of innate gift augmented by hard work. As in music, poetry, painting, sculpture, and perhaps dance, you have either got it or not. No amount of hard work in and of itself is going to make up for that fine cutting edge of the sublime gift with which some are endowed.

Yes, but that is not the same thing as saying we cannot learn to appreciate elementary areas of mathematics and see how we might use them. As teachers, we must realize that many areas of mathematics hold the potential to open up and describe areas of human experience that would otherwise remain hidden. This is because these "languages" (the quotation marks are deliberate and important here) form the natural expression for many important ideas, and in some cases the only practical and feasible expression. While ideas expressed naturally in mathematics may be partially translated into more familiar verbal and graphical forms, there are some areas of human knowledge and discourse that are extremely difficult to translate into these more familiar expressions. If we are to understand them and really appreciate these ideas in the original, we must take our coats off and get to work. There is no other way.

For example, the simple graphics of catastrophe theory may give us a glimmer of the most elementary forms, such as the cusp catastrophe, but the four-dimensional swallowtail, and the five-dimensional hyperbolic and elliptic umbilic catastrophes are not *realizable* in words and pictures—and the dictionary says that means they cannot be presented as real, or converted into fact. The problem, of course, is that the closed and circular system called a dictionary is telling us that we cannot talk about these things in words, the very thing dictionaries are all about, and that we cannot even draw pictures of them. Yet they certainly exist in the language of algebraic topology.

Can we not encourage the teaching of mathematics as a mode of expression, and motivate people to explore and learn various forms by showing them how we can use them to express things geographical? Most geographers are bored stiff by much formal mathematics (I certainly am myself), but equally most students of geography—and one hopes that includes all of us—can see the utility of using such abstract constructions to express and advance geographical ideas and thought. But we, the geographers, are responsible for generating the motivation: perhaps we can then find mathematicians, such as those who composed the original Foundation Course for the Open University,[1] to help us along. Otherwise important areas of applied, contemporary geography such as optimization, stability and control theory, network analysis, simulation, spatial statistics, entropy and information modeling, topological structure, and so on, will remain forever closed. And since these are some of the growing conceptual areas, rather than just more factual bulk, more and more people are going to be increasingly isolated from the truly exciting ideas of the field. Let me suggest that there are too many specialized and increasingly isolated people around already to encourage the production of more. Rather, our task is to encourage liberal education in the best eighteenth-century sense of a liberating education that throws open windows and unlocks doors. We cannot teach everything in a three- or four-year period, but we can try to teach efficiently, and we must try to say: "Look. There . . . and there . . . and over there"—and so give people a vision of intellectual challenge and joy for the rest of their lives.

This is the reason I place such an emphasis upon teaching methods of inquiry—*mere* methods, they are called by some. For how can a person learn to inquire without tools, and how can a person today read about the inquiries of others without a knowledge of the basic approaches? Learning about methodological tools from books is possible, it can be done with great perseverance and diligence, but for most people this is not an efficient way. Methodology really *is* worth teaching to help people acquire the tools of inquiry efficiently and early. Also the tools themselves, whether simple regression methods, goal programming, or Monte Carlo models, quickly open up new lines of thinking about other problems. Having thought about how one thing varies with changes in others, how things can constrain us from reaching an optimal solution, how the rules of a spatial game may be written and their effects simulated, people are prepared to think in imaginative ways about other, analogous problems.

What is fascinating to observe over the past fifteen years is the way in which geographers with solid, contemporary methodological backgrounds have dominated the applied areas of planning, consulting, and sponsored research. Those who have such backgrounds have a flexibility to move from one problem area to another, and an ability to make both practical and intellectual contributions unmatched by conventionally trained people. This is because they are used to simplifying and modeling complex problems to the point of comprehensibility, and because their training allows them to work effectively with others from allied fields because they share many of the same languages.

Furthermore, I have the strong suspicion that as geographers we have something very special in our traditional concern for maps and mapping. I do not want to define spatial thinking—I am not sure I can—but I think we have the ability to teach people something of great value in that it will help them to think more creatively about the world around them. But to do this effectively we shall have to enlarge greatly our own vision of the map. We should see the traditional map as a model, a filtered picture of reality, and see the act of mapping as a particular form of the mathematician's more general and powerful definition.

I think it helps to define a map simply, but quite profoundly, as a representation of relationships between things, for in this way we break loose from traditional earth space and see that many of the spaces where people desperately need maps today are conceptual rather than terrestrial. By all means give people who want to learn about maps the once-over-lightly with the traditional *portolani* (old maps of sailing directions) and the Hereford map; certainly it is genuinely interesting to see how our ability to draw an accurate picture of the world developed. But people can read about such developments, we do not have to drag them out in lectures, and you cannot *think* with these historical examples. Rather, we should expose people to the most imaginative examples of mapping we can find: maps of relationships between married couples in various emotional states, maps of the South Pacific in geobotanical space, maps of New Zealand in changing aircost space, sound maps of moths terrified by bat squeaks, maps of intellectual winds blowing through psychology journal space, maps of the world in priority press telegram space, maps of phase spaces showing trajectories of systems, maps of influenza epidemics in cartograms of demographic space, projections of four-dimensional spaces on to three-dimensional maps of diffusion where a hierarchical process appears to be contagious, maps of Shakespearean tragedies based on the verbal interaction of the players, maps of the trajectories of characters in Bergman space, and so on.[2]

After all, who else is going to do this in the Academy? Who else has such a deep tradition of mapping, of spatial thinking, of mental visualization? This is not what we usually think of as cartography? No, thank Heaven, it is not. Instead, let us get some intellectual challenge and excitement into this fundamental field, not by spending hours pushing the pen and applying the dot patterns,[3] and certainly not by trying to make a profound problem out of the trivial question of where to draw the choropleths (lines breaking up a distribution for mapping). Let us consider what is really worth teaching about maps, and show people how relative spaces can be created and mapped, how multidimensional scaling can be used imaginatively by first-year students, how computer mapping programs can generate patterns of any reflectance we like, how Euclidean regression between spatial patterns in two-on-up dimensional spaces gives us vector residuals and exciting fields of forces, . . . and so on. If you took people in their first year and exposed them to a sequence of imaginative examples, if you showed them how a computer could do the human-benumbing dirty work, would you not be giving them something to think *with*, something that gives them a new ability to think about the spatial dimensions of the world they live in? And not just the conventional world of the atlas and wall map, but new, transformed, and more relevant worlds?

Finally, it is worth teaching people to see that geographic problems are just particular examples of more general cases. We *must* teach, or have others teach for us with our blessing and abetment, the intellectual perspective of systems analysis. This is a higher order discipline, an algebraic cover set for geography in an institution laughingly called a *university*, which does its best to partition knowledge into little, and frequently jealous, fiefdoms. We teach today that geography has some precious integrity of its own, but in fact it is a base, one of many bases, from which we can help people to grow and enlarge their capacities for intellectual experience and inquiry. Can you imagine what it would be like to teach people in geography who have taken the first-year systems analysis course of the Open University,[4] and who have followed it up with the second? Of course, we may have to pull our own socks up and take the courses first; otherwise our roles may be reversed. But with such a background, people in geography would have rich sources of analogy, and excellent problem-solving experiences that would make them far more effective in dealing with practical, applied problems.

Analogy, no matter how rich the source, is dangerous, you say? Of course it is; all creative and imaginative thinking is dangerous, simply being an intelligent human being is dangerous, but analogical thought remains one of the most powerful forms of ordering and structuring our experiences. As for practical applied geography, what is it fundamentally but examining one's values, choosing goals, and then steering very complex systems toward them, while at the same time modifying the values, altering the goals, monitoring constantly the state of the system and its present trajectory, examining the stability of the system—in brief, a mixture of optimization, control theory, sensitivity to spatial consequences, systems simulation, and ethical concern?

The last is difficult, and I have said nothing about ethics, or teaching people ethical conduct in their professional lives. Certainly I think it is worth exposing people to the writing of those who have thought deeply about such questions, people who have pointed out how regional planning can lock a spatial system into the *status quo* with distressing and unforeseen human consequences—the counterintuitive effects of Jay Forrester.[5] But this is all we can do outside of dogma, because I do not believe rational discourse about ethical matters is possible. Like Wittgenstein refusing to sit at high table, or pointing to a map as a proposition, we can quietly set examples as teachers by the quality of our instruction and the professional work we do. I doubt that we should go further.

Teaching and Learning as Filtering

As I try to think through what is really worth teaching in geography, it becomes obvious that good teaching involves a great deal of selection. This means we are really filters, people who constantly examine and scan as much of the enormous body of human knowledge as we can, and then filter out bits and pieces to shape our courses, lectures and teaching. The idea of a filter is a very general and useful one, for it raises the question of our own selection process. What we teach we have filtered out, but how often do we examine the nuggets in the sieve? How often do we toss back our favorites and dip in again? And how often do we examine the porosity of the filter, the mesh of the sieve? It seems to me that one of the things the physical sciences are very good at, and some areas of modern literary criticism are getting much better at, is examining what is of real value and intellectual worth in their fields. It is not necessary, for example, to recreate the whole history of physics in a contemporary physics course today. Decisions (that is, *selections*) have been made, and much that was formerly taught has been greatly compressed, thrown out, or shown to be a special example of a more general, and intellectually more challenging and worthwhile, idea.

An Example: All You Wanted to Know about Linear Systems but Were Afraid to Ask

As for teaching, let me try to give you an actual example of what I mean. The example is extremely simple, and I use it as a point of departure for other things in a course with beginning students in a large American university. The example owes a great deal to a former colleague, Conrad Strack, with whom I had the privilege of teaching a few years ago. I also had the pleasure of teaching a course recently to second-year Canadian students, and I will draw upon that experience, too.

Suppose we have four towns connected by a railroad (fig. 23a)—and let us stop right here for a minute. This is a pretty boring and abstract sort of diagram, and we have got to liven it up. Why? Because people who are learning about networks and movements deserve some human color. Lines joining points, railway tracks joining towns, are *vital*—ask the people of Malaya, Nigeria, Guatemala, and Gabon who live on them, and do not be so insensitive and blasé about taking the 8:05 commuter to town. We have become so inured to the exciting space-shrinking and space-warping properties of transportation that we forget how human lives are changed by them. I often use Ghana as an example, and talk about the impact of the new railway upon the cocoa farmers, how one train was worth 20,000 people carrying goods on their heads, how in cost-space the towns suddenly moved toward each other, how towns were able to reach out like magnets and draw people from the countryside to them, and so on. But almost any place, at any time, will do.

Then why not take the real world of Ghana? Because it is too difficult and detailed; for most of our teaching we should throw the real world out of our examples and exercises, and make up stories about greatly simplified situations and examples that illustrate the concepts, *not* the actual facts. Then the person learning can get such a clear view of the basic *ideas* that he or she will never be able to see a map of transportation routes in quite the same way again.

So let us take Ghageria Leone as our country, and label the points (the names are also imaginary but plausible), put in the tracks, and make Dunkweh a seaport on the coast (fig. 23b). Now suppose we have some cocoa at each town

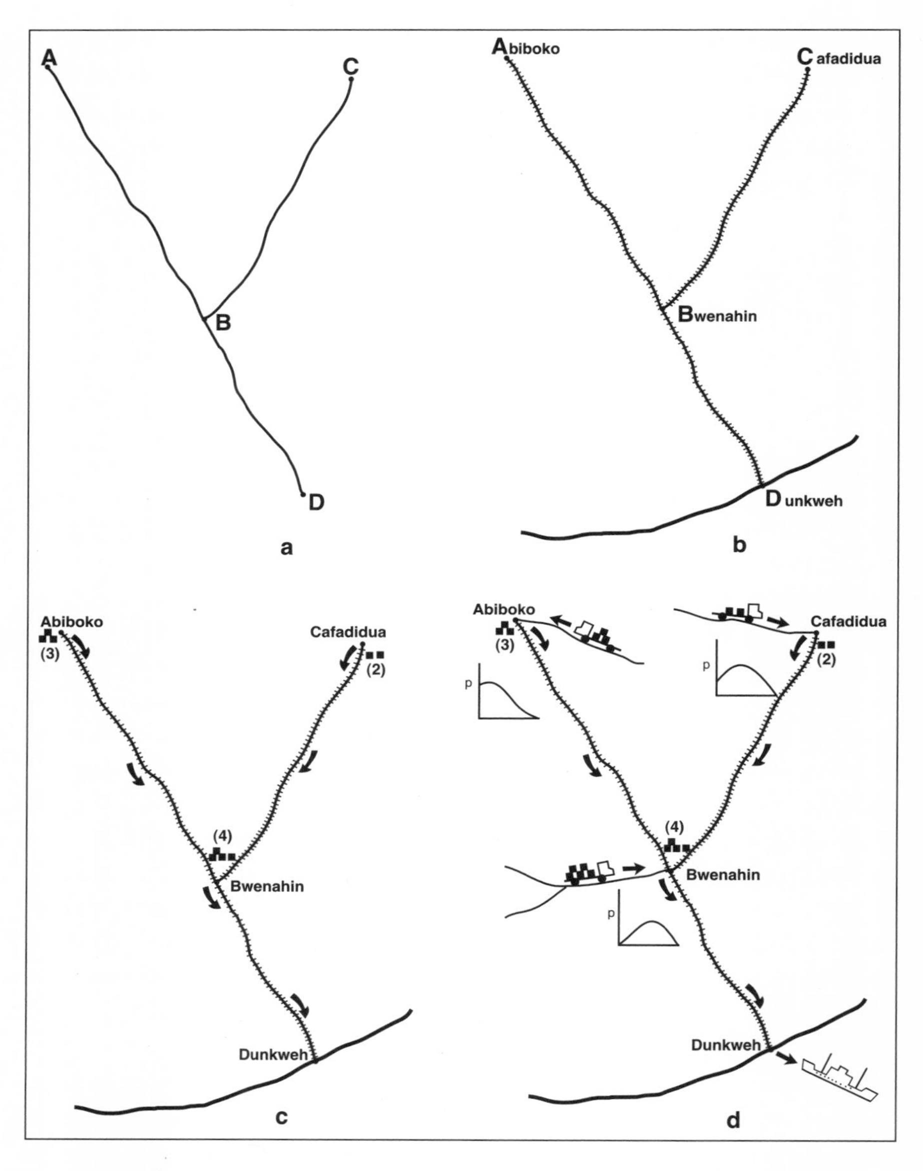

Fig. 23. Developing an example of a dynamic geographical system: (a) the bare "graph theoretic" bones; (b) a railway line in Ghageria Leone; (c) tons of cocoa waiting to be shipped to Dunkweh for export; and (d) daily arrivals of cocoa at the stations described by probability distributions.

waiting to be sent down to the port for shipment overseas (fig. 23c). Since we are interested at this point only in export flows, we can think of the towns as being connected by one-way links. This gives us another way of describing the network; by making a square table, or matrix, and simply recording the presence or absence of a link. Like this:

$$\begin{matrix} & A & B & C & D \\ & \downarrow & \downarrow & \downarrow & \downarrow \\ A \leftarrow & & & & \\ B \leftarrow & & & & \\ C \leftarrow & & & & \\ D \leftarrow & & & & \end{matrix} \begin{bmatrix} 0 & 0 & 0 & 0 \\ 1 & 0 & 1 & 0 \\ 0 & 0 & 0 & 0 \\ 0 & 1 & 0 & 0 \end{bmatrix},$$

a 4×4 matrix that we shall call **B**.

This directed connectivity matrix **B** could not be simpler, but notice how we have already used three languages in our regional description: the verbal, the graphic, and the algebraic (the connectivity matrix **B**). And although the example is simple, the idea is important; we must try to teach people that there are alternative ways of describing things, each with its own strengths and weaknesses. We need words to give human color and life, to provide a setting of reasonable verisimilitude. The maps give us a visual picture, a different perspective on a particular spatial proposition that we must deal with. But the matrix gives us (as we shall see in a minute), a *manipulatable* description, a description with which we can actually explore further. Obviously, since we are dealing with first-year people, I have chosen a situation where a familiar and important problem *is* expressible in all three languages, and translations between them are quite easy and straightforward. But if we can get people to see how alternative descriptions are possible, and why they are important, we can then lead them later into areas where translations may become exceedingly difficult—so difficult that everything may be lost by the translation, and the ideas *realized* only in the algebraic languages.

So here we have a little geographic system: three towns and a port, some connections between them, and some cocoa waiting to be shipped. What is the *state* of the system at this time (say t) as we first observe it? Well, we have our map (fig. 23c), and this is a perfectly good description of the system's state. But there is not much we can do with it, except frame it and hang it on the wall. Suppose, however, we make a list of the towns, and the amounts of cocoa at each, like this:

$$\begin{array}{lll} \text{Abiboko} & A = 3 \\ \text{Bwenahin} & B = 4 \\ \text{Cafadidua} & C = 2 \\ \text{Dunkweh} & D = 0 \end{array} \quad \begin{bmatrix} 3 \\ 4 \\ 2 \\ 0 \end{bmatrix} = \mathbf{X}_t.$$

We can call this list our state vector, and give it the shorthand notation $\mathbf{X}_t$. We do not do this to be pretentious or jargony, but simply because it brings this particular description of a geographic system under the umbrella of the more general language of systems analysis. By using more general words from geography's cover set, we can teach people that all sorts of systems, which appear initially and quite superficially to be totally different, are really identical because they share a similar, deeper structure.

What happens to our system after a short period? Abiboko and Cafadidua have shipped their cocoa to Bwenahin, while Bwenahin has shipped its cocoa to the port of Dunkweh. We could draw another little map showing the state of the system at $t+1$, or we could say the same thing algebraically by noting that

$$B_{t+1} = A_t + C_t,$$
$$D_{t+1} = B_t.$$

This means that our system is now described by the state vector

$$\begin{bmatrix} 0 \\ 5 \\ 0 \\ 4 \end{bmatrix} = \mathbf{X}_{t+1}.$$

In systems terms, $\mathbf{X}_t$ was our *operand*, while $\mathbf{X}_{t+1}$ is our *transform*. To send our operand $\mathbf{X}_t$ into our transform $\mathbf{X}_{t+1}$ we need an *operator*, which we can think of as a little engine powering our system along. Actually, we already have it: it is nothing more than our directed connectivity matrix **B**. If we multiply the state vector $\mathbf{X}_t$ by the matrix **B**, we shall get the new state vector $\mathbf{X}_{t+1}$, or $\mathbf{X}_{t+1} = \mathbf{B} \, . \, \mathbf{X}_t$.

Of course, we need the matrix operation of multiplication to do this, but most people have had some sort of contact with elementary matrix algebra today, and even if they have not, then the basic operations of addition, subtraction, and multiplication can be taught very quickly—literally five minutes on the blackboard and a few homework exercises for practice. The point is that people now see why they are being asked to learn this new type of algebra, with its rather odd way of multiplying, rather than being asked to make great leaps of faith that "one day it will be useful."

At the moment, our genuinely dynamic system is not very interesting—although stop for a moment and ask yourself, how many geographic descriptions have you come across that explicitly take time as well as space into account? At $t+2$ everything has been cleared out of the up-country towns, and only Dunkweh has five units waiting at the dockside to be loaded and exported. In algebra:

$$\mathbf{X}_{t+2} = \mathbf{B} \cdot \mathbf{X}_{t+1} = \mathbf{B}(\mathbf{B} \cdot \mathbf{X}_t) = \mathbf{B}^2 \cdot X_t,$$

and if you want to give people some practice multiplying matrices, let them confirm for themselves that these statements are true. The exercise is not the world's greatest eureka experience, but many people feel secretly rather pleased the first time they confirm these expressions for themselves.

Suppose now (fig. 23d), that during each time period the farmers bring more cocoa by local trucks (mammy wagons) to the railhead. In other words, our system has regular inputs to it that we can describe by another vector:

$$\begin{matrix} A = 3 \\ B = 5 \\ C = 2 \\ D = 0 \end{matrix} \quad \begin{bmatrix} 3 \\ 5 \\ 2 \\ 0 \end{bmatrix} = \mathbf{A}.$$

Now the state of our system will be given not just by **B . X**, but requires the addition of **A** each time. We now have a complete description, which looks like this:

$$\begin{bmatrix} 3 \\ 10 \\ 2 \\ 4 \end{bmatrix} = \begin{bmatrix} 3 \\ 5 \\ 2 \\ 0 \end{bmatrix} + \begin{bmatrix} 0 & 0 & 0 & 0 \\ 1 & 0 & 1 & 0 \\ 0 & 0 & 0 & 0 \\ 0 & 1 & 0 & 0 \end{bmatrix} \cdot \begin{bmatrix} 3 \\ 4 \\ 2 \\ 0 \end{bmatrix}.$$

This is precisely the same form as the usual expression of a straight line:

$$\mathbf{X}_{t+1} = \mathbf{A} + \mathbf{B} \cdot \mathbf{X}_t,$$
$$\mathbf{Y} = \mathbf{A} + \mathbf{B} \cdot \mathbf{X},$$

and for the first time many people see why we call certain systems *linear.*

What is the equilibrium, the steady state of this system? As a teacher you now have a natural opportunity to talk simply and effectively about the whole concept of equilibrium: what it means, the different sorts of equilibria, and how systems do or do not return to a particular equilibrium state when something dislocates them. You can also point out that perturbations can come only from two sources: **A**, the vector telling us what is coming into the system; and **B**, the matrix telling us what the internal *structure* of the system is. If either of these changes, a new equilibrium may result. Our present system is so simple (and converges so rapidly to a steady rate—which might give us the opportunity to talk about relaxation times in spatial systems?) that anyone can play with a map for a few time periods and work out the steady state vector. Then you can show people that the same thing can be done algebraically, because obviously at equilibrium $\mathbf{X}_{t+1} = \mathbf{X}_t = \mathbf{X}_e$: after all, equilibrium means not changing from one time period to the next. So we can write:

$$\mathbf{X}_e = \mathbf{A} + \mathbf{B} \cdot \mathbf{X}_e$$

and with a bit of simple rearranging:

$$\mathbf{X}_e - \mathbf{B} \cdot \mathbf{X}_e = \mathbf{A},$$
$$(\mathbf{I} - \mathbf{B})\mathbf{X}_e = \mathbf{A}.$$

Suppose we multiply each side by $1/(\mathbf{I} - \mathbf{B})$, so that

$$\frac{1}{(\mathbf{I} - \mathbf{B})} (\mathbf{I} - \mathbf{B}) \cdot \mathbf{X}_e = \frac{1}{(\mathbf{I} - \mathbf{B})} \cdot \mathbf{A}$$

or

$$\mathbf{X}_e = (\mathbf{I} - \mathbf{B})^{-1} \cdot \mathbf{A}.$$

If you want to, you can bring in the idea of an identity matrix (**I**) and an inverse $(\mathbf{I} - \mathbf{B})^{-1}$, noting casually that this is called a Leontief inverse and links up with something rather useful called input-output analysis—which also needs matrix algebra as a language of expression and exploration. Or you can tell people that it is

$$\begin{bmatrix} 1 & 0 & 0 & 0 \\ 1 & 1 & 1 & 0 \\ 0 & 0 & 1 & 0 \\ 1 & 1 & 1 & 1 \end{bmatrix}$$

and get them to check that (1) you are not pulling the wool over their eyes, and (2) by multiplying $(\mathbf{I} - \mathbf{B})^{-1} \cdot \mathbf{A}$ they really do get $\mathbf{X}_e$. You might also point out that when **A** is zero, and no farmers are bringing any more cocoa in, the equilibrium state is also zero—which was obviously what was happening in the early stages of building the model.

Do you see what I mean when I say we can show people how the different languages of our investigations can be used, how we can slip in the algebra naturally as an obvious requirement to describe and model things geographic, and how we can enlarge their perspective to see geographic systems as only special cases of a larger, richer, and more powerful body of ideas? And can you see how this simple little example can be used as a takeoff point for a whole series of other exciting things that will help people *think* about dynamic systems?[6]

I do not want to exaggerate, and certainly some of the things that follow may be too complicated to deal with properly in the first year, but we should not be afraid of giving people views of more difficult, but intellectually more challenging and exciting, horizons. After all, how many of our geography courses really do give people a sense that there are still lots of exciting ideas ahead? Not more mounds of factual detail and busywork to plod through, but whole areas of organizing ideas and concepts that can be used to learn and think with?

For example, what happens if we cannot specify the inputs of **A** exactly each time—a much more realistic notion, since the farmers themselves are subject to random events. Ah! Random events! We have not even thought about these, but as soon as we do we are in an intellectual quagmire. I have drawn in little probability distributions (fig. 23d) to represent the notion that we may be able only to specify the probability of a certain size of load appearing at each town. Our model is no longer a state *determined* system, but a nasty, if more realistic, probabilistic system. So because of a host of unique events (the wife of a farmer is ill, a mammy wagon breaks its axle, a driver drinks too much *apoteshi* [a distillation of palm wine] the night before, it rains and we have to wait for the roads to dry out, etc.) that we can only treat as random, our inputs now have an uncertain arrival date. Queuing theory? Yes, obviously, and I have had first-year students rolling dice, playing at random arrival rates and constant service times, building queues, seeing what happens as the arrival to service time changes, and so on. Why not? The *idea* of things arriving at random, and waiting to be serviced, is terribly important, general, and quite simple (like many powerful and important ideas). That the mathematics of advanced multichannel queuing theory in networks is so intractable that Monte Carlo simulation is the only way out should not prevent us from exposing people to the basic ideas.

Monte Carlo simulation? What does that mean? And off we go on another tack, explaining how we can write rules of a spatial game, and get a computer to play it time and time again to give us a spread of outcomes, so that we can see which are more likely than others.

Which are more likely? Ah! The most likely state of a system, when many, perhaps astronomically many, states are possible. And off we go into entropy-maximization models. Not deeply—not in the first year, when people probably do not have the language of the calculus to enter effectively into an area of such conceptual richness that one's whole view of the world may be wrenched about and changed. But we can convey some intuitive feel for the sorts of questions asked in this area of geographic inquiry, so that people are no longer afraid of the literature.

Talking about the most likely state vector and the distribution of other states of the system should raise the question of how *constraints* mold and shape the possibilities. And off we go again, showing people how there might be loading constraints at the harbor of Dunkweh (very realistic in view of the many ships lined up at ports with inadequate loading facilities)and capacity constraints on the railway lines.[7] Cocoa may back up at the towns and have to be stored. Stored? There is a whole theory of storage, a body of ideas not only of great relevance to human and physical geographers (rivers, dams, soil capacities, runoff, etc.), but one that links directly with inventory control, management science, and operations research. Nor is the human color and relevance incidental. In many developing countries the storage of food and export produce is a critical consideration: great losses are experienced in storage because of rodents and fungi, the quality of exports (and therefore foreign exchange for imports) declines, and we really have to think about scheduling arrivals and trying to optimize the whole system. Optimize? With constraints? And off we go again into linear programming, with simple problems of maximization and minimization treated in our verbal and graphic languages. As for the mathematics here, there is no need to present the linear algebra; simply build up the problem slowly, using the tabular expression of the Simplex Tableau. It can be done easily, and if you have a simplex algorithm in your computer laboratory, why not make up some exercises for homework? People enjoy "playing" with their own networks, changing the capacities of the roads, having tropical storms cut links, and seeing how their optimal solutions change as more and more constraints are added to the problem. They are not, of course, "playing" at all, but learning to think about ways of approaching these very general sorts of problems in human and geographical affairs.

And describing a network as a linear programming problem raises other questions in turn. What about the empty trains returning up-country from Dunkweh? How do we minimize the wasted time and empty carriages? This problem should perhaps wait for a more advanced course, but at least we can give people a sense of real, and humanly important, problems. Put the exercise in terms of the drought in the Sahel, and the brave, but often futile and wasteful, attempts to get food to hundreds of thousands of stricken people, and you have an excellent example of what happens when we do not have sufficient information to model a real and complex network.[8] And in such a situation, perhaps it is not too early to expose people to the ideas of goal programming, describing how scarce resources and priority constraints based upon actual human values can alter solutions, so that people die or are saved according to the value constraints placed upon a problem. Such explanations can point out that goal programming is just a more general approach, of which linear programming is only a particular example. The advantage of simple linear programming in teaching is that we can show, through small Simplex Tableaux, that apparently quite different problems (minimizing the cost of a nutritious diet for children, maximizing the output of a small farm, shipping goods at least cost) have an *identical* structure. And such examples teach the very valuable idea that there are deeper structures and analogies linking problems that superficially, literally on the surface, appear quite different.

Enough: I have already overstayed my welcome in this new public meeting place, but I hope this example has put some meaning into the earlier opinions, so that you can see fairly precisely why I emphasize the languages of our inquiries, maps and spatial thinking, and the cover set of systems analysis. I am convinced that these three ideas can help us to teach people to learn more effectively, so that we can produce more competent, imaginative, and joyously creative people who have been shown areas of such intellectual excitement that they can continue to explore for the rest of their lives.

And perhaps, if we choose our examples carefully, to show that the structures of these problems appear again and again, we shall also create more thoughtful and compassionate citizens—not of Country X or Country Y, those archaic, ethnocentric remnants of earlier days, but of the world. After all, from space satellites and platforms we can see ourselves as a system: so why not teach from that thoroughly geographic perspective?

Perspectives and Sensitivities

Teaching as the Creation of Conditions of Possibility for Geographic Thinking

> Being . . . is the totality of all meaning . . . the lighting-process in and by which beings are illuminated as what they are.
>
> —Joseph Kockelmans
> *On the Truth of Being*

Creating Possibilities

My apologies for the rather complex and cumbersome title, but it refuses to compress to something short and snappy. Anyone who has taught conscientiously knows that teaching well is one of the most difficult things in the world, and we all recognize the difference between teaching that is anecdotal entertainment, and that which opens up new ways of seeing. Perhaps the two need not be totally distinct; perhaps the entertaining anecdote can produce that "Oh, I see!"—one of those moments of small eureka that are our most cherished rewards—but my concern is with the latter. Martin Heidegger wrote about the "horizon of our world," the conditions of possibility that our "world" gives us to allow thinking to expand to new ways of seeing and illuminating. In the domain of geography, these ways may be new perspectives, allowing us to approach something of geographic interest in a manner that was not a possibility before; they may involve the choice of a new methodology, literally a new way *(hodos)* to approach an analytical task; they may be heightened sensitivities, allowing a caring concern to enter and inform our inquiries and the choices we make. Good teaching—I am tempted to say genuine teaching—always attempts to open the horizon, to create new conditions of possibility for thinking.

First published in *The Journal of Geography in Higher Education* 18 (1994): 177–189.

The Teacher as Fraud

So why, after more than thirty years of teaching, do I remind myself and others of these transparently obvious things? Because, after a sabbatical year of intensive and varied reading, I experienced an uncomfortable feeling this semester (spring 1994), heightened by a worry that has been gnawing at me for many years. The uncomfortable feeling was a sense of fraudulence brought on by teaching a course of the history and philosophy of geography variety, built around David Livingstone's splendid book *The Geographical Tradition.*[1] For the first time in decades, I found myself teaching a factual course, something I had not done since my earliest days of teaching, a time when I was responsible for a regional course on Africa. Yet once again I found myself saying, "Why should I compose lectures culled from other peoples' publications, and provide a sort of predigested pablum for bright and intelligent students who are perfectly capable of reading and thinking for themselves, or for lazy and unorganized students who are sitting back waiting to be entertained, and anticipating summaries that surely will form the factual foundation for the factual exam that evaluates their feats of nocturnal memorization in a factual course?" The whole situation seemed a long way from the lector, the master reader, surrounded by attentive students, reading, commenting on, and literally explicating in the sense of unfolding, a single precious copy of Aristotle at the University of Paris in the late twelfth century.

Now obviously these same feelings of fraud were not generated by David's book. His leitmotif of geography as a continually contested enterprise appealed to me greatly, not simply because I had experienced such contestation myself, but because I believe agreement produces those smug silences that are the death of any discipline. Once we all agree, it is easy to recognize "deviations from the course of historical development," or to make "great retreats" as we back away from the true geography that has been revealed to us.[2] Smug agreement, even within esoteric coteries, produces nothing but adoring acolytes who seem to spend their own professional lives creating hagiographies for Maître.[3] The temptation to guruism is always there, of course, but why actually encourage *dependent* thinking? So to reduce my own sense of fraudulence, I put a lid on the course enrollment, and decided we would treat each session as a seminar discussion. What I would not do was stand up on my hind legs lecturing in the traditional fashion, essentially stealing lectures by summarizing the chapters in David's book. The students would be perfectly capable of doing that for themselves.

The Teacher as . . . "Realist"?

Things did not turn out that way. A book that I had read easily and with great pleasure after thirty or more years as a professional geographer, and from which I learned many new things, and recalled many old things long forgotten, was tough and challenging work for most of the students, including those in their first year of graduate work. With rather lost and blank faces looking at me, I recalled Derek Gregory's comment some years ago, when he visited us at Penn State in our Distinguished Visitors Program, that "The most difficult thing that we face at Cambridge is teaching people how to read," something I resonated to immediately, because I only really learnt to read myself at the age of forty-seven.[4] What I tried constantly to remember when I taught methodological courses, and what I had clearly forgotten in approaching this teaching task, was that a text is demanding to those without much of a context in which to place it. What I can only call the "historical sense" was weak in many of the students, the very contextual matrix within which David was placing and relating the geography of those contested episodes. It turned out that he had written a demanding work for some. Well, good: anything is better than talking down to people; we should reach and stretch, . . . or what's a heaven for?

So I recalled the style of late twelfth-century Paris, and the example of contemporary philosophers, and spent roughly half the semester trying to unfold, that *explicare,* the text. We took one session for each chapter up to 1940, and then splurged two each on "The Regionalising Ritual" and "Statistics Don't Bleed." After all, the Age of Adoration was getting closer to home; even the students born in the early 1970s were aware of intriguing undercurrents. Then, with this shared experience behind us, the second half of the semester moved to contemporary contestations with their more immediate implications for those reading about them. Discussions tended to be much more vigorous (and contested!), for these were ideas to be thought about, not facts to be memorized for an exam. That underlying threnody of David's, "controlling the discourse," was recognized again and again, and so were issues of power. Once they have been pointed out to you, you begin to recognize them all over the place. Were geographers today independent scholars with open horizons to new ideas? Or were they marionettes jerking to the strings of research grants reflecting the Establishment's "flavor of the week" down in Washington, or even their own unexamined discourses of ideology?

Paths to Possibilities or Dead-End Traps?

But I said my initial feelings of fraudulence were heightened by a long-standing and unresolved worry about teaching as purveying facts to be memorized, or teaching as opening up ways of thinking. In the last couple of years, I have found myself touching upon it a number of times, yet always backing away from a full attempt to think it through. It is a worry of some delicacy today, open to misinterpretation, accusations of unreflective nostalgia, positivist insensitivity to the human condition, and charges of inhumane reification. Could I find myself being charged as an old lag in yet another, but updated, complicity?[5] The worry was amplified further by rereading my favorite author, one with whom I frequently (but not always) see eye to eye.[6] It was the old question of how to give people things to think *with*, rather than facts to think *about*. Methodological teaching, helping people to a knowledge *(logos)* of a way *(hodos)* to approach problems, has the potential to enlarge the conditions of possibility for thinking. It also has the potential for clamping on intellectual blinkers, for setting peoples' feet on one path, which sometimes becomes *the* path. We all know of geographers who were intellectually potty-trained in their postgraduate days, and have clung to their particular methodological pot ever since. Which simply says that teaching is a dangerous game, and we have the responsibility to open up as many paths as we can for thinking in quite different contexts. People put not only a lot of time, but also a lot of themselves, into acquiring the technical skills to make particular methodological approaches operational. Of course, subsequent declarations of faith in these most fruitful perspectives, or even in "only perspectives," are not confined to the technical-methodological realm. However, and given the alternative of flabby uncommitment, I suggest we tend to respect the dedicated enthusiast, even if we think she is misguided, over the languid detachment and studied pretensions of the uncommitted dilettante. After all, why should we care about someone who does not care? Students certainly do not.

Rethinking in a Nexus of Possibilities

Let me suggest that any thinking about methodological teaching has to recognize that it is taking place in a "context of juxtapositions," that a number of technical advances and intellectual reappraisals have come together, and that this nexus of new developments has forced the question of rethinking the teaching task to the fore. Two clarifications first: some geographers, those at the analytical frontiers, may say that these developments have been around for quite a while, and "where have *you* been?" Carefully qualified, my response might be one of grudging agreement, while arguing at the same time that there is not much evidence around that this conjunction of events has been thought through in its full potential for teaching. Second, I shall deliberately confine this discussion to the quantitative methodologies, using "quantitative" as a shorthand to acknowledge other "qualitative" approaches. As may become clear with some concrete examples later, I am reluctant to label the latter as "humanistic," because human, not to say humane, caring is neither captured nor confined by these artificially partitioned categories. If the postmodern discourse can do anything, it can surely help us eschew a single perspective, and so allow us to acknowledge the potential for genuine illumination within and from multiple perspectives. And notice that this is not the relativism that so many seem to view with a sense of shocked, almost Victorian, morality, but an acknowledgment that in our human freedom we can make different choices in different contexts, or even take these up deliberately in the same context.

What developments have come together to form the nexus of possibilities for rethinking quantitative methodological teaching? So many that perhaps we have a nice example here of what Julie Graham called, in a very different context, "overdetermination."[7] But following her antiessentialist lead, let us cut into this tangle of juxtapositions at four places. First, the *ideological,* recognizing the tensions within prominent historical materialist positions between the shocked recognition of reification anytime the human world is approached analytically with quantitative tools, and the desire to remain logically rigorous, analytically penetrating, scientifically respectable and, thus, capable of enhancing the power of rhetorically persuasive narratives by the marshaling of empirical evidence, that *videre,* that "seeing is believing." The tension has been felt for at least a couple of decades; from Harvey's Marxian awakening on a Damascan road from explanation to justice and beyond; to bizarre humanist claims equating the "spatializing" of the quantitative turn with logical positivism; to the labeling, not to say damning, of any scientific approach as yet another example of the hegemonic phallocentric gaze. There are all sorts of reasons around for *not* bothering to rethink quantitative methodological teaching, while forgetting that *any* perspective on problem solving is always a two-edged sword, capable of moral or immoral use, and always illuminating and concealing at the same time. Inquiry, any inquiry, is delicate, even dangerous, work.

A second, *philosophical* cut can be made, using the term as a shorthand to mean that we find ourselves in a geographical discipline today of vastly expanded horizons. The conditions of possibility for doing geography have exploded in the past fifteen to twenty years. This means we are prepared to ask very different questions: of an ontological, epistemological, and historical nature; of substantive focus, particularly from the renewal of social and cultural geography; of increased subtlety and sensitivity from perspectives ranging from the feminist to the environmental. The recognition that any research program contains within it problems of representation, from initial conceptualization to final publication, simply was not in sight, in other words in geographic *thinking*, twenty years ago. And if the question of representation will not go away, as it never can, at least acknowledging this fundamental human problem—of re-presenting to others that which comes to presence in us—makes us more sensitive to what we are doing, and raises the moral question of why we are doing it in *this*, rather than *that*, way. There are far fewer taken-for-granteds and "unexamined discourses" today as a result of such discomforting questions, even if there are still plenty of examples around. After all, geographers are doing too much philosophy these days: those discomforting, but of course irrelevant, -isms will soon be -wasms, and it will be Establishment "business as usual" once again. Who cares about Aristotle's virtues? One more Dead White European Male, and he was not even a geographer.[8]

A third cut into the complexity of modern geography acknowledges a *technical* revolution, a world of personal computers, workstations, geographical information systems, computer cartography, libraries of standardized software, interactive capabilities for visualization (not the least, animation), all backed up by storage and access developments that have essentially shrunk some of the largest teaching machines of the early 1980s into the shoeboxes of the 1990s. Let me suggest that, for *most* teaching tasks, speed and capacity are no longer a problem. This is not to say that research projects cannot generate very quickly computing requirements still taxing the largest machines, but only that if these concrete problems were to form the basis for meaningful teaching exercises, then most of them could be shrunk down to exhibit their underlying principles. It is hard to see where technical capabilities are even close to binding constraints for the teaching task.

The final cut acknowledges that we sit at the edge of an *analytical* revolution that itself is composed of many intertwined threads. The origins of some of these go back a long way, to the variation-absorbing traditions of regression, and its many multiple, trend, and Euclidian variations; to partitioning and assigning variation to orthogonal dimensions, and using such components or factors as the basis for other approaches, perhaps taxonomic. All of these contained within them old concepts of optimization (minimizing or maximizing sums of squares, etc.), and they form special cases of more general optimization approaches (linear programming and its variations, location-allocation models, entropy-maximizing under increasingly severe constraints, and so on). But out of these traditional approaches have come greatly enhanced and expanded analytical capabilities. Spatiotemporal "cubes" of data, often displayed as map sequences showing changes over time of a spatial series, invite exploration for their nonrandom information by transformation, expansion, and spatial adaptive filtering methods. The enhanced speed and capacity of computers have made quite practical permutational approaches to generating statistical distributions appropriate to *particular* geographical configurations, and provide yet another approach to identifying significant spatial clustering. Neural computing in geography has made significant strides, and has contributed to such various topics as precipitation controls, self-organizing maps, classification, and map prediction.[9] Imaginative graphics allow us not only to display, but to understand, the spatial patterns, structures, and implications contained in relatively small, but still overwhelming, numerical data sets.

Any Space for "Spatial" Science?[10]

What are the implications for this juxtaposition of ideological, philosophical, technical, and analytical developments for rethinking the teaching task in the "quantitative" area? Is it really possible to wipe the blackboard of thirty years clean and start again? Once an institutional framework is in place, and staffed (not to say guarded) by Old Boys and Old Girls who "did it this way in our time," that anarchistic act of wiping everything out and starting again is very difficult to achieve. How can we get ourselves to think, right from the beginning, about (a) the disciplinary, and potential interdisciplinary, needs of geographers, and (b) the way such methodological perspectives can constitute conditions that allow possible ways of seeing to come to a geographer's thinking? The task of thinking *this* through is so complex and overlapping that I am tempted to draw Venn diagrams—except that they would need more dimensions than this page allows. What I can only offer here is a sort of potpourri of elements that might be *some* of the things entering and informing a much more knowledgeable and complete discussion.

Quantitative methods use numbers, or symbols standing for numbers or sets of numbers. This is so obvious that we usually forget about such a basic fact, but it should remind us of two things. First, *all* mathematical and logical operations on a digital computer are conducted in binary arithmetic as it is transformed from the set of integers augmented by zero. This claim makes many people uncomfortable the first time they meet it. "What about 6.147 meters per second, or −212/423?" they say. "These are hardly integers!" But simple scaling gives me 6,147 millimeters per second, or whatever integer unit I require to my limit of significant measurement, and my negative rational fraction is −5,011,820 microparts of a whole pig's whistle, again to the limits of my ability to measure. Like the mortal beings who conduct them, all *computations* are finite, and most accumulate error, as Lorenz discovered to his consternation and delight in stumbling across the beginnings of what we now call chaos theory.[11] So all quantitative science is creating scales, counting integers on them, and relating the scales. That is all; really a rather simple task, and not nearly as difficult as writing a memorable sonnet.

Second, sets of elements, some of which can contain numbers (vectors, matrices, etc.), can have various binary operations defined on them to create particular algebraic structures. For example, the binary operation $a \cdot b = b \cdot a$ is well defined in simple algebra (and others that commute), but it is often absurd, or undefined, in linear algebra. The *structures* are different, as nineteenth-century mathematicians in London knew very well as they ridiculed these absurd ideas from across the Irish Sea. Today, we realize that these aberrations of Hamilton and Boole can prove quite useful for describing things. But we must always remember that *we* create such structures, and *we* impose them on observations, if *we* decide they are useful simplifications, and *we* can persuade others of their utility. But *we* should never forget that *we* are looking at our own fabrications—our own, literally, *fictions*, the things we have posited and made. In a post-Enlightenment world, how could it be otherwise, unless you think God is a mathematician, and we are simply searching for what she has created? Many physical scientists still believe this, totally unaware of their implicit theological underpinning, which would probably embarrass many of them if they cared to think about it. So what is "quantitative geography"? It is creating information that comes to us as a light signal (penciled digits on a sheet of paper, digitized map coordinates, readings of a voltmeter measuring decibels or spontaneous radioactive decay or depth of water or albedo reflectance or . . . you name it), and mapping those sets of digits in some way to the elements of other sets. Good: as geographers we start with mappings.

Mappings come in many varieties, and are nothing more than rules relating the elements of one set to another. Ordinary regression maps a set of X's to a set of Y's with a simple scalar ($Y = b \cdot X$); multiple regression just uses more information and more scalars, and so do all component- and factor-based methods; Euclidean regression maps one set of coordinates to another; dy/dx is just a rule mapping $f(x)$ to $g(x)$; $d_{ij} / P_i \cdot P_j$ maps health districts in Iceland, or counties in Ohio, to measles and AIDS spaces, respectively;[12] multidimensional scaling maps measures of (dis)similarity between things to distances between points; the black box of a trained neural net maps one set of numbers (say pressure) to another (say precipitation) . . . and so on—mappings all, superficially different, but really the same.

Many mappings partition amounts of variation and assign it to different boxes—explained and unexplained, Factor I, II, III, . . . this or that taxonomic box—one-to-many mappings all, superficially different, but really the same. It is difficult to think of any analytical procedure (quantitative or qualitative, for that matter) that is *not* a mapping. Those A_i's and B_j's in entropy maximization weight sets of origins and destinations to numbers of trips between them, and the mapping will vary depending upon the squeeze you put on it, until the constraints take you to the minimization of linear programming, or infeasibility. Clustering algorithms compare points per quadrat with expected values or spatially distributed values with a distribution specifically computed by permutations for each point;[13] but what are such comparisons if not mappings of sets of numbers to others? Commuter flows around New York are transformed (mapped by the binary operations of + and ÷) to probabilities in a Markov process; take the fixed-point vector (map the probabilities to the elements of the eigenvector) and you have a global measure of access to the "system." Another mapping (regression) tells us that it soaks up about 90 percent of the variation in the AIDS rates in the region.[14] Searching for spatiotemporal structure in those (x,y,t) cubes to predict the next maps with spatial adaptive filtering and other expansion methods is nothing more (and nothing less) than finding mappings based on the spatiotemporal information in a pile of maps to predict the next ones.

Do these examples of geographical analysis exhaust the idea of mapping? Not a bit. Every location-allocation model maps a set of elements (say hospitals) from one set of coordinates to another set that the algorithm specifies as "most efficient." Algorithms are mappings, rules of transformation. Perhaps you are tempted to say that efficiency, measured as the minimization of cost or distance, may not be humanly relevant, that this is just managerial reification and manipulation? Then ask people dying of AIDS in a hospital if they would like to be close to loved ones and see what answer you get. Let every horrified cliché about reification be examined and thought through. Context, not cliché; a culture of caring, not precious posturing, will decide on the morality of an analytical act.

You say some algorithmic mappings are difficult to find? You bet they are; many are heuristic, unsupported by proof, but so what? Some researchers may be inclined to devise proofs for hill-climbing procedures taking us to our analytical Everest, rather than getting us mired in the foothills of local maxima, but meanwhile the rest of us have work to do, much of it highly practical and humane in societies trying to renew the social contract. Given today's computing speed, most heuristics seem to work pretty well if they are run again and again from different initial conditions. The fact that many multidimensionally scaled configurations in the literature are probably stuck on local minima should not deter us from teaching these approaches allowing people to create other sorts of spaces and then *think* in them. And that means to *see*, visualize, to comprehend in those spaces. I confess I am getting tired of little Lacano-Freudians labeling every act of scientific inquiry as an erotic penetrating, of intellectual transvestites disparaging the "gaze" as phallocentric, of the visual being labeled masculine voyeurism. Ask each of them which of the five senses they would least like to lose, and then listen carefully to their replies.

Hands-On Visualization

But understanding *as* visualization, truth as an unconcealing and a bringing into the light of human thinking, places severe responsibilities on the teaching task. This is a task radically changed by interactive computer graphics that have, in turn, a remedial bite for one of the most difficult problems we face in this computerized "press button" age. It is the problem of staying somehow close to the data, close to what is actually going on—so close that analytical procedures become part of us, ways of thinking, ways of enlarging those conditions of possibility we all strive for as teachers. Somehow the enormously enhanced potential for "visualized analysis" must be directed to the task of genuine understanding, the sort you can think *with*, rather than "If you push REG you get a number between −1.0 and +1.0." To take the simplest idea: if people have never actually added up a set of numbers, computed the mean, taken it away from each number, satisfied themselves that the deviations sum to zero, *seen* why squaring is necessary to get the total variation, divided it by N to get the average variation, taken the square root to get back "out of squares,"

. . . how can they understand—I mean really understand—what a standard deviation is? I know that there is a button with a lowercase sigma on every pocket calculator, and that by pressing it you get the same number. But that is precisely the point; you get the same number, but not the same understanding. And the same goes for least-squares, residuals above and below the plane, the location of an eigenvector, the starting set of locations for MDS, the best guesses for location-allocation models, the search for clusters, the whole process of mapping intuitive hunches onto sets of numbers, relationships, maps, equations, and so on—the whole *visual* paraphernalia that lets people say "I see!" (that adrenaline jolt of eureka that is not gendered but simply human). Phallocentric gaze or not, those black blobs of Stan Openshaw's confounded the professions of medicine and epidemiology, and raised questions about leukemic children once again, instead of shutting them down and burying them in an Establishment report.[15]

Yes, I know, visualization is not everything, seeing may be illusion as well as believing (or perhaps believing in illusion), because we tend to see what we want to see. But as geographers we seldom hear, smell, touch, or listen to a map. And so that is all there is to geography? Maps? No, of course not. But I dare you to take them, and their graphic relations, away. They are frequently the sources of our hunches, they are often our analytical tools, and they form some of the most powerful and convincing ways of presenting our conclusions. Graphical rhetoric? You bet. That old and honorable art of persuasion, with the visual augmenting the traditional spoken or written form.

How can we combine the speed of interactive computer graphics, the power of analytical methods, and the problem of staying close to the data to inform the teaching task? Obviously anything in science that can be done with pencil and paper can be done on a computer screen. So let us do it. Where are the interactive programs that let people plot an X variable against a Y variable and then fit *by eye* a "best-fitting" line, all the while computing the sums of the vertical deviations, the sums of the perpendicular deviations, the sums of squared deviations, and so on? Are you willing to let people see *why* least-squares is unique and stable, or do you want to shove it down their throats by pedagogic authority? Where are the interactive programs that let students divide two spatial distributions as they will, and then compute the correlation coefficient? After a few such hands-on exercises, you do not need a "million correlation coefficients" because you have experienced the problem yourself.[16] What software lets a person gradually move an eigenvector around in a space of unit-variable vectors, while computing the sum of the squared orthogonal projections that will become the associated eigenvalue when the maximum is reached? Who is going to accept factor outputs on faith (just because they are published?), when you have tried yourself, visually on a screen, to rotate a structure to a varimax position on an evenly spread fan of vectors? Mathematically you can do it, but having experienced what is actually going on, having seen what is happening with your own eyes, you know it may be meaningless, simply the thoughtless imposition of an orthogonal structure upon which some, almost any, interpretation is desperately impressed.

Who has a chance to play with—and that means *work* with—a small entropy maximization problem, and see how patterns of flows change as origin and destination constraints, and costs of movement, are altered? It is only when you can change the constraints, and see immediately the shift in spatial pattern, that you get a sense of the closed, even *tautological*, nature of such systems. Why cannot people experience location-allocation problems guessing the best one, two, three, . . . positions, and then see how close they are? Only such immediate, visual experience lets you understand why the addition of a new facility produces a sudden, and totally different, optimal configuration. Where can a student design a small ecological system, and see almost immediately the dynamics it produces?[17] Where is the software that will let a person see the connectivity between a set of things, and how some taxonomic algorithm chops it up willy-nilly, arbitrarily, but always "scientifically"! People have to be able to move from Euclidian regression and asymmetric interaction matrices to vector fields of residuals and "winds of influence" in something like "graphic interaction time" so they can see how changes alter the appearance of such vector fields summarizing perhaps thousands of numerical bits of information.[18] You have to be able to *feel* what is happening; otherwise how can you *think* with these approaches, meaning that they have become a part of yourself, a new condition of possibility for doing geography?

Such interactive, essentially pedagogically orientated, developments clearly need the best approaches characteristic of Geographical Information Systems (GIS). Indeed, some might claim that they are GIS approaches. But the flow is not one-way, from GIS to teaching. Engaging in such tasks, scaling real and tough problems to exercise proportions for teaching, could integrate analytical methods into GIS to a degree not yet seen. In this way, GIS might become involved in something useful and practical, instead of constantly worrying about the latest bell and whistle to hang on the Mark XXV version of an algorithm that has accreted over such a long time that no one can follow it through anymore except on faith. This is an increasingly common experience in large systems algorithms, where modification has followed modification, one more bell hung on yet another previous whistle, often employing different computer languages for subprograms.[19]

Teaching Integrated Analytical Approaches

Where can we go from here? And I mean that literally. How can we get together in front of that freshly cleaned blackboard and think through the integration of quantitative methods, analytical cartography, and GIS in the context of greatly enhanced computing power, and for the cause of teaching? Real problems abound, from both the physical and the human realms of geography, and increasingly from the interface between them—or should it be "joining them"? All social problems with a history have a geography, and Zeus knows we have enough of those to sensitize all exercises devised for teaching. Simply recall that AIDS rates around New York City are very closely predicted by an index of accessibility to the commuting system. But the new social-cultural geography has already highlighted scores of other examples.

What would you think of an international conference, reasonably small and compact (thirty-plus people, not too many more), for which the purpose is not to display the latest analytical twist, but how to enhance the teaching of numerically based analytical approaches to geographic problem solving? It would need, above all, people with experience in the teaching task who had kept abreast of, and even pioneered in, analytical developments. It would not need methodological pot clingers, those who saw geographical problems through a single analytical lens, but those open to a spectrum of approaches.

Is this a possibility? Many science foundations at the national level have set aside modest funds for the enhancement of undergraduate teaching. It would not take much from the half-dozen countries or so that might support participants. After all, in science is there anything more basic than the next generation?

A Petition to Judas Gould

Oh, Judas Gould, methinks you're wrong
I plan to study all week long.
Dear Mom and Dad, they miss me so,
I haven't been home since long ago.
I'm sure it's not all feasting and fun
(They'd like to see some homework done),
But spend Sunday night at the Computer Center?
Will Apollo[1] allow me then to enter?
With cold towel in hand, stat facts in my bones,
Will I be forced to wish I'd never gone home?
There's no game there, no players in cleats,
Football will not find me on the streets.
Weekend computers are your bread and butter,
But they only put me in the gutter.
I'm not alone—take a sample or poll—
This weekend homewards the wheels will roll.
Monday morning we'll all be in trances
On postponed homework—what are the chances?

Judas Gould Replies

Season of mists and mellow fruitfulness,
Close-bosomed friend of the maturing sun,[2]
Fall terms when tired and shattered people
Back to Mom and Dad do run;
When midterms, projects, homework, papers
Fall like leaves so bright and gold,
And drudgeries leave no room for capers,
And for relief petitions bold
Exhort the Judas Gould to pity,
Forgetting the treacherous of that band

Did not betray his charge and master
Without bright silver on his hand.
Silver's no good . . . but beer will do,
A bottle's a thing of pulchritude,
But bribes for matters academic
Are gross, immoral turpitude.
Bribe me no bribes—on the next exam
Hand not in sheets so blank, pathetic,
No matter what belief, recall
The Protestant without an ethic.
And towels wrung out in glacial streams,
And wrapped 'round fevered, aching heads,
Do more to justify one's life
At eight o'clock, than slugabeds.
Judas? My name is Petrus, Madame,
Not on *this* rock shall learning bow,
We're on this earth because we're sinful,
Only hard work will save us now.
But, Father William, I am old,
The hard steel heart is cracked and rusted,
And even I have got behind,
And oft and many a gut have busted.
All right then, but on one condition,
No prisoner homeward on the lam,
At three o'clock, in blue and white,
You're cheering there 'gainst Alabam.[3]
Let your cheering ring to southward,
Help the Lions on every play,
As they fight for Alma Mater . . .

. . . then homework's due on Wodensday.

Thinking about Learning

I AM SURE many of my students over the years have done a "there-he-goes-again" when I have pointed out that the verb "to teach" is usually transitive, while the verb "to learn" is generally used intransitively. That is, of course, if anyone these days knows the difference. So I tell them, "I can teach you, but I can't learn you," and get the point across that way. Which says nothing really about the symbiotic relation between the two, or the way both of them invoke a shared responsibility. Am I that far off when I say that the life of a teacher is an angst-ridden life? Not the angst of hollow eye sockets and gaunt cheeks, but an always present and mild angst of caring, whether faced with the avid student, to whom you never seem to be able to give enough, or confronted with the "teach-me-I-dare-you's" in the back row, awake only enough to be resentful at eight in the morning in your required course. "You little bastards," I have thought on occasion, "if I can dig my way out on a cold and snowy winter's morning ready to teach my guts out, couldn't you at least pretend you care too?"

At some point, we have all suffered bad teachers, those who go through the motions, or simply do not have the aptitude to share their own intellectual passions. Perhaps because they have no passions left? I am skeptical about "Teacher Improvement Seminars," with all the paraphernalia of TV instant feedback, and "skilled pedagogues" hovering nearby to elicit a more positive response and dialogue with your companions in learning. I mean, come on: teaching is a calling, and no made-to-measure delivery system crafted by a pedagogue is going to elicit a response from your "companions in learning" if they do not want to learn, and do not believe you want to teach. Anyway, pedagogue, if you're so damned good, why don't you go and . . . pedagogue? Euphemisms should be banned in the university.

Over time, our views of our teachers change. With any luck, the bad ones fade to forgotten pentimentos, while the good ones are remembered with heartfelt thanks. Teaching can be a lonely task: sometimes, if you are lucky, a letter will arrive, always when you least expect it, saying "Thanks for what you did," even as you may have no recollection of what was a simple, caring gesture you made in the midst of a busy day long ago. Such letters are worth more than all honors. I had my share of good and bad teachers, the latter either lazy to the point of unbelief, or so incapable of empathizing with another's struggles to understand that they produced only despair. Mathematics seems to be full of them, perhaps because many mathematicians find the elements so easy and obvious that they simply cannot comprehend another's difficulties, at which point they become exasperated, impatient, and dismissive.

But of the good teachers I had, two from my school days stand out, and "A Lasting Legacy" was my attempt to recall and give thanks—in one case too late. And now that the fine Irishman (what is it about the Irish and language?) Fred Doyley can no longer read these words, let me confess that I never could share his enthusiasm for George Meredith. I found his work head-noddingly boring, second only to Alexander Kinglake, whose *Eothen* we had to trudge through as one of the books required that year for Higher School Certificate. While usually absorbed by good travel books, from Mary Kingsley to Patrick Leigh Fermor and Bruce Chatwin, I still find *Eothen* unreadable. On the other hand, I took Winston Churchill's advice seriously, when he was asked by a young man how to write well. "Read Kinglake's *History of the Invasion of the Crimea*," said Churchill. Somewhat taken aback, the young man tentatively inquired further if there was anything else he might do to improve his style. "More Kinglake," replied Churchill.

For many years, I tried to obtain a copy, and had completely forgotten a previous request, obviously hidden away in some old file at Foyle's on the Charing Cross Road. Then, one day, out of the blue, came a letter: they had obtained a copy; did I still want it? I can still remember the thrill of unpacking all six volumes, volumes previously belonging to Collegium Corporis Christi in Academia Oxon—"Deleted by order of the library committee," read the stamp on the bookplate. Well, ole Corpus, your throwaways are my delight! And what splendid volumes they are, with their numerous foldout maps, the "thin red line" marked in bright red, the French allies in somewhat faded blue, and the tyrannical Ruskies in a rather bilious green. And tucked away in the back, the publisher, William Blackwood of Edinburgh, placed numerous advertisements: *The Diary of a Late Physician* (did patients survive his tardiness?), *Lays of the Scottish Cavaliers* (not what you think), *The Great Problem, Can It Be Solved?* ("likely to be of great service to a vast number of wavering and unstable minds," said the reviewer), *A Manual of Modern Geography; Mathematical* ("the result of many years' unremitting application"), *Quedah; or Stray*

Leaves From a Journal in Malayan Waters, and *The Geology of Pennsylvania*, and many others. Advertised in 1876, taken together they capture a totally different world of over one hundred years ago. I started to read my way through Kinglake's account, starting dutifully with a slight bow in Churchill's direction, but continuing to the end with great pleasure and excitement. His prose is, yes, Victorian, but of the very best—crystal clear, beautifully structured, leading you on effortlessly to see what happens on the next page. It was an era that produced mountains of somewhat mannered but always clear narratives, before fads, before jargons, before sociologeze. Some of the best writing is to be found in the scientific journals of the time, straightforward accounts of why this topic was interesting, what had been discovered, and what remained puzzling to the author.

Which leads to the world of the second teacher I had, this time in quite classical, Newtonian physics. Ernest Beet was a graduate of nearby Reading University, and a respected Fellow of the Royal Astronomical Society. Given an eclipse of the moon, a partial or full eclipse of the sun, or a year with many sunspots, he would have boys crowding around his telescope for a look. He believed the task of physics was to explicate this wonderful world, and that meant that you *did* physics, as well as reading about it. I remember, many summers ago, meeting two young Frenchmen who were studying for the *bachot* physics exams coming up during the next autumn. For them, physics was one saying to the other, "Fifty-two!" or some such number between one and a hundred. "Fifty-two . . . fifty-two . . . ah, yes! One calorimeter, one thermometer, one cylinder of copper, one Bunsen burner," and so on. "Experiments" in physics were feats of rote memorization helped by mnemonic devices. Neither of them had actually determined the specific heat of copper, or done any other hands-on experiments in their lives. No wonder French physics has so few recruits of distinction, and only eight Nobel Prizes out of 137, the last one in 1970.

My own belief in hands-on geography, *doing* exercises, always map oriented, no matter whether the computations are done on a pocket calculator or computer, stems from Ernest Beet's own meticulously designed series of experiments in light, heat, sound, magnetism, and electricity. Once he had explained the principles of calorimetry, shown you how *this* side of the equation had to balance *that* side, and guided you through your own experiment, the fog of the textbook cleared and it all seemed so simple and obvious. Intransitive or not, I have to confess "He learned me"!

The essay about August Lösch brings quite different thoughts about learning. Nearly thirty years after reading his marvelous book *The Economics of Lo-*

cation,[1] written and revised just before and during the Second World War, I was asked to contribute to a volume in his memory, for in 1945 he had died of malnutrition and scarlet fever in a garret in Kiel, after sending his wife and child, with all their available food, to southern Germany for greater safety. More than a quarter of a century on, I was a very different person from the graduate student who had read Lösch with such fascination. The invitation also came at a time when, for five years previously, I had read text after text of Martin Heidegger under the tutelage of one of the finest scholars it has ever been my privilege to meet. The philosopher Joseph Kockelmans always worked with parallel German and English texts, and with the Greek text too if it were appropriate, and *slowly* opened up the thinking that lay behind Heidegger's tenacious quest. If anyone had told me we would spend sixteen weeks on the forty-five pages of Heidegger's "Anaximander Fragment," I would have laughed,[2] but at the age of forty-seven, I began to read properly.

I had a small room in the library that semester, and after an eight o'clock class I would go there and read, *slowly,* the work of August Lösch again, realizing, again and again, what a very different person was now coming to the text, what a very different child of a different time was now trying to see what was there. How the essay honored August Lösch is for others to say: for me it was one of the most revealing learning experiences of my life.

Plate 1. The Salcombe estuary and its "bicycle ride" surroundings, the teenage world of Peter Gould from 1945 to 1951. Taken from the Ordnance Survey Sheet 202 Torbay and Dartmoor at 1:50,000, revised 1976. By permission of the Ordnance Survey.

Plate 2. The area identical to that of plate 1, made up from Sheets 23 and 24 at 1:31,680 of the Old Series (1809) Ordnance Survey, engraved from surveyors' drawings completed in 1803–1804. Nearly a quarter of its place names are no longer to be found on the modern map.

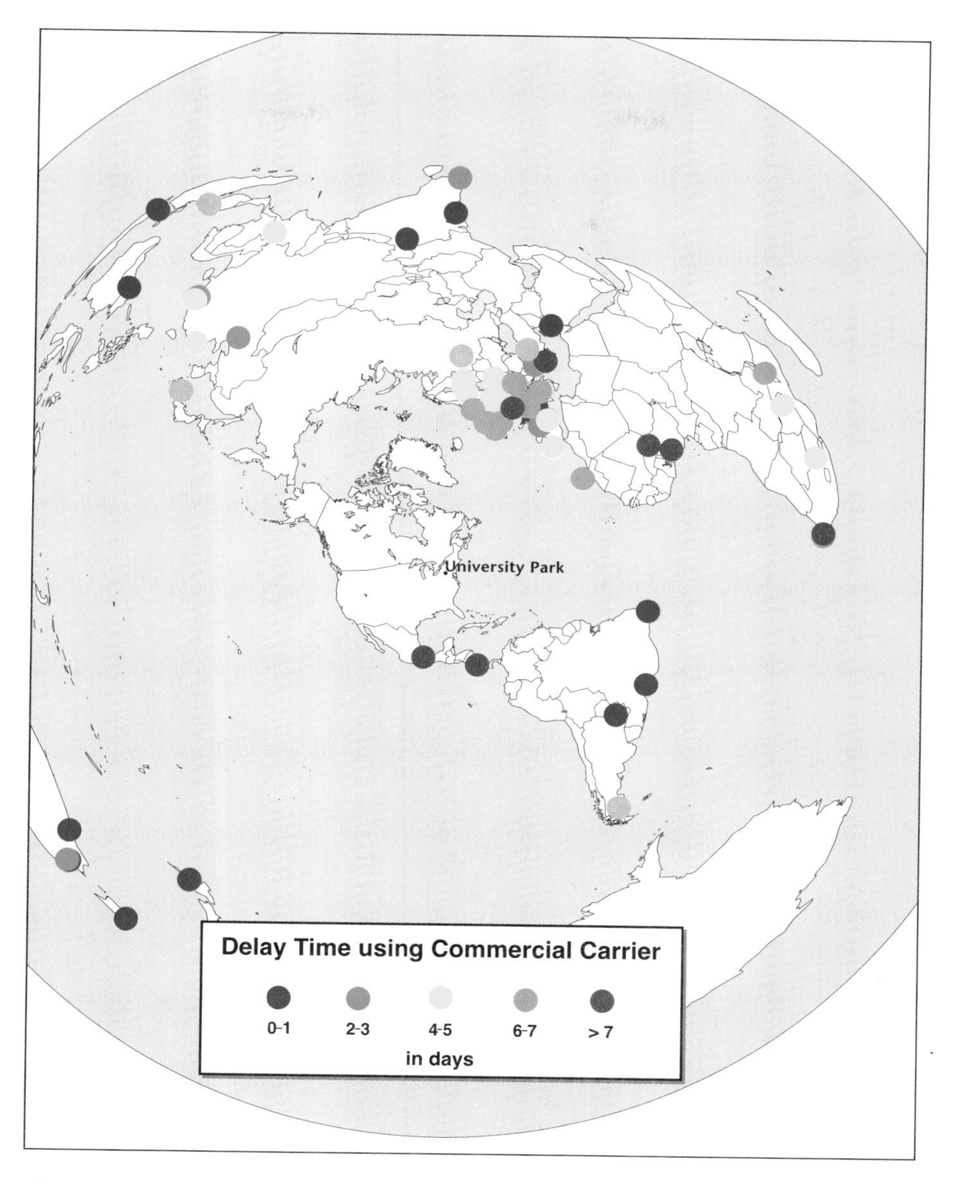

Plate 3. Penn State (University Park) at the center of postal delay space resulting from the use of a commercial carrier rather than the United States Postal Service.

Plate 4. The oblique "picture-map" of the course as it appears on the small brochure available to skiers at Beaver Creek, Colorado. Copyright ©Vail Trademarks Inc.

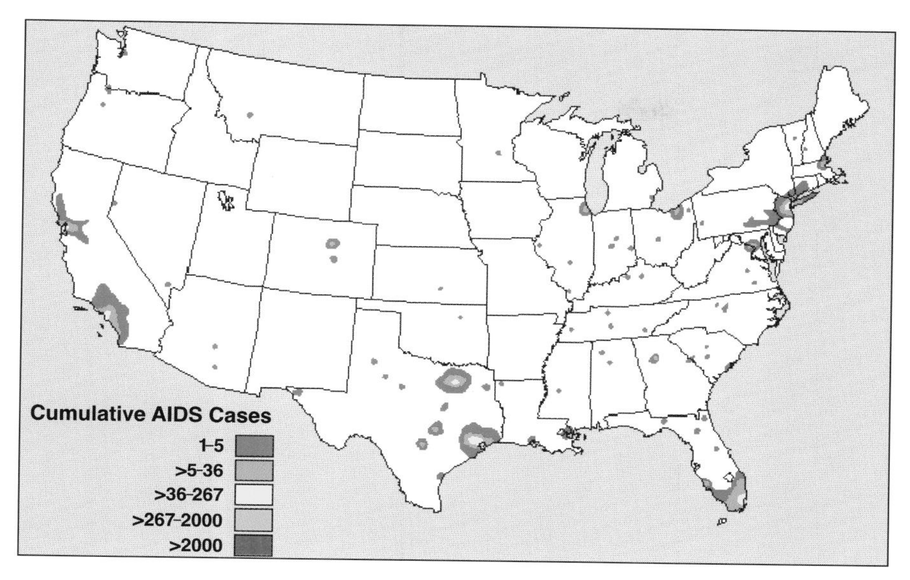

Plate 5. The intensity of the AIDS epidemic in 1982. The color-contour interval is geometric, each change multiplying the previous value by 7.5. The effects of hierarchical diffusion are already evident.

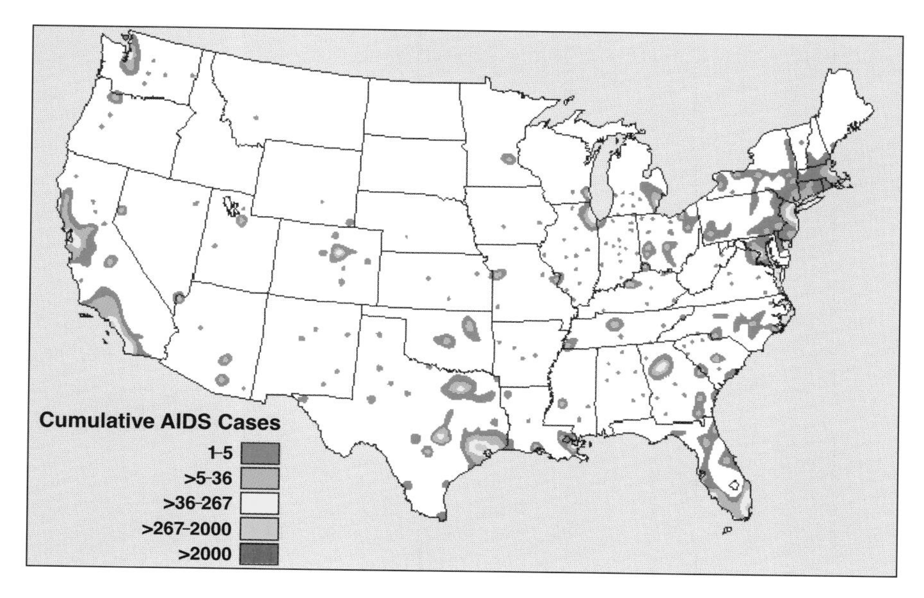

Plate 6. AIDS in 1984: hierarchical diffusion has allowed the virus to move down the hierarchy to smaller centers.

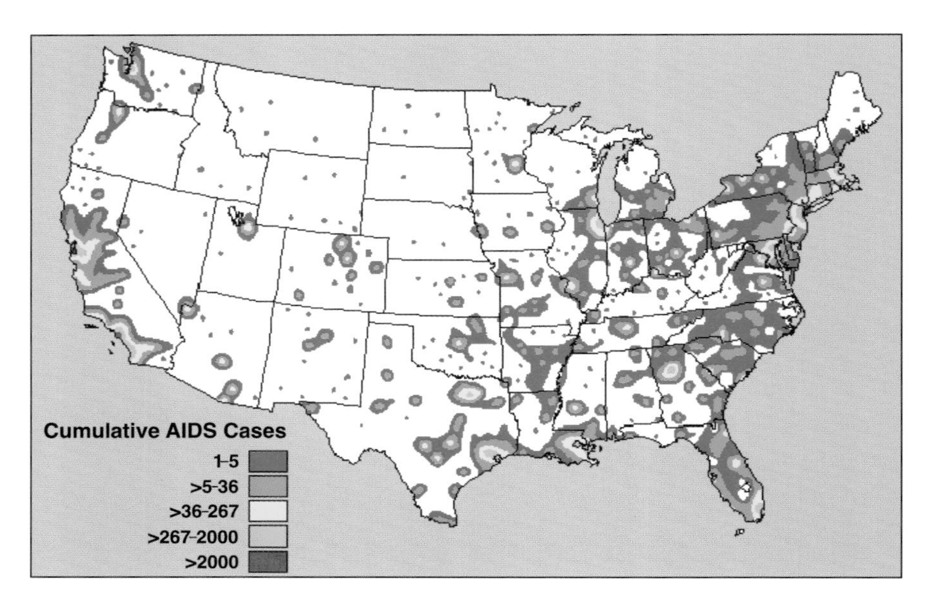

Plate 7. AIDS in 1986: spatially contagious diffusion now becomes more prominent as urban commuter fields pump the virus into the suburbs of regional epicenters.

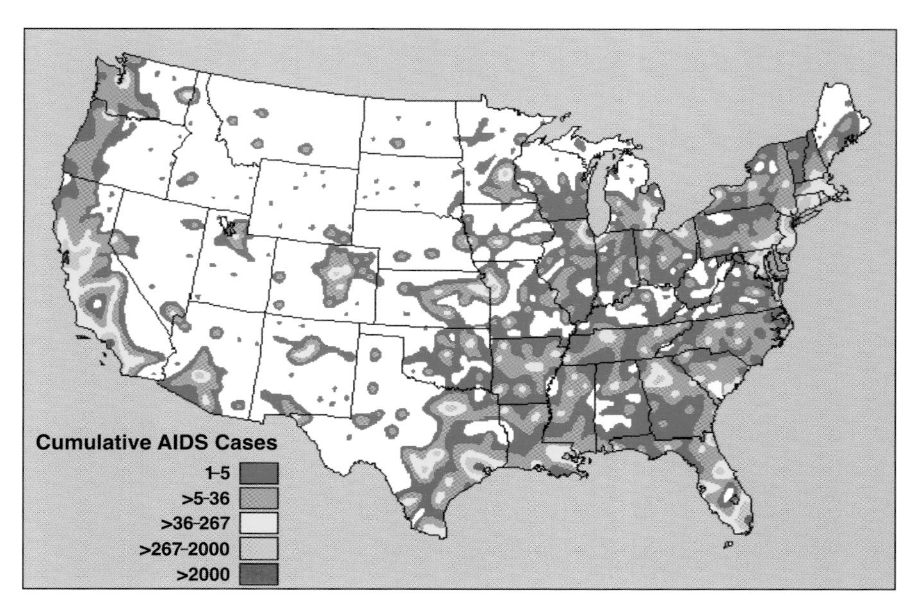

Plate 8. AIDS in 1988: clear alignments along interstate highway systems are intensifying, and earlier urban "beads on a string" are reaching out to coalesce.

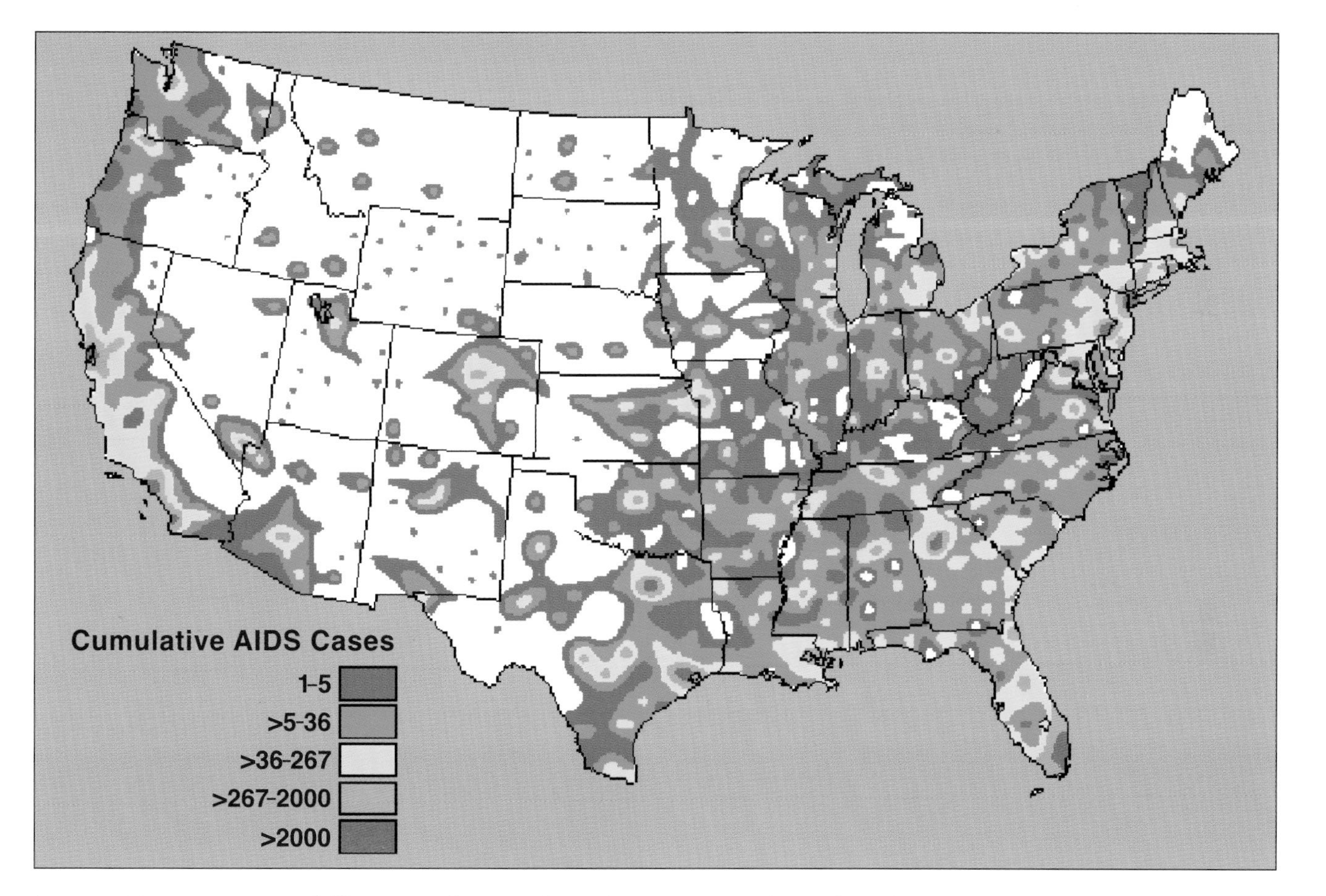

Plate 9. AIDS in 1990: interstitial filling intensifies in rural areas between major urban nodes and national transport magistrals.

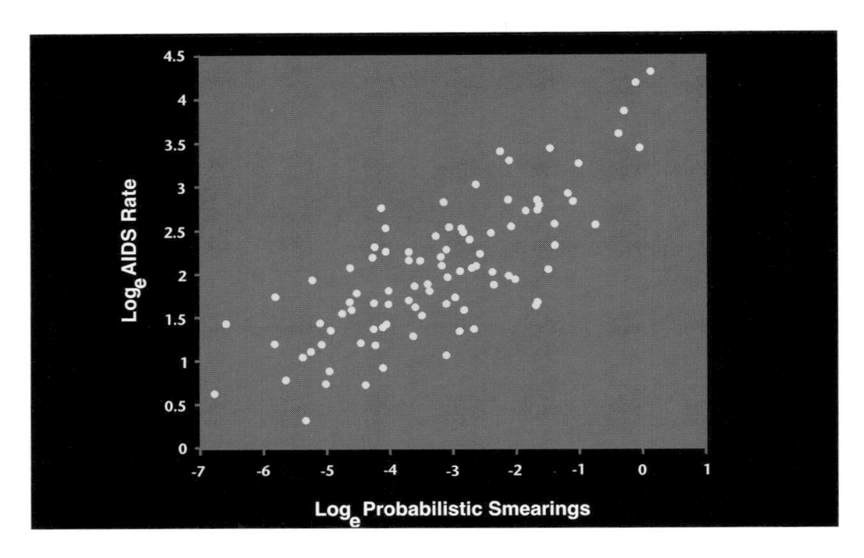

Plate 10. The "iterative smearings" produced by our air passenger operator, after injecting the HIV at New York, Los Angeles, San Francisco, Miami, and Houston, predict AIDS rates in 1986 and subsequent years about 80 percent correctly.

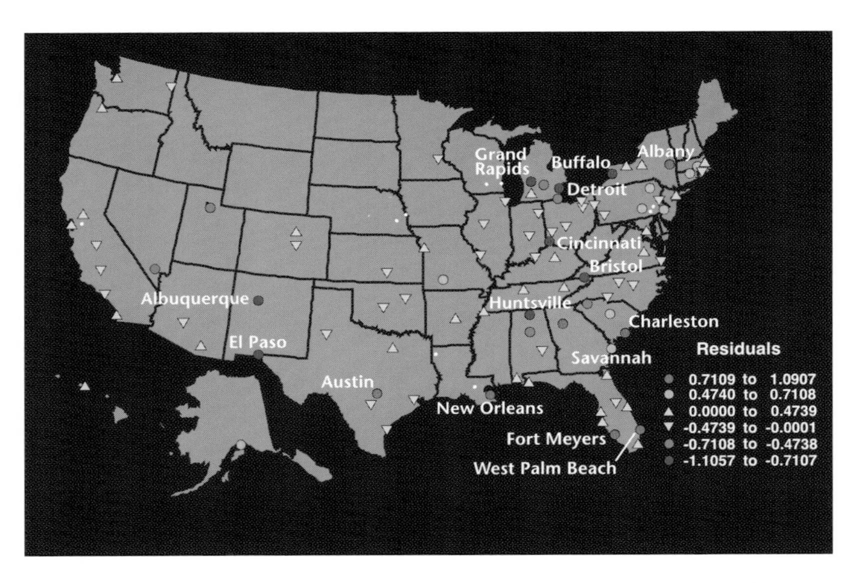

Plate 11. The residuals from the AIDS rates versus "smearings" relation, defining cities (red and orange) with AIDS rates greater than we would expect, and those (blue and green) with rates less than we would expect. Cities with rates well predicted are shown in yellow, the upward pointed triangles *slightly* greater than expected, the downward pointed ones slightly lower.

August Lösch as a Child of His Time

Glükseeling Suevien, meine Mutter
—Friedrich Hölderlin
Die Wanderung

Hommage d'Auteur

THE SLATE GRAY BOOK has been on my shelf for nearly thirty years. "Peter Gould, January, 1957" it says on the flyleaf, and the back cover still tells me it cost $7.50. I took it to that first class of Northwestern's winter term, glad to get out of the raw wind that swept flurries of snow across Lake Michigan. Charlie Tiebout arrived a few minutes later, and with a seemingly casual style that marked the sharp clarity of all his lectures, he started to explicate, to unfold and lay out for all of us to see, the great wealth of thinking that lay within *The Economics of Location* by August Lösch.[1]

Today, it is very difficult to convey to others the jolt of excitement that even a casual leafing through the book generated. For someone called to geography, it was the renewal of hope at a time when economic aspects of geography had long been divorced from economics, and when all the rich and informing connections to other areas of geographic inquiry had been severed by the sort of partitional thinking that still shreds geography into the disconnected boxes of tidy, but unimaginative, scholars and administrators. To return to the standard texts, and many of the professional articles of economic geography in those days is not to recall, honor, and respect the always limited and historically contingent insights of another time, but to recollect, and feel once again, the numbing banality of unstructured compilations, and the sense of shame that came from being associated with them. Lösch, with his unusual breadth of learning, honored the spatial, economic, and geographic insights of Chamberlin, Christaller, Haberler, Hoover, Launhardt, Ohlin, Palander, Predöhl and von

Originally published in *Space, Structure, Economy*, ed. R. Funck and A. Kuklinski (Karlsruhe: von Löper, 1986), 7–19

Thünen. A geographic work of the same era, drawing heavily upon the same German tradition, noted one of these (Palander) as a critic of geography, perfunctorily dismissed von Thünen with a cursory footnote, and totally ignored the rest.[2] That was the difference between two approaches to the common task, one defensive and dismissive, the other open and acknowledging. It was upon such a difference that a new perspective in geography was to arise.

Preliminary Clarifications

August Lösch's *Economics of Location* is an enormously rich book, rich in thinking, speculations, assertions, insights, and anticipations. No short, and necessarily compressed, presentation can do justice to a book whose stature is measured precisely by its ability to generate other thoughtful works about it. This essay cannot even be counted as an exploration in intellectual history, but simply as a scouting expedition. Even as a preliminary foray, the temptation to quote *in extenso* must be resisted, although Lösch, even in translation, invariably says it better. For this reason, an air of superficiality cannot be avoided, but neither does it need an excuse if "superficiality" is read properly and etymologically as "lying at the surface." My only purpose here is to scratch that surface deeply enough so that others may catch a hint of what lies beneath, and so turn to the work itself.

In reading this essay, three points of misunderstanding may arise from the traditions informing the writing. I owe it to you, the reader, and to the memory of August Lösch, the man, to try to clear up such possible misunderstandings immediately. First, I wish to undertake an exercise in *critique*, and to note clearly that in the philosophical tradition this does *not* constitute *criticism*. I wish only to lay out a few claims for inspection, and to try to understand them in the context of the informing and shaping "worlds" out of which they arose. Secondly, no hint of denigration should be construed in the title. We are *always* children of our time, whether we accept, reject, or try to enlarge the horizon of the tradition into which we have been thrown, and in which we always stand. It is important to point to this to induce that self-reflective concern and thinking that has become ever more crucial since "the individual sciences broke away . . . and philosophy set them free."[3] The price the human sciences have paid for such liberation has been a terrible one. Few practitioners of any of them demonstrate a capacity for that truly questioning critique that generates understanding of origins and informed humility towards future possibilities.

Finally, certain passages of this essay will stand in a tradition of retrieve, in a hermeneutic tradition that recognizes that the act of interpretation is constantly informed by that which the interpreter brings to the text, that which the interpreter *is*. In rereading August Lösch, I was constantly and acutely aware that I was not the same person who read him during the winter of 1957. Such a tradition of retrieve implies that we are not only trying to be just in our attempt to understand what an author meant in his time—undeniably important though this may be—but trying to retrieve from the text that which may become a condition of possibility for generating new insights for ourselves in our own time. To use a text in the mode of retrieval is not to put words into the mouth of the author, or to impute to the author meanings that could not possibly have been meant, but rather to allow the caring tradition in which the author stood to touch us, to inform us upon the way of our own thinking.[4]

Creating a "World" from a Text

A work of art, music, architecture, literature, or science cannot be divorced from its time, even as it seeks to confirm, reject, or enlarge the vision of that time. Confirming, rejecting, or enlarging, it still relates to the tradition in which it stands. What was that tradition in which August Lösch stood, that world of the highly educated and intelligent German of the first half of this century? Is it possible to retrieve a sense of that informing world that stands behind a text that seeks constantly to bring the spatial, social, political, and geographic to the fore, to make these perspectives copresent with the economic? One thing is certain: it is a world that seems much more richly informed by a sense of the historical, a sense that allows Lösch to draw upon one historical example after another, and to assume a shared sense of meaning with his reader. It seems to me that this constant presence of the historical points to an understanding, or at least an implicit assumption, that all within the Western tradition is connected into a whole, that all, in a sense, is "grist for the mill," that nothing learned is wasted, that the separate pieces are always there waiting to be connected into a larger fabric. In fact, is it not the enrichment of the knowledge and understanding of this heritage that we understand by *education*, as opposed to the increasing sense of *training*, or indoctrination to correct ideologies, that epitomizes trends in so many universities today?

It is the philosophers and poets who provide the most thoughtful and penetrating comments upon their "worlds," and who thus shape in turn the thoughts of those who listen to them. We can only guess at the possible texts read by a German scholar born in 1906, writing the first edition of his work in the late 1930s, and revising the second in the middle of the terrible war years. But such a speculation is not necessarily idle and without meaning. Even though there are no philosophers named in the index, their voices are still heard in the footnotes. Though subsidiary in a literal sense, these constitute over 30 percent of the book, as Lösch uses them to explicate, elaborate, and comment upon points made more sparely in the text itself. And it is here, in this accompanying chorus of commentary, that we hear the informing philosophical tradition. Hegel is explicitly noted twice,[5] the first time to recall his "self-confidence of thinking," the second to emphasize the power of a rationally based system, and the disintegration of life that may follow upon its demise. Such a pointing to the unraveling of the fabric of a coherent life means that Kant is there too, for Hegel's voice is not heard alone. Schelling is also there, struggling with the problem of human freedom, and behind all three lies Descartes. Lösch is acutely aware that with Descartes the human being placed itself at the center, to establish an order through reason and rationality mediating the freedom to produce the consistently logical. This is precisely man as the copula, the *medius terminus* between what absolutely *is* (system) and what comes to be (reality). And more distantly, more faintly, yet still firmly implicit, the voices of the Greeks are heard to inform the long tradition of that Western world in which Lösch stands. For one who writes "man acts according to his *idea* of laws," and "What matters is that statesmen shall act correctly . . . ,"[6] Plato is present, whether *The Republic* is explicitly referenced or not.

That informing presence of the philosophical tradition is, in its essence, a presence of thinking struggling to illuminate the ever recurring tensions of the human condition—at least as that condition has unfolded in the Western world of the last two and a half millennia. It is a struggle of truth versus certainty, reality versus theory, and freedom versus system. It is the struggle to find a break in the circle of argument that leads thinking to reason and rationality, and so directly to logic—logic that becomes explicitly calculating in turn, and so transmutes the hope of reason into the determining tyranny of mechanism that contradicts the freedom that forms the condition of possibility for thinking itself. This same sense of struggle permeates the text of Lösch, not simply as a sense of intellectual struggle, as he attempts to widen the horizon of thinking in a limited domain of inquiry, but as a moral struggle that feels the contradictions even as they are being set down. It is in this deep sense that he is a child of his time, and his text may be read as a commentary upon that time—which is still ours.

On one side of the contradiction (if one may be allowed to speak this way), we have a man steeped in the tradition of German Idealism, a tradition in which knowledge must be absolute knowledge, a tradition in which certainty is sought through reason. But to be certain, to be in a realm in which certainty reigns, the structure of that realm of thinking must always be self-consistent, rational, and logical. At this point there is no other choice: the realm that forms the "world" of such thinking toward certainty must be, can only be, mathematical. A *system,* and this is precisely what Lösch is trying to construct, a "system is a mathematical system of reason. . . . [T]his system is *not* derived from experience; but rather set up *for* it."[7] It is in such an embedding ethos of Idealism that we find perhaps some of the most famous phrases of Lösch: "Comparison now has to be drawn no longer to test the theory, but to test the reality! Now it must be determined whether reality is rational,"[8] a passage that follows his comments on "capitulation to reality,"[9] and his reference to that "Reality of reason upon which comparably more depends in the long run than upon the reality of the factual."[10] Let me suggest that these are statements to which no thoughtful physicist could possibly subscribe. Can we imagine a Bohr, a Schrödinger, or a Heisenberg maintaining, as a general principle, that the result of an experiment in high-energy physics was not *rational*? It is true that there are a number of examples in the history of science where the results of a particular experiment have been discarded or ignored on deeply intuitive and insightful grounds. But even an Einstein, confident that astronomical observations during an eclipse would confirm his theory of relativity, was prepared to acknowledge the sense of humor of *der Alte* to give the mortal physicist a surprise. "In the field of natural science . . . nature immediately takes its revenge upon a wrongheaded approach."[11] Thinking driven in this way quickly comes upon the mathematical, and, therefore, upon the essential mechanism that underpins any notion of general equilibrium.

It is here, in the midst of his concern for general equilibrium and theory, that it is worth pointing briefly, but explicitly, to the structure of the mathematics employed by Lösch. It is above all a functional mathematics, grounded upon those structures that were devised from the seventeenth through the nineteenth centuries to describe the physical world, a world that permeates by analogy aspects of the human world that it is Lösch's goal to illuminate through such mathematical structures. At many points the analogies are stated forthrightly: price waves of spatiotemporal succession are like physical waves, and so invoke reference to Huygens's principle;[12] routes that minimized costs invoke the law of refraction, and point to a "significant identity between a law of nature and a law based on reason."[13] But such analogies are, in a sense, peripheral. The core of such thinking is grounded in the functional mathematics of mechanism, in which a general theory of location is cast as a system of simultaneous equations whose solution will lead not to partial, but general, equilibrium.

Yet this path only discloses another facet of the struggle: even in his discussion of industrial location theory, he notes that "algebraic treatment leads to equations of insoluble degree,"[14] and this irreducible fact that a multitude of local minima may conceal a global solution also plagues the discussions of agricultural and central place landscapes. Today, when we have allowed the mathematical possibility of the one-to-many mapping to enter the realm of scientific description, we know that the parameters of nonlinear systems may reach critical values to produce bifurcation. This is important, not because we could ever marshal concrete historical evidence for such a change in system trajectory,[15] but because such an essentially nonlinear view leads to symmetry breaking.[16] For Lösch, in his mood or disposition towards reason, rationality, logic, and mechanism, the breaking of geometric symmetry appears, in the quite literal sense, to be unthinkable. The advantages of geometric representation are noted,[17] and geometry is constantly invoked in the same breath as *analysis*,[18] by which term it seems legitimate to imply that Lösch meant a unique solution cleanly derived by pure analytical means, rather than by methods grounded in numerical approximation or combinatorial search. Indeed, and this is crucial to obtain the intellectual flavor of those precomputer days, combinatorial search is equated with "empirical testing" and "only trial and error,"[19] and this prevents, in turn, a "scientific solution."

In agricultural location, the same drive to a geometrical, analytical, and *therefore* (in Lösch's thinking) scientific solution, must also be abandoned, and the same mischievous combinatorial spirit plagues him when the production schedules for the individual farm are examined.[20] "Even the production system that would be most advantageous for the farm cannot be scientifically unequivocal . . . since innumerable combinations are possible."[21] Today, we give the name of linear programming to such a problem of combinatorial search over a large number of vertices of a convex set, produced by the simplifying assumptions of linear constraints that may approximate, as closely as we like, any nonlinearities we may be capable of specifying. Nevertheless, "science can still be of service," he maintains doggedly, and proceeds to the geometricization of landscapes by a mechanism of rotated lattices of varying central place meshes that preserve geometrical symmetry by a means that only a god twirling vast geometric roulette wheels could achieve. But even such a celestial method to achieve rationality is ultimately grounded upon what we might legitimately call a "geographic tautography"; namely, that any weighted point distribution in the plane may be divided into "rich" and "poor" sectors, when "rich," "poor," and even "sectors" itself are only weakly defined by an ordering relation. By definition, if I draw lines to make one sector as poor as possible, the adjacent sector can only appear richer, and so on around to close the disc. Thus, thinking reduces a human landscape to order (?) via an identically closed, and therefore ultimately tautological, system, as physics orders the physical world.[22]

But it is out of these struggles to create and preserve a system, a system of reason and logical mechanism founded upon the "unreasonable effectiveness"[23] of mathematical structures devised for purely mechanical systems (celestial, continuum, quantum, etc.), that we see an August Lösch in another mood and disposition. It is a disposition that allows a trap to be seen, even as the trap itself, granted and arising from the conditions of possibility of that "world," is being constructed. Perhaps it is not really too far-fetched to draw the parallel to Hegel's own sense of universal freedom in the midst of his drive to the systematic. And the fact that unfreedom increasingly appeared in this day, and is still amplified in our own, does not deny the principle. "So much the worse for the facts, he might say," says Gadamer, interpreting Hegel.[24] "So much the worse for reality," might have echoed Lösch. And *yet*, right there, at the very center of his concern, stand undeniably the same human beings who shape and give meaning to the economic landscapes that must be ordered by human thought directed *freely*. Even as he acknowledges that analytical solutions are "impossible,"[25] that trial and error (combinatorial search) is "unscientific," that "irregularity of conditions . . . prevents a scientific solution," and that "a systematic theory is impossible,"[26] the necessity of human decision intrudes, and with it that freedom that underpins the moral dimension. "Nature *must*, man *may*, act correctly" (original italics); the humanly created landscape differs from the physical world of nature "as the moral differs from the mechanical."[27] Planning enters, as a judicious control of the *free* market, and "levers" (Lösch's own word) are sought by which economic regulation can be carried out. Even as the human being demonstrates his freedom to choose, he reaches for the mechanical lever to secure his choice. The struggle between the moral and the mechanical, between individuals freely deciding and a totally planned order by "Them," is not over. On the contrary, it still permeates the discourse, as Lösch is torn between free market forces straight from Smith, Ricardo, and Malthus on the one hand, with a need for a higher-order economic policy grounded in spatiotemporal awareness on the other. Yet it is here, in these thoughtful irruptions of freedom into the logical drive, that we sense the grounding humanity and humane concern of the man himself.

There is, first, the sense that we are in the presence of thinking that displays theoretical concern rooted in real problems. In the same way that ontological concern takes meaning only as it seeks to illuminate the ontic realm, so a concern for theory only becomes meaningful to the degree that it illuminates the empirical. It is here that Lösch obliges us to take a hard look at ourselves, and reflect upon the degree to which theoretical concern may becoming increasingly disconnected today from any problematic other than the mathematical intricacies that it generates for itself. Lösch was unequivocal on this question: "Actual knowledge of a real area is the mainspring of research. It is the best guarantee that investigation will be directed toward actual possibilities instead of losing itself in a mere play of the intellect."[28] And when we come to his beloved Swabia, all doubt vanishes: an imposition of reason may disclose logical order, but no one is going to understand the Swabian economy, and its ability to weather economic storms, "without an understanding of Swabian philosophy."[29] By which it is quite clear that he means the Swabian "world," a "world" constantly maintained by those who are thrown into it, and who pass it on, refurbished and renewed, to their children in turn.[30] Such a sense of structural constancy and integrity in the social realm should resonate deeply in those who are concerned today to understand the maintenance and influence of social structures in their spatiotemporal settings.[31]

It is precisely this willingness, even this necessity, to bring into the economic realm an older tradition and meaning of the "philosophical"—the cultural, political, and thoroughly human social aspects of a region—that characterizes his thinking. Or, to put it another way, it is his refusal to partition the human complexity of real regions into separate, nonoverlapping boxes, boxes in which we and our students still sit today in the course catalogues of universities, that allows him to temper even the most severe geometrical constructions to generate a human illumination of a concrete situation. The landscape is not just a geometry, but is inhabited by people joined by a complexity of structural relations, not the least of which may be a deep sense of rootedness, of *Bodenständigkeit*, in the region itself.[32] Few had a deeper sense of such rootedness than Lösch himself, who constantly uses his Swabian homeland as an intellectual touchstone to test the truth of a humane assertion. It is this constant interweaving of the abstract and the concrete, the irruption of humane concern and insight into the most constrained geometry, that shapes his writing. His most strident statements of faith in deductively derived theoretical constructions immediately precede his most concrete examples. Theorems, whose human meaning can only be mechanistic determination, lead to his most undetermined, empirical, and freely generated North American examples. And along such a way, his own struggles in the midst of these apparent contradictions enlarge the horizon within which the thinking of others in this realm can move freely—although still within bounds that are always there. But thinking that moves in a newly opened and enlarged horizon may begin to glimpse things not seen before. These are anticipations.

Anticipations

Any work that genuinely enlarges the condition of possibility of thinking generates anticipations, things once concealed that now emerge, at least at the edge of the clearing. When Lösch writes "it is not easy to stand before this rich harvest with hands tied,"[33] we sense his own mood of anticipation, and wonder what possibilities he saw to which we may still be blind. Nevertheless, his text stands, and so allows us, in the act of retrieve, to think toward possibilities of our own, whether those were directly anticipated by him or not. Always we stand at the edge of a tradition. Hegel, Fichte, and Schelling may have disputed Kant, but they were the first to acknowledge that Kant placed them where they were. Lösch takes issue with the Webers and von Thünens of his day, but acknowledges willingly the issues they raised. In the same way, we can become more sensitive to the fact that even as we may disagree with Lösch, it is his thinking, work, and anticipations that allow us to stand where we do today.

In this sense, the text is very rich, although only a few points can be used here as illustrations. Long before ideas of entropy and relaxation times entered geographical descriptions of the human world, Lösch anticipated the effects of joining into a common union two areas that were previously "closed systems."[34] It is an anticipation of the approaches of Alan Wilson and Peter Allen,[35] as we live in the midst of a world today in which many territories have been rearranged and separated by changes in political boundaries. Both the tribes of Africa and the tribes of Europe have been severed and joined without thought for the human consequences or the spatially explicit dynamic effects. And if there are new centers today, there are also new peripheries, even as the centers, political as well as economic, may move, as Brazilians and Belizians well know. But all such rearrangements are the result of human decisions, and arise in the context of power and planning, and the policy implications that follow. But for

Lösch, policy must be both rational, based upon conscious and careful planning, and humane. One must be sensitive to the delicate fabric of the social condition in its spatiotemporal setting, to the unanticipated results of planning that Jay Forrester, thirty years later, would call "counter-intuitive effects."[36] This means that the fabric, in its social, political, economic, spatial, and *ecological* dimensions, has to be well described, and here we see Lösch anticipate explicitly the hierarchical nature of such descriptions. Long before Christopher Alexander declared that a "city was not a tree,"[37] Lösch knew perfectly well that a partitional hierarchy could not possibly describe the intricate relations that truly structure a genuinely informed planning process. Forty year later, we would realize that he was pointing to a hierarchy of cover sets, to be shown in all its graphic complexity, rather than in the obscuring tabular form he was forced to employ, a form that effectively concealed that which he had seen. In fact, he gave us a remarkable prescription for geographic planning at a comprehensive level, even if he lacked a mathematical form, and a high-speed computer, to make it operational.

Finally, on that last page entitled *Epilogue* (and here I wish to speculate openly), is it possible that we can hear another voice that informs and undergirds the caring and concern as Lösch leaves his reader? Is it not remarkable that upon this last single page, a "verbal finger" pointing to the unfolding of human society across space and through time, that we find the words *world, finiteness, horizon,* and *existence,* and a rejoicing that "geographical and cultural roots" may yet be preserved, that the sense of *Bodenständigkeit,* marking his own deep sense of Swabian being, may yet give meaning to a world uprooted?[38] No man who dedicates his greatest work to "the land of my birth, the land that I love" could have been untouched by his fellow Swabians, that same trio that touched a later thinker of Lösch's own time.

And so we see August Lösch as the child of Hegel, the great systematizer; of Schelling, who struggled to prevent the denial of freedom that system seemed to imply; and of Hölderlin, whose unbearable sense of the human dilemma led him to gentle madness, and to the poetry of his *vaterländische Gesänge* that illuminate a deeper meaning of the human condition.

Swabians all.

Floreat Swabia! Whose children, among them August Lösch, gave us so much to think about, and to be thankful for.

A Lasting Legacy

And gladly wolde he lerne, and gladly teche.
—Geoffrey Chaucer
Canterbury Tales

You have been running a section of memories entitled "I Think I Remember But I Could Be Wrong" for several issues now, but perhaps you can find room for a small *postscriptum* of commentary from someone who possibly holds the record for the length of time he was a cadet. I entered in January 1946, the youngest in the College, and left in December 1950, the oldest. As someone who has spent the past thirty-one years on the other side of the Atlantic River, I view many things rather differently today, and cannot, in all honesty, give my wholehearted approval to private educational institutions in a country that so desperately needs to unify its people, rather than produce even deeper distinctions between us and them. But these personal convictions about education aside, let me add a couple of thoughts to those contained in the last Headmaster's Report, and in the process acknowledge and honor the deep influence of two devoted teachers who served the College with great dedication and care during the years I was there.

To take the Headmaster's points: Yes, we do indeed "stand at a time when the advent of several new technologies . . . have begun to revolutionize industrial society," and we have the responsibility to help young people thrown into such a world cope with it and use it to human, and above all, to humane ends. So, good for computer studies. Yes, of course, we must learn to work with one another, and to common ends; that is what it means to be a member of a human society. So good for team efforts, on or off the playing fields, in classrooms or laboratories, and in all other areas of college life. Yes, we desperately need exposure to creative leisure activities, interests that will engage the heart and thinking, and enhance and encourage latent aesthetic sensibilities, to provide continuing joy in lives increasingly marked by glazed and passive eyes

Originally a letter to the editor of the *Pangbournian*, 1984.

staring at TV sets. So, good for drama rehearsals, musical productions, and engagement in all the many facets of our rich artistic heritage. But it seems to me that there are two higher priorities, two gifts that the college must try to give each boy and girl who passes through.

The first gift is the gift of language—the English language. It is the most priceless gift of all, the gift of clear and articulate speech and writing. We think in language, and in a very deep sense we are, as human beings, our language. No one knew that better than Fred Doyley (I have never dared to call him Fred before!), who read our weekly essays with such an eye of disciplined concern that you could mark your garbled syntax and split infinitives by the rhythmic pursings of his lips. Reading a boy's essay was not a matter of quick perusal, jotting down some marginal comments, and handing it back the next time. Your essay was read on Friday, the "reading day," and while others were immersed in Meredith (or other "sound" writers he approved of), you went to his desk and stood beside him. "Come, boy, this is not an English sentence. . . . Shouldn't this thought have come before that one? . . . What were you trying to say? . . . Well, why didn't you say it then!" The red ink flowed, and you returned clutching your bloodstained manuscript, wondering how all those awkward phrases and misspelled words had escaped you. Yet with luck and care, the day eventually came when his pen made only a few quick strokes (mere grazing flesh wounds), and with a chuckle he gave it back with a "not bad, . . . not bad at all." Those were the days when you floated back to your desk, and thought "Move over, Meredith, . . . here I come!"

What kept Fred Doyley going? Have you ever tried to read hundreds of badly written essays in a term? Term after term? Year after year? In such situations and walks of life only one thing can keep you going: you have to care, deeply and passionately, for the English language, and see it as a priceless heritage that you can, with goodwill on either side, pass on to the next generation. Fred Doyley cared, and I bless his name and his caring. We need such caring now.

The second gift is the gift of disciplined inquiry, and this was Ernest Beet's world, the world of classical physics, a human endeavor carefully constructed, and then tested and modified by evidence and experience over the centuries. Books were, of course, necessary, and you were wise if you read the assignments on time, and did the problems to hand in when they were due. But entrance to this world of rational exploration was not to be gained simply through books, but by seeing and doing. Ernest Beet's performance with sticks of glass and amber and small sheets of gold foil in glass bubbles was magical, and I often wondered if he practiced before a mirror—like the magician he was. But insight also came from doing, and as a teacher I marvel at the care, devotion, and thought he put into the series of carefully graduated experiments he designed. I did not become a physicist, but in an amateur way I have tried to keep up with those relativistic and quantum worlds in which thinking the unthinkable constitutes normal, everyday, but always disciplined inquiry. I can also feel his influence in my own work, and perhaps the best geography today really is disciplined exploration in another realm of human curiosity.

To weather rough and changing seas, ships need firm but sensitive hands on the wheel. Sensitive, because a new and sudden breeze can be used to good effect; firm, because the home port of an education remains constant. Those who teach at a college with a nautical tradition know this better than most. Fred Doyley and Ernest Beet certainly did.

Thinking about Thinking

THINKING about thinking is the most difficult thing in the world. It means, in some very strange and quite contradictory sense, that we have to get outside of ourselves and think about what we are thinking about as we think we are thinking about our thinking. Aaaagh! The trouble is that as soon as some philosophical masochist raises the question, you find that it itches so much that at some point you have to scratch it. Or putting it more conventionally, until you start to reflect upon it seriously. Of course, you do not get very far, particularly at the deepest and most difficult levels, but sometimes the exercise turns out to be worth it in more mundane ways.

Classification, and the sort of thinking that lies behind it, is a case in point. From Aristotle's *katagoria* onwards we have known that we have to name things, assigning some things to the category "trees," others to a box we label "fishes," and so on. Carl von Linné, better known as Linnaeus, is rightly respected to this day for sorting out the living world into taxonomic boxes, distinguishing at ever-finer levels plants and animals until the twigs on his taxonomic tree stop at the point where only the plants and animals there can sustain their existence by reproduction. Mules are a bit of a problem, and ligers and tions are possible, but generally things look pretty orderly, although experts still wrangle over particular bugs, and DNA analysis has illuminated differences and similarities in ways not possible for Linné.

The problem is that the whole question of classification, the act of assigning things to separate boxes, has become mechanized with computer algorithms, and in a very real sense has become thought-*less*. By this I mean you can put any set of jumbled-up things into a computer, define some measures of similarity, and the algorithm will, *must,* partition the things into subsets that can only be a product of the particular clustering algorithm you chose to use. No matter how tightly related or similar some things are, all taxonomic algorithms will hack them apart and classify them for you. And the classification is quite "objective"; indeed, it is untouched by human hands, let alone thinking, because it was done numerically, and therefore scientifically, by a machine. I mean, how objective can you be? The result is that thoughtless people, far removed from the sort of concern shown by Linné, shove anything through the machines these days to produce nonsense all beautifully and scientifically laid out on the computer sheet. Foraminifera, little tiny animals who leave their skeletons of calcium carbonate on the sea floor, are not the most exciting things for many people, although they are important to geologists, particularly petrologists. But they do illustrate, in a quite concrete way, how much conceptual damage can be done by machines that cannot think.

Machines do not think, and all sorts of problems arise when thinks begin to machine, when human beings start acting mechanically without thinking. This is especially so when the objects of concern are people—who are *not* things. Many of the early approaches to psychoanalysis constructed quite rigid frameworks out of the nineteenth-century heritage of mechanical biology to use in treating a patient. Too frequently a "cure" was effected when the patient agreed with the blinkered diagnosis of the analyst; and frameworks, devised out of a highly limited set of patients, were then applied in totally different circumstances. Classifying human beings in need is a difficult and dangerous game, particularly when shock, drug, and other treatments are predicated upon the taxonomic box to which a person is assigned.

And then there is another form of thinking, much more everyday stuff, much closer to home. It appears when someone says, "Well, what do you think?"—meaning what is your *opinion* about some situation, person, country, policy, academic trend, and so on (there are lots of things to have an opinion about, to think about in this limited sense). A university should try to produce independent, self-confident thinkers—or what's a university for? A democratic society, one that really works despite all the difficulties, requires independent thinkers, people who can weigh up this opinion and that, and then come to what they feel is a defensible, although perhaps always open, position. This is what I tried to say to the new students at Clark University during the convocation ceremonies that opened their first academic year. It was not a particularly popular address with everyone in the audience, but independent thinking, thinking for yourself, does not mean thinking like everyone else.

Do Foraminifera Assemblages Exist—At Least in the Persian Gulf?

A ballet dance of bloodless categories.

—Francis Bradley
Principles and Logic

Partitions and Kategoriai

WHENEVER we approach a set of things (animals, archaeological artifacts, rocks, poems, cities, clouds, books, . . . what you will), in what I can only call a "classificatory state of thinking," we tend to be oblivious to the way we accept unthinkingly a lot of assumptions from the world into which we have been thrown. We take it for granted that when we have to classify things (all too often because we cannot think of anything more imaginative to do), we are going to put similar things together in one box, and different things in other boxes. In other words, we are going to take some initial unsorted jumble, and sort everything out into quite separate piles, a process a mathematician would call formally "creating a partition on the set with an equivalence relation." This means that each thing is safely tucked away in its box with similar companions, and quite separated from different things in all the other boxes. This is so naïvely obvious that few of those who have been intellectually potty-trained can see what the problem is. After all, what *is* classification if not dividing things up and separating them into similar and dissimilar groups?

For much more reflective reasons, this type of partitional thinking has a long history in philosophy, particularly from the time of Aristotle's *Kategoriai* onwards. Much human meaning depends on it, and in different ways it is reflected in all human languages. We categorize things into "trees," "clouds," "people" (who should never be thought of as "things"), and so on, for we could

---

Written for this volume, but drawing upon the scientific example in my essay "A Structural Language of Relations," in *Future Trends in Geomathematics*, ed. Richard Craig and Mark Labovitz (London: Pion, 1981), 281–31

not have human communication without such general terms or categories. Incidentally, these are not given by the gods; we create and come to accept these as we need them. The Ojibway hunter has lots of names distinguishing different types of things that we lump together as "rapids" on a river, the Eskimo's propensity for naming lots of different types of snow is well known, and the Kalam of central Papua–New Guinea refuse to classify the cassowary as a bird.[1] When we come across unfamiliar things, we tend to call them "thingamajigs" until we learn that they are sitars or quarks. Scientists especially are continually discovering or making strange new things, and they either make up new names for them or tuck them away in ready-made boxes with more familiar, often higher level, names—subatomic particles, galaxies, beetles, various sorts of viruses, and so on. Occasionally, a new thing does not fit a previous category because it is not similar enough to things we already know, so to keep things tidy a new box is created for it.

Linnaeus, Darwin, and Computers

Three historical events of the past two hundred years have encouraged this rather strong human propensity to divide the world up into lots of nonoverlapping boxes. The first was the extraordinary ordering of the living world by the Swedish naturalist Linnaeus (Linné), choosing as his criteria morphological things he could observe—the shapes of bones, teeth, and other anatomical details for animals, and a rather strong focus on sexual organs for plants.[2] Very general categories or partitions like "birds" were broken down and partitioned again into orders like *Strigiformes* (owls), *Falconiformes* (falcons), *Passeriformes*, and so on. These were partitioned again into families like *Tytonidae* (barn owls), *Vulturidae* (vultures), and *Corvidae* (crows); and the latter into ravens, jackdaws, jays; and the latter into the European jay, blue jay, Stellar's jay, scrub jay, pinyon jay, Mexican jay, . . . until observing ornithologists could agree on no more significant distinctions, or quit out of sheer exhaustion. You created a hierarchy with very general terms or categories at the top, and these were split or partitioned more and more as you went down to the most specific level. After Linnaeus and his followers had done it, it all seemed perfectly "natural." In fact, Linnaeus thought he had discovered the order and beauty in God's handiwork.

Then along came Darwin, well versed in Linnaeus's partitional type of thinking, and the hierarchical splitting got turned upside down into the familiar tree-like diagram everyone knows today. In biological and evolutionary terms, it made perfectly good sense, because your finest categories, the boxes that were attached to the twigs at the ends of the branches, contained things that needed each other to reproduce and so maintain their very existence. Mules were a bit difficult, existing in a rather strange box of their own, but needing things from the rather blurred donkey and horse boxes next door. Ligers and tions were unknown at the time. The fundamental rule seemed to be that you kept splitting and partitioning until you came to a point where the things in your box could not reproduce themselves faithfully without each other. Well, fair enough: in biological and evolutionary terms, the familiar taxonomic tree made good sense. Even today, viral geneticists and molecular biologists use such tree diagrams to show how one sort of influenza or immunodeficiency virus differs from another, although at the molecular level we know now that the RNA strands can sometimes exchange bits and pieces all by themselves, and so produce new, somewhat different types.[3] Some of the viruses seem to be evolutionary adaptive machines; unlike the leopard, they constantly change their spots to find a niche somewhere in immune systems in order to survive.

But the virologists and molecular biologists today rely almost entirely on the third historical event shaping the partitional way we think about classification—the modern computer. With its huge capacity and enormous speed, it allows us to transform a table of things (rows) and their characteristics (columns) into spaces of various dimensions in which the things we want to classify have *locations* specified by coordinates $(x,y,z, \ldots)$.[4] And once things have locations in a space, they each have *distances* to all the other things. And once they have distances, then things close together are perceived to be more similar than things far apart. Unfortunately for "objective" science, the spaces, distances, and similarities all depend on what characteristics you choose to record about the things in the first place, but assuming you can make a persuasive argument for these (rhetoric is never far away in science), you can start grouping things together that are close to each other. The grouping process, based on distances measured in a multidimensional space, is usually represented by a taxonomic tree in a two-dimensional space—the usual flat piece of paper. Nobody asks whether a tree ought to be in a three- or four- or higher dimensional space. Such questions allow thinking to escape from the accepted paradigm—what Thomas Kuhn called "normal science." Clearly, people who rock scientific boats should be thrown overboard.

But it is precisely here, in the so-called clustering algorithms, that things start getting tricky: once two or more things get clustered together, you have to decide what location will stand for the whole cluster. And notice that this location will determine its distance to all the other things, some of them already clustered together, others waiting patiently by themselves to be clustered, like shy people waiting to be asked to dance. And what happens if your rule about assigning new locations and distances (and algorithms are precisely that, a set of absolutely consistent rules that a machine can follow), produces long, snake-like chains in the space, or even brings some things closer as a result of one step in the clustering process than they were before the step was taken? That would make for very awkward "trees," with branches bending backwards to catch the things that were suddenly brought closer as a result of the last step in the algorithm.

And so more rules are added, more bells and whistles attached to the algorithm, in order to *force* a partition on the set of things. Or rather, all sorts of partitions, because where you choose to cut horizontally across your taxonomic tree will depend on how many partitions you get. Too often, various clustering algorithms are tried on a set of data until the result more or less resembles what you wanted in the first place. Except that when you publish your results everything has been done "scientifically" and "objectively." The problem, of course, is that no one asks whether it is appropriate to force a partition in the first place on the set of things to be classified, because classification itself has come to be defined as a partition on a set. Once you are willing to enter that tautological loop, it is difficult for thinking to escape. Perhaps, in the biological world, where the question of self-reproduction can arise as a basic criterion to stop the partitioning process, this sort of unthinking, unreflective, but always "scientific" training does little damage, but its effects can be devastating when it moves sideways into totally different topics.

Only think of the problem of classifying a book, something librarians have to do every day, often with key words that might help a person search for it and the subject it is about. The trouble is that librarians cannot read every book that comes to the library, and often titles are not very informative about the real contents, so librarians have to make a guess at a few words. Sometimes they are right; sometimes, inevitably, they are not. But each of the words they choose can be thought of as a link or connection to other books on the same topic *or set of topics connected together by the author.* It is precisely the *connections* between topics, and how they connect to other books, that you want to preserve, and it is precisely these sorts of connections that are severed whenever partitional thinking forces some books into some taxonomic box and other books into other boxes. Is there not some way we can keep the connections so that potential readers can explore the structures these create, moving around the interconnected pieces until they find what they are looking for, or, even better, until they stumble across a piece of the interconnected structure they never suspected? This is one of the great advantages of browsing in the more conventional sense along library shelves; serendipity is often the result.

So the question arises: If we really want to describe things as faithfully as we can, and see how they connect to other things, is there any way we can do this in order to preserve the useful "structure of connection" that partitional classificatory thinking tears apart to satisfy what appears increasingly to be an archaic "tidiness complex"? Before the days of computers, the answer would probably have been "No," except for the very smallest of problems; in other words, just a few things with a few characteristics. But now computers allow us to deal with essentially any number of things described by any number of characteristics. The trouble is that instead of using the computer to free ourselves from old-fashioned partitional thinking, we use it to do what we always did, only faster. Once you sit on that tautological pot, it is very difficult to conceive of other ways of hygienic clarification.

Keeping the Connections

In the early 1970s, a mathematician at Essex University, Ron Atkin, tried to develop a mathematics for structural description out of his work in algebraic topology.[5] In other words, if you were prepared to talk about such things as "social structure," "regional structure," "the structure of a curriculum," "the architectural structure of a town square," and so on and so on, then you owed it to yourself and others to make that intuitive notion of structure as precise, well defined, and useful as possible. Otherwise, you reduced your perfectly good intuitive sense of structure, and why and how it might be important, to pretentious cocktail party chatter and jargon. All the details and ramifications of his "language of structure" are not important here. What is important is that you were obliged to define the set of things you were talking about in a clear and unambiguous way. Equally, you had to define the descriptive terms with great care, often sorting them out into more general terms at high hierarchical levels that Atkin called "cover sets," with more specific terms lower down. A simple, intuitive example would be a set like {algebra, arithmetic, mathematics, geometry}. Obviously "mathematics" belongs at a higher level than the other three, and in that superior position it "covers" them.

Once you had your set of things well defined as the rows of a matrix, with the columns the descriptive terms carefully specified at the same level of generality in your hierarchy of cover sets, you could describe each element of the set as a little geometric figure (a line, a triangle, a tetrahedron, etc.) called a *simplex*, whose vertices were the descriptive terms. If two things shared, say, three descriptive terms, they were joined together by the triangle formed by them, or a two-dimensional "face." Taken altogether, your set of things (simplices) formed a structure, called by mathematicians a "simplicial complex." Obviously, if they shared lots of descriptive terms, they could form up into a single structure, although one that could take many different shapes and dimensions, and even have holes in it. Atkin called such a well-defined structure a "backcloth," while other things, which might need such a backcloth in order to exist, he called "traffic." Clearly, if the traffic was going to be transmitted from one part of the structure to another, you had to have sufficient connections in place.

Suppose we think about books in a library as a well-defined set of things, and the key words (carefully sorted out into a well-defined hierarchy of specificity) as the descriptive vertices. It is easy to see that all our simplex-books are going to be connected together by the key words they share into a simplicial complex, our backcloth. At this point, we could think of ourselves as traffic needing such a structure to exist as scholars, while our browsing is simply an exploratory transmission of our thinking as the traffic. Notice that if lots of scholars tend to end up as traffic on one piece of the backcloth, then the citation indices tend to increase as yet another form of traffic. Moreover, if a subset of books on some, literally esoteric, topic is tightly connected together, with little or no connective descriptive "tissue" to other books in the library, then they will form a disconnected, i.e. severed and partitioned, piece of the structural backcloth. With no structural connectivity in place to help you jump over the gap to that disconnected bit, you better be part of that small group initiated into that rather rare subject—which is exactly what *esoteric*, as opposed to *exoteric*, means. What is important for our purposes here is that such a description certainly allows the backcloth to fall into separate pieces, or, in traditional terms, partitions severed from the rest. But it does not *force* them on the set by doing all sorts of dreadful and undefined algebraic operations, squeezing perfectly nice and honest topological spaces down into Euclidian metric spaces, in which totally artificial "distances" are then used to cut a connected whole up into little pieces for conventional box stuffing.

Foraminifera Assemblages[6]

To see what dreadful things conventional partitional thinking can do, I want to take a rather esoteric topic that will illustrate a rather exoteric conceptual lesson. Foraminifera are minute organisms living in the sea, and when they die they sink to the bottom, where their skeletons of calcium carbonate, some as beautiful as snowflakes, accumulate. Years later, in seabed samples, or in chalk or limestone cores taken from the seabed, geologists think the various types of foraminifera are useful indicators of the conditions under which the ooze or sedimentary rocks formed. They also speak of *assemblages*, groups or clusters of foraminifera appearing together, and they interpret these as being even more specific indicators of seabed and other environmental conditions, like temperature and depth of water.

In the 1950s, the geologist Jacob Houbolt took 137 grab samples from the floor of the Persian Gulf,[7] and analyzed each of them for the various amounts of foraminifera they contained, labeling them on a scale of 1 to 6, depending on whether a foraminifera type was single or rare (1), or whether it formed up to a quarter of the sample (6). Houbolt claimed to have seen various distinct assemblages all over the place. A decade later, another geologist, Graeme Bonham-Carter, took Houbolt's detailed monograph as a concrete example of how to use the new computerized clustering algorithm that was then all the rage in numerical taxonomy.[8] For some reason, perhaps because only a small pedagogic example was required, he chose only 70 out of the 137 grab samples, and then squashed down the scale of six intensities of foraminifera to three. He wanted to simplify this already reduced ordinal scale still further in order to use a coding scheme that let him measure the degree of relationship between each grab sample with something called a Jacquard coefficient, a value ranging from 0.0 to 1.0. So eventually a 70×70 matrix of coefficients was fed into the machine, and the cluster algorithm made several "passes" of the matrix, each time grouping together the most closely associated samples, and then substituting a weighted mean to represent the clusters formed. The conventional taxonomic tree was produced, and the branches sliced across at an association value between 0.34 and 0.41 to produce seven clusters or partitions, A, . . . G.

Before we examine these clusters more closely, a number of things are worth commenting upon. First, and although the cluster analysis was based on a simple ordinal scale reduced from an original, more discriminating one, calculating a coefficient of association immediately produced a number that was combined with others to compute, for example, a "weighted mean." However, this involved arithmetic operations like division, which is defined in a metric space but has no meaning in the ordinal space grounding the analysis. Secondly, the algorithm is a deterministic partitional machine, which means that any set of numbers can be substituted, including those simply chosen at random, and a taxonomic tree partitioning the set of grab samples will always be produced. Thirdly, taking the 70 samples and subjecting them to such an abstract analysis tears them out of the specific context of their locations in the Persian Gulf. Location, the "where" of the grab samples, is completely obliterated by moving from geographic to "computer" space.

Now suppose instead we start a structural analysis with a matrix of 70 rows of grab sample observations, and 20 columns of various types of foraminifera, allowing the ordinal scale 1 to 3 to designate, in a rather crude way, the intensity of their presence in each sample. We can "slice" this matrix at 1, 2, or 3, allowing each grab sample to be defined as a simplex over the set of foraminifera vertices, depending on whether we want a single specimen to form a vertex (1) and so a possible shared connection, or whether we require more than 100 specimens (3). If we slice at 1, it means that even the single or rare foraminifera in a sample will form one of the defining vertices in its simplex representation. To analyze this complexity, we can think of putting on a pair of spectacles that allows us to see sample-simplices of varying dimensionality. For example, if we look through lenses of 19-dimensional power, we see nothing at all, because no grab sample contained all 20 foraminifera to form a 19-dimensional simplex. But as we gradually lower the dimensional power of our lenses, at last a grab sample called RC400(A) comes into view at dimension 13. The (A) in RC400(A) simply means that it is eventually going to end up in the algorithm's A-group.

The trouble is that as we continue to lower our dimensional lenses, by the time we reach those of 9-power, eight of the eventual 15 samples in the algorithm's A-group can already be seen, but they are in two separate "clusters," while six of the 15 samples in the algorithm's D-group are joined together. But we have only to put on 8-power lenses to see that these all join up into one group, meaning that they are all connected by at least nine types of foraminifera in common. Yet these highly connected sample-simplices are torn asunder by the mechanism of the taxonomic algorithm, even though it means destroying all their connective tissue to force a partition where none exists.

In fact, it does not really require any dimensional or structural analysis to see this. The geologist arranged his taxonomic groups in such a way that they implied that they were controlled by the depth of water (fig. 24), but just scanning the diagram by eye, we can see the total overlap between A and D, E and F, and G and C, and a considerable overlap (connection!) between all the others—which is exactly what our structural analysis tells us. In fact, it does not matter where we slice and analyze, there is not a single shred of evidence of any distinct "assemblage" at all. Even when we go back to Houbolt's original monograph, and analyze all 137 grab samples over all six of his "intensities," no assemblage partitions appear, although at one point a very interesting and provocative piece of structure does pop out. When we slice at 3, and put on our 6-power lenses, a subset of five sample-simplices comes into view, linked together by seven, and even eight, different types of foraminifera. Back in geographic space, it turns out that four of these were taken only 2–3 kilometers apart on the Shah Allum Shoal, and there is evidence at other dimensional levels that samples taken near the Halul, Cable Bank, and Shah Allum salt domes tend to cluster together, perhaps because these areas represent specialized environments. Yet the computer algorithm tears these apart and puts them into A and D partitions! It also takes the two samples, numbered 539 and 540, taken next to each other in 10–11 fathoms, and assigns them to clusters A and E. One is reminded strongly of a numerical taxonomy to produce a hominoid classification in which female chimpanzees, orangutans, and gorillas were clustered with male and female human beings before joining the male partners of their own genera.[9] Of course, one must be broad-minded these days, but clearly the partitional mechanics of some algorithms can produce strange bedfellows.

As for the relationship to the depth of water, when a plot is made of all twenty foraminifera found in all 137 samples (fig. 25), it is difficult to see any

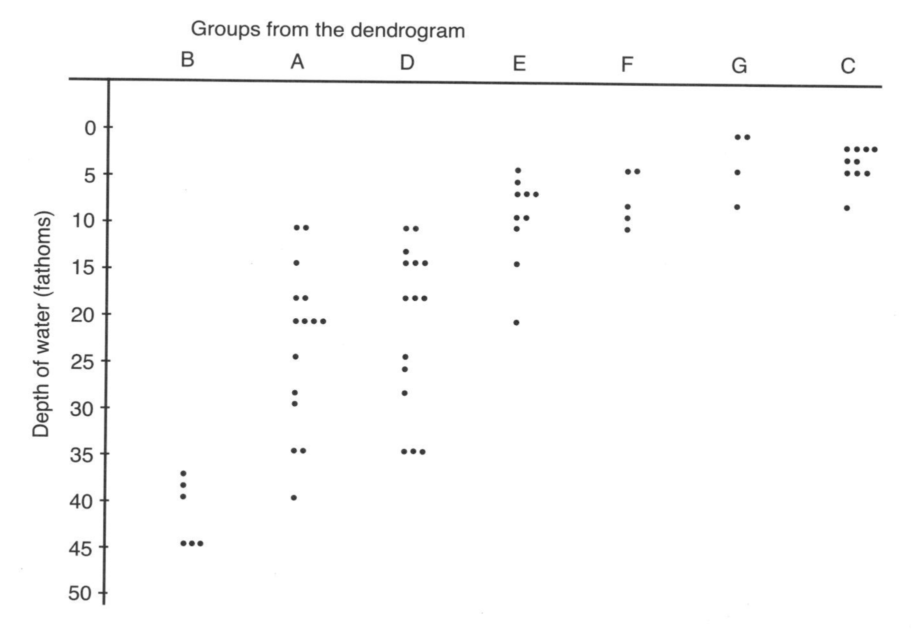

Fig. 24. The plot of the A to G clusters produced by the computer algorithm against the depth of water.

implied functional relationship to anything. The depth ranges are large; rough upper and lower limits can be drawn for about half the foraminifera, but even these are wide (10–40 fathoms); samples are not equal at each depth; and densities do not appear to vary significantly. In other words, what we have here is a piece of algorithmic partitional nonsense. The process reminds one of a lunatic hacking apart a pumpkin with a broadaxe, and notice how intellectually seductive the results are. Points that are close together in the pieces hacked apart were definitely close together in the pumpkin, no matter *where* the cuts were made. Because one fact is tautologically indisputable, given our original, tightly connected "foraminifera pumpkin": No matter what clustering routine is applied, points close together in the space (pumpkin) will often appear in the same groups (pieces hacked apart). But the real question is whether we should be hacking around with pumpkins in the first place.

Let us recap, and take any of the typical problems for which numerical taxonomic methods are used. Information is recorded about a set of objects over another set of variables. Mathematically, and often using undefined operations that would turn the hair of any competent mathematician white overnight, an orthogonal, metric space is created in which the objects are assigned locational coordinates. Then some sort of rule is applied to group or cluster the objects that are close together. If the objects really do form small and distinct groups in the space, which in all the textbooks they do so obviously that a child of five could point to them, then we may have evidence for a genuine partition on the set. And if this is truly the *structure* of our data set, a structural analysis keeping careful track of all the connections will certainly tell us this. In somewhat formal terms, a partition can be considered as a special case of a cover. It is certainly possible that the tree-like diagrams, with continuously splitting branches at each level of the hierarchy, are truthful descriptions, and we do not rule them out. But in fact, most of the time they are imposed by thoughtless partitional thinkers. Not surprisingly, they litter the organizational charts of companies and other authoritarian institutions—the private reports to the sergeant who reports to the lieutenant who reports to the major, and so on—but anyone in such an organization knows that such a chart has little bearing on how the organization really works and gets the important jobs done. Perhaps we should pay more attention to what actually lies in front of us, and approach it respectfully, with a sense that if we are open, and do our best to describe faithfully, we may get a much more truthful, and so richer, view of the world. Then the word "scientific" may once again have some real meaning.

And we might even find the book we were really looking for.

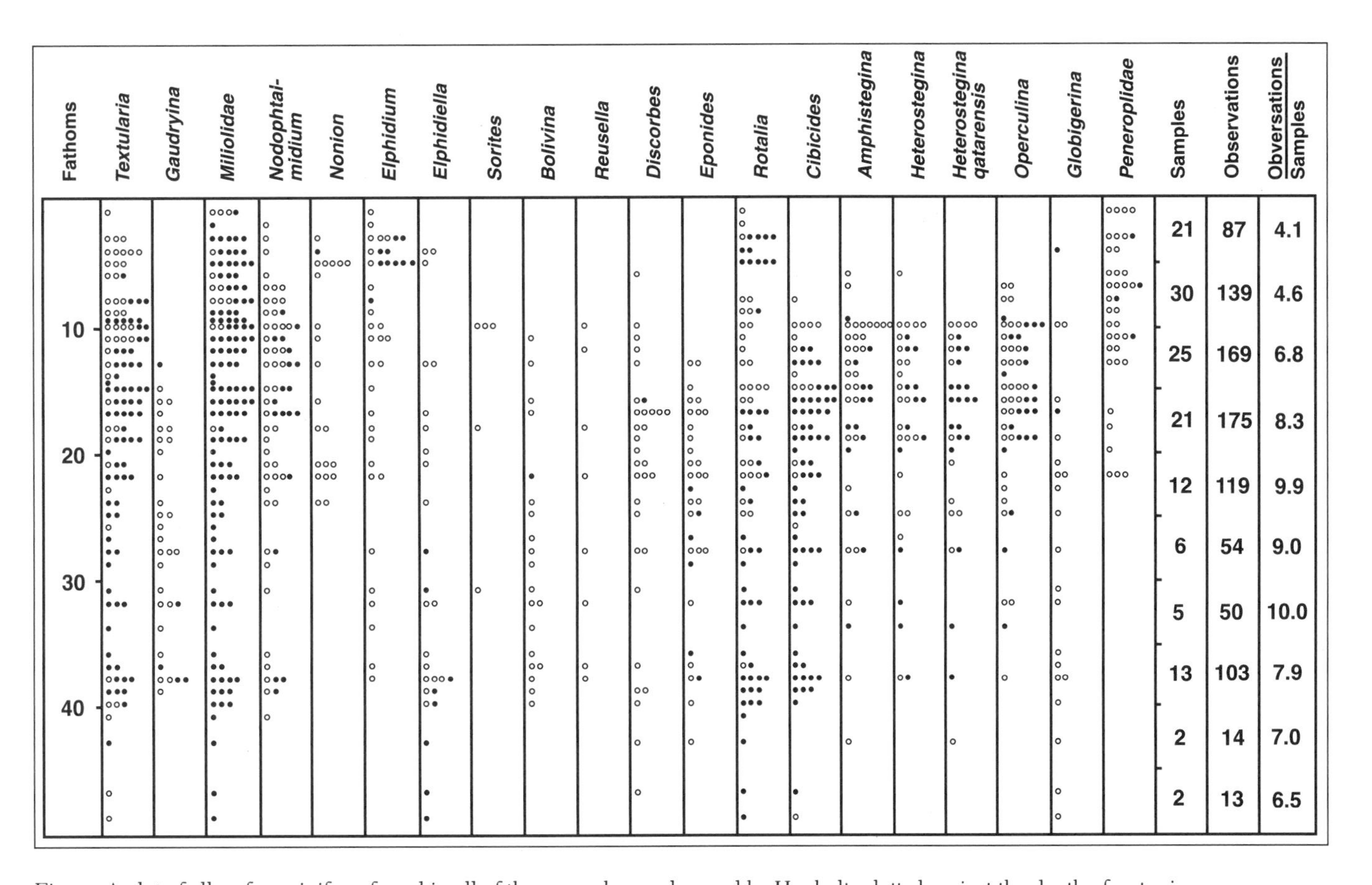

Fig. 25. A plot of all 20 foraminifera, found in all of the 137 grab samples used by Houbolt, plotted against the depth of water in which they were found. The reader is invited to find *any* functional relationship implied by fig. 24.

Thinks That Machine

The psychiatrist is too preoccupied with categorizing the behavior of the patient to notice that the "psychotic" utterances of the latter are reasonable, if disguised, objections to being merely classified and not treated as a person.

—Peter Lomas
Psychoanalysis Observed

LIKE ALL PEOPLE at all times, we find ourselves today thrown into a world that is not of our own making. We are no more responsible for our robot production lines, computers, or atomic bombs than a person cast into the world of the fourteenth century was responsible for the shipyard, the astrolabe, or the crossbow. But even if historians trace the faint beginnings of our modern world from that earlier age, it is clear that the two are radically different. Our world today is permeated almost to the point of saturation by technology, and that implies, in a most fundamental sense, that it is characterized by the machine. It is, in its essence, a mechanistic world, and that means a world that tends to direct our attention and thinking toward mechanistic descriptions, logics, and solutions—what we sometimes call the "technical fix." Thrown into and made human by this world, how could it be otherwise? And even if it becomes such an anathema that we reject it, our very posture of rejection discloses our presence in the world of technology, the fate given to us to deal with as best we can. After all, we have no other.

But loose and imprecise talk about responsibility and fate can quickly become irresponsible and fatalistic. To say we are not responsible for the mechanistic world that is our fate does not imply at all that we do not care, that we turn away from it, or that we think of ourselves simply as puppets pulled by its strings. When the philosopher Friedrich Nietzsche wrote joyously of *amor fati*—the love of fate or destiny—he was declaring no sense of resigned impotence or Lear-like railing, even if there were moments in his own life when

Written at the invitation of the editors of *Integrative Psychiatry* 3 (1985): 229–232.

he knew both. Rather, he was advocating a thinking, reflective, and therefore thoroughly human engagement with the world in which we find ourselves, an engagement characterized by that sense of caring that is constitutive of being human.

What does it mean to reflect upon the world of mechanism when we are prepared to think within the realm of caring for others who have severe difficulties coping with that world? This is not an easy question to answer directly and immediately, for how can thinking schooled in mechanism find an Archimedean point outside of the world upon which it tries to reflect? Our technological world is a scientific world, and as each physical, biological, and human science has successively hived off from its origins in philosophy, so it has become more removed from the tradition of reflecting upon itself—what it is doing and where it is going.

The framework adopted or inherited from the sciences today remains a mechanistic framework, one that has its origins in the earthly and celestial mechanics of Galileo, Newton, and many of the other great figures forming the scientific pantheon that is one of the glories of human endeavor in the Western world. But it is always and essentially a framework for thinking about things, and we must remember that all the theorizing and all the mathematics were originally directed to understanding those nonsentient and nonconscious things for which self-reflection and meaning has . . . no meaning! So familiar is this mechanistic world of science today that we sometimes forget how theorizing in mathematical ways came to be unthinkingly accepted. Theorizing is generally considered a contemplative activity, and in the old Latin root of *contemplari* we find a clue. For this is how the Romans translated the Greek *theōrien*, and in the process distorted its meaning almost beyond recognition.[1] For *contemplari* means to cut something off, to separate it, to partition it, and we still hear the older meaning in our English word *template*. But what is a template but something we use to force the same shape, to make things conform to the framework we have previously brought to them? Our theories are a priori frameworks to which things must conform.

As for the mathematics brought along with this thinking, much of it is simply a child of the same templating approaches that require conformity. It is basically a functional mathematics that requires everything to balance around that equality sign. Write a check on the right-hand side, and the balance on the left must also change in order to conform. Push the lever down on the right, and the load on the left-hand side rises. Turn the cogwheel on the right, and the gear on the left turns, too. So much of our mathematics has its origins in the same mechanical world, the world of the machine. But what happens to people when we try to describe them with the mechanical mathematics of the function? Will they too begin to look like machines? As the psychotherapist Medard Boss noted, "People who live in the technological age are at home with machines and even prone to draw inferences about human nature from them."[2]

Naturally, some of these a priori theoretical frameworks are very useful and we cannot live without them. But the real question is, can we live decently within them? Listen to the way our technological, scientific, and mechanical dilemma is poignantly impaled on the double meaning of the French word *grille.* It can mean a framework for thinking, but it also has the connotation of bars over a prison window. To what extent are the frameworks we create for thinking also our prisons? Of course, we realize these frameworks are not immutable. Of course, as scientists, we recognize that eventually they break down, only to be replaced by more secure viewpoints. After all, science is always open—at least, so the myth of science asserts. Yet even as we nod our heads gravely in public assent and approval, we know in the quiet of the night the enormous power that such a priori constructions have over us, how they represent a "pre-structuring" of our thinking, and how difficult it is sometimes to let go of them. We all become *involved* in these preset ways of looking at the world, and I mean that in the literal sense. Volvos are not just machines from Sweden: we are all in-volvos, all rolled into, wrapped up in, the a priori structures that have such personal meaning for us that they become a part of ourselves and our way of seeing. Thrown into a mechanical world not of our own making, to what extent do we become machines, preconditioned, preprogrammed to think mechanically along certain lines and not others?

"Yes, yes, that's all very well," said Candide in an exasperated tone to a pontificating Dr. Pangloss, "but I think it's about time we went out to dig the garden." Fate, technology, machines, and mathematics; Greek, Latin, and French words; philosophers and a priori frameworks . . . what has all this got to do with psychiatry—integrated or dissipated? The answer is that even psychiatrists are thrown into worlds, and as children of those worlds they are not immune to their influence.

We only have to think of Freud himself. Coming from the tradition of nineteenth-century medicine, with its almost total admiration for a classical mechanics of causes and effects, forces and fields, Sigmund Freud felt constantly obliged to emulate this powerful example, even as his thinking toward new realms of inquiry marks him as one of the major figures of his century. But sometimes his templates were cast in reinforced concrete, so that even the difficulties of little five-year-old children were forced to conform to the death wishes conjured from an adult world.[3] Apostates, whose thinking led out of the natural scientific framework with its mechanical requirements toward the truly human experience, were dealt with harshly, even as acolytes tried to reinforce the model of mechanism, postulating a physical mind of parts and pieces where forces acted, memories were stored, and thoughts and experiences pushed underneath to be repressed. Today, a Jacques Lacan declares that any denial or criticism of the "Unconscious" is itself proof of an unconscious, which is required as a place to which things can be consigned to be repressed. Criticism of an initially tentative theoretical construct, for which there is not a shred of evidence,[4] is now interpreted as proof of the existence of the construct itself. It is difficult to conceive of a more impregnable tautologic defense, except that we must remember that defenses designed to keep a world at bay also lock us into the fortresses we have constructed. Ironically, the image of the defensive ramparts protecting the mechanistic theory is Freud's own, one he used frequently to characterize the psychiatrist's assault to break through a patient's resistance.

It would, of course, be grossly misleading to characterize all psychotherapeutic practice today in these terms, for there are clear indications of movement away from the machine-like image. In the open accounts of an Anthony Storr,[5] who recognizes the influences of men such as Laing and Szasz, we can acknowledge much more sensitive attempts to think through to the human reality rather than the machine surrogate. And from the Zollikon seminars of Martin Heidegger,[6] and the tradition of existential and Daseinanalysis, we see a grounding that constitutes a radical replanting of the psychiatric encounter directly into the realm of the human being, rather than in the natural scientific world of things.[7] Not all the confrontational and authoritarian aspects disappear immediately as a result, but released from the necessity of forcing a patient's experiences onto a preconstructed theoretical framework, a psychiatric encounter approaches more nearly the authentic dialogue that allows genuinely meaningful interpretation.

It seems to me legitimate to take a step back, and view such reappraisals and reflections as purposeful attempts to demechanize psychiatry, even as we acknowledge that this movement takes place partly as a reaction to an increasing trend in the opposite direction. This opposite tendency is also a child of its time, one more indication of the steady blurring of the lines between the mechanistic world of things and the world of human beings. We see it everywhere. For example, the way teenagers plug themselves into TV games is causing great concern, but this is only one of many instances where the machine becomes an integral part of what is euphemistically called re-creation. Snowmobiles and jet skis shatter the silence and pollute those tranquil environments where genuine recreation might strengthen and heal. In athletics, the human being is treated increasingly as a physiologic machine to be fine-tuned,[8] while a film like *Chariots of Fire* only emphasizes how far we have come in sixty years. Amateur Scots are all very romantic, but we are obviously not going to excel at the Olympics with a short warm-up and a prayer—although let us make sure we stop the steroid injections early enough to get through the screening tests. In a recent analysis of international TV, we had to create a descriptive category *Man-Machine Ambiguity* to capture all the bionic women, men with five-million-dollar mechanical insides, robots of all sorts, and intelligent computers.[9] It was when Joseph Weizenbaum wrote the computer program ELIZA to simulate the responses of a psychotherapist that the full implications of the computer hit him.[10] Not only was he astonished that people forgot they were disclosing their most private thoughts and fantasies to a machine, but the misplaced enthusiasms of some practicing psychiatrists raised deep ethical questions.

It is the ethical questions that lie between the demechanization and remechanization of psychotherapeutic practice, and nowhere are these encountered with such immediacy as in the realm of the computer. Perhaps more than any other machine in history, it has already altered our world, and it is clear that the revolution has only just begun. And let me scotch immediately any mistaken impression that this essay is simply a tiresome tirade against computers as machines. It is nothing of the sort, for like any other aspect of our technology they can be used for good or ill. If twenty physiologic variables of mine have to be carefully monitored as I lie recovering in a postoperative room, I would rather have a computer watching those limits and sounding the alarm than a tired, overworked, and distracted nurse. As a matter of fact, my next-door-neighbor's life was recently saved by the diagnostic program of a computer capable of scanning millions of combinations of symptoms to raise the possibility of a very rare disease recorded only six times before in medical history. In any case, tirades do not lead to thinking, and it is our very best thinking, informed by ethical concern, that we have to direct toward the presence of the computer in our world.

There are obviously legitimate ends to which computers are the proper means, not the least the daily humdrum tasks of keeping readily accessible accounts, records of reimbursements and treatments, and so on. To make them accessible means we have to categorize and classify them in some way, and it is here that problems begin—not with the computers, but with us. The problem is that we bring to these taxonomic tasks a framework within which we are forced to define classification in a particular and quite rigid way. Our traditional ideas of classification force us to take a set of things and break it up in order to stuff the pieces into separate little boxes that are not allowed to overlap in any way. Such thinking is an old intellectual legacy, part of our scientific heritage going back to Linnaeus, and one still reinforced by work in biology and genetics.[11]

The problem is that these principles of partitional thinking are not only carried over into other, quite different realms, but also increasingly incorporated into the algorithms—the step-by-step instructions and procedures—that are written for computers to mechanize the whole process of classification. And this is where the real difficulties arise. Just think of classifying a book: we know intuitively that many of the most interesting books lie between the conventional boxes. It is hardly necessary to point this out in *Integrative Psychiatry*! But synthesis, integration, and the emergence of problems between past categories mean that connections are being made that structure our ideas in a different way. And when these connections are torn apart by nonsentient, unreflective, deterministic partitional machines, and the "things" being destructured and disconnected are people, do we face an ethical problem?

Suppose, for example, that a set of people are characterized by things that purport to record various aspects of neuroses. If we were to give the table of data, with the people as rows and the neurotic symptoms as columns, to a computer it would certainly partition the people into boxes. That is why I called such algorithms deterministic partitional machines: no matter what the structure, no matter what the connective tissue of the relation between the people and the symptoms, the computer will always sever the connections, tear them apart, and put the people into boxes. And then, depending upon the boxes into which they are stuffed, the treatment they receive will be completely determined.[12] Yet we know that in the realm of neurosis, and in many other common psychiatric categories, our definitions are often distressingly imprecise, and the whole history of psychiatry over the past one hundred years should make us extremely cautious about affixing verbal labels too quickly. Such labels are words, and because we think in language they trap our thoughts along certain lines. Equally, they close off others.

In one sense we have no choice, for language is constitutive of us as human beings, but at least we can stand open and be sensitively aware of our too-ready propensity to categorize in verbal concrete. To take this imprecise language that expresses uncertain and changing knowledge, and then harden it into mechanical devices that unthinkingly assign human beings to treatments, is to make a move from the human world where thinking is possible to the mechanical world where thinking cannot exist. For in the mechanical world of things there is no meaning.

Think What You Like, but Think for Yourself

> A foolish consistency is the hobgoblin of little minds. . . . Speak what you think to-day in words as hard as cannon-balls, and to-morrow speak what to-morrow thinks in hard words again, though it contradicts every thing you said today.
>
> —Ralph Waldo Emerson
> *Self-Reliance*

PRESIDENT TRAINA, Geographer and Provost Kasperson, Dean Krefetz, members of the faculty, and, most especially, students of Clark University. On this day of *con-vocatio*, a day of "calling together" that marks the beginning of another academic year, it is a great privilege to address you all. But I am sure that those of you who are not new arrivals will understand if some of my remarks are addressed specifically to the class that will be the last of the "Nineteens"—the Class of 1999.

Nineteen ninety-nine seems a long way off; looking back, it will have come all too quickly. And at that moment, when Commencement replaces Convocation, you too will be robed in formal fashion, and will wear the gown, hood and hat, the traditional marks, since medieval times, of the educated, reasoned and responsible person, whether that person goes on to a life of scholarship, or takes his or her place in all the possible ways society has to offer. And robe and hat, worn upon that day, will signify full, justified, and irrevocable membership in an institution with one of the oldest and most honorable heritages of the Western world.

The university is a dangerous institution, and for that reason it is often in well-justified trouble. If you have totalitarian leanings, it is one of the first institutions that you try to control, after seizing the TV and radio stations, and shooting the poets. Until quite recently, Plato's *Republic* was much too dan-

---

Modified very slightly from the convocation address given at Clark University, August 29, 1995. The invitation was extended after the incoming freshmen, during the orientation week seminar, had read selected chapters from my book *The Slow Plague* (Oxford: Blackwell, 1993). Hence the references to the HIV/AIDS pandemic.

gerous to teach in Prague, and secret police took the names of students attending a seminar held in a philosopher's private home. It is an institution essential, that is to say it is "of the essence," to a democratic society, and to the extent that a university in our society is placed in jeopardy by external controls, so the democratic nature of the society itself is gravely threatened. In our country, to make my point more immediate, there are very few institutions left where ideas are fully free to be explored, unfettered by fears of power exercised by an Establishment—or, better still, Establishments, because they come in various sorts and guises.

Let me give you two examples, which you may feel are far removed from your concern at the moment, but whose effects can work their way down to the level of everyday university life quicker than you think. One of the great glories of the university is its privileged opportunity to extend knowledge, to push out even further the bounds of our present knowing. Research is a critically important part of university life, and of individual scholarly lives within the university. It sometimes appears to be, and occasionally is, in conflict with teaching: at its best, it is in creative tension, informing—I am tempted to say infusing—the teaching task. Over the next few years, you will know instinctively how to identify the *teachers*, precisely because they are involved in the excitement of pushing back the frontiers of their discipline and learning themselves, as opposed to the *pedagogues*, who assume you are trapped with them for the semester to plod along well-worn ways. If you find yourself in lectures given by pedagogues instead of real teachers, drop them like hot bricks. Remember that the origin of the "lecture" is the *lector*, Latin for "the reader." It comes from the earliest days of the university, long before printing, when there may have been only a few, precious, handwritten manuscripts of a text, and students hired a learned "lector" to read and explicate—*explicare*, literally "unfold"—a text for them. "Lectors" were paid directly by the students: the good ones ate and drank moderately well; the poor ones starved, or looked elsewhere. I hope you never feel that such a system might have had certain merits!

The problem is that much research today is funded by institutions external to the university itself, the great majority of them government agencies exercising the enormous power of the purse. You should know that many faculty at many universities are under tremendous pressure to get external funding, and sometimes such pressures, particularly on young and untenured faculty, mean that people are being judged mainly by the dollars they raise, rather than how they learn and teach. Unfortunately, funded research is too often Establishment research, what the historian and philosopher of science, Thomas Kuhn, called "normal science." Nothing necessarily wrong with it—most science is built slowly and patiently from careful, steady research within a well-accepted paradigm—but when funding decisions about normal science are made by peer groups of "normal," read "Establishment," scientists, there may be little room for innovation, new perspectives, or new research directions and topics. It can be the same in the humanities and social sciences, where schools and fads produce the same effects. The long-term results may be insidious, because faculty, especially young faculty, may feel great pressure to put aside their own interests and intellectual passions, and direct their research towards whatever happens to be the "flavor of the week" down in Washington. Too many faculty in universities these days resemble marionettes, jerking to strings pulled by funding agencies, rather than truly independent scholars. As undergraduates, you may not feel these pressures in any personal, immediate sense, but be aware of them—and talk to the graduate students!

I felt these pressures strongly in the research on the diffusion of the AIDS epidemic, because doctors and epidemiologists had forgotten that every epidemic always has a geography as well as a history. In the same way that we can extract valuable scientific information from time series, say the growing number of cases each year, so we can extract enormously valuable scientific information from spatial series, what we call, more prosaically, map distributions. Unfortunately, epidemiologists, with doctoral degrees from the most distinguished medical schools, had been trained to think only with their differential equations down the single dimension of time, to produce numbers that no one believed in, and numbers that no one could do anything with. They had forgotten that by putting time and space together, we could predict not just numbers, but what the next maps would look like at all sorts of different scales, from the horror of the Bronx, to the metropolitan region,[1] all the way up to the nation.[2] Today with a supercomputer, we could predict the "Million Map" of the United States—when it will come and what it will look like. As a nation we are just about halfway there. Remember, every planning decision to extend health care always involves the geographer's *where*, in addition to the historian's *when*. And when you link actual and predicted maps together with animation for television, the educational impact may be startling. And education is all we have to combat the HIV.

But power is not simply exercised externally to the university; it can also be exercised within. In an institution committed to open, reasoned debate about ideas of all sorts, the pressures to think uniformly, to conform, to be "politically correct" may be very great. And here we enter truly treacherous ground, not the least because ideas may be held, and motivations pursued, with the utmost sincerity—as they often are when people feel deeply ethical issues are at stake, and when their sense of justice has been aroused. It is difficult to find and keep your own sense of balance when these questions arise, questions that appear when there are wrongs to be righted, when instinctively you feel that fairness and decency must replace injustice. In these situations, it is difficult to think for yourself, to be committed, and yet at the same time to maintain just that slight sense of distancing in your perspective, of being aware that you might have a little chink in your own armor of righteousness and truth that will let you listen to another side, to another view, to an alternative or extended argument. These situations often arise when ways to right previous wrongs are pushed to such extremes that they themselves create injustices once more. I know of fine scholars and excellent teachers, who wish to be judged only as scholars and teachers and *not* as women, whose further academic careers have been blocked because they would not join the radical women's group on their campus. Some of them have been labeled as "traitors." This is not just, and the university should stand for justice. I know of a man in an equally prestigious university who was taken quietly aside by two vice-provosts and told that a tenure track position was his if he declared himself homosexual. The position had been reserved for either a lesbian woman or a homosexual man, and since the former would have been chromosomally impossible, he was offered the latter. He refused, and a more "suitable" candidate was found. This is not just, and the university should stand for justice. And if these examples seem extreme, there will be other, different questions and moments in your life here over the next four years, when you may have to think through such questions of what is truly right and wrong, fair and unfair, just and unjust. They will not be easy questions—they never are—and some of them may leave you with a feeling that no matter how you resolve them you are partly damned if you do, and damned if you don't. The important thing is that you think these matters through for yourself, trying to be as just and fair-minded as only you know how as you lie there in the quiet of the night thinking.

Because that is what the university is about—thinking. And this university is your university now, and you are a part, and a most important part, of it. Without you, the student body, there would be no university in any true sense, just as there would be no university without the faculty. You are in it together, teacher and student, in a symbiotic relationship that can be one of the great glories of the human condition. As a student, there is nothing so rewarding as that eureka-like moment when, quite suddenly, understanding breaks through and you say "I see," meaning that something which seemed dark and difficult suddenly becomes clear and falls into place. It is no accident that we use metaphors of light, of seeing, when we speak of the truth. It is one more of those many points at which the Greek heritage of the Western world touches us in our daily scholarly lives. Because the Greek for "truth" was *aletheia* . . . and in it you can hear the *lethe*, the underworld, the region of darkness. But in the Greek that beginning "a" negates the *lethe* (*a-letheia*), it negates the darkness, to make truth a lighting up, an unconcealing, that lets us say that marvelous "Oh, I see!"

And as a faculty member, I can assure you that there is nothing so rewarding as helping another person to that point, that point of seeing and understanding. The great German philosopher Martin Heidegger once wrote that the task of the teacher was *to bring one another to learning.*[3] Remember, no one can "learn you": I can always try to teach you, but I can never "learn you." Learning is that lonely, solitary task you have to undertake yourself. Sometimes with excitement and joy; sometimes with an exercise of will and the gritting of teeth, knowing that you have to "get through this stuff," because only then will other ideas become unconcealed, will other conditions of possibility for thinking enter your own horizon, your own "world."

Let me give you a tip: the best way to learn is to place yourself in the role of the teacher. Believe me, teachers learn very quickly, not the least because their own pride is on the line tomorrow in front of all those students! As you read, as you go over lecture notes (and always go over them before the next lecture), get away by yourself and pretend you are teaching this material to someone else. Pretend that someone else in the class comes up to you and says, "What on earth was that guy saying in class yesterday? What is this stuff? I don't get it!" And you, because you have made that material a part of yourself as a teacher, you say patiently, "Oh, it's not that difficult; you see, it's just this and this . . . and so on," going slowly, explaining carefully, remembering the bits where you had difficulty. Because good teachers are always those who once had difficulties themselves, and who remember what it was like. You do not need an actual person to ask you: pretend to yourself that you are teaching. Talk to yourself, and if anyone else sees you whispering and gesticulating, well . . . let them think you are crazy. And do not say anything when you get an A, and they get a C, at the end of the semester. Remember talking to yourself is not a sign of madness. After all, what is thinking if not having a conversation with yourself? And what is the university if not a place for thinking?

And remember, too, that if you take the 30 weeks of two semesters away from 52, that still leaves you 22 weeks each year to attend to your social calendar. So while you are here, give the university all you have, push yourself in an institution that will make demands on you to the degree that ultimately you make demands upon yourself. And then, when Commencement replaces Convocation four years from now, you will not look back on those years with regret for opportunities lost, but look back for the disciplined joy they have given you, a discipline and a joy that will be with you all your lives.

Ladies and gentlemen, welcome to the most dangerous and exciting institution in the world.

Epilogue
The Arrival of a Spatial Century

DESPITE the millennial madness that will be displayed one year before the millennium,[1] it is still worth pointing out that if we had twelve fingers our approaching millennium would only be the duodecimal year 1669, nothing particularly noteworthy, with postmodernism nowhere in sight. Still, we are what we are, and like waiting for the odometer on our car to finally roll that sixth wheel to 100,000,[2] so we anticipate the arrival of the year 2000. And why not? Anticipating the future, wondering what it will bring, is a very human thing to do. And long-range forecasts are usually harmless: if you have the good fortune to be right, nobody else remembers; and if you have the misfortune to be wrong, nobody else remembers. Which is why long-range demographic forecasting is so popular. Fifty years from now, who will remember, and who will care?

But it is tempting, and I am weak. Forgive, Dear Reader, our trespasses, but lead us once in a while into temptation. Here goes. I have the strong gut feeling that we are rapidly approaching a century where—yes, a "century where," time and place—spatial thinking will emerge very strongly; or, perhaps better still, spatiotemporal thinking will mark many areas of human endeavor. In some ways it is already here, you can feel it in your bones and all around you, but it is ready to blossom and expand. It has many strands, but in one way or another it has always seemed to me to be a very geographic way of thinking. You find it in David Harvey's concerted effort to force traditional Marxist perspectives to acknowledge geography and space, augmenting the nineteenth-century emphasis on history and time. You find it in the verbally expressed dynamics of historical geographic writing, and in the numerically expressed dynamics of analytic, not to say algorithmic, geography. You find it in the acknowledgment that with trillions of bits of information monitoring our planetary home (let alone the entire solar-planetary system), data compression and visualization are essential to understanding. You find it in the thoroughly practical, ongoing world of urban and regional planning, no matter how inadequately it ensures a humane environment in some areas, or how successful its achievements are in others. You find it, and the vocabulary it brings with it, as an increasingly rich source of analogy in the humanities and human sciences, and recall that we think in and with those "spatial" words like "distance," "region," "boundary," "place," "zone," "transformation," and many others, including "space" itself. From time to time, some of it becomes mantra chanting, which usually means fruitful analogical thinking has disappeared, but at its best it has the capacity to light up some aspects of the human condition from a new and quite different perspective.

There is a geographer in most people. If you have a certain fascination for maps, if you have caught yourself daydreaming about faraway places, wondering what they and their peoples are like, then you know intuitively what "spatial thinking" means as you cast your thinking like a fisherman's net across the world. If you are becoming a professional geographer, you know what led you to this point: the map again, the clusters, the patterns, the linkages and connections, the juxtapositions, the coming-to-be sequences, the relationships in space, the rules-of-the-game, the warpings and transformations, and the spatial structures, whether guiding water in a river system or diseases in a population. The world needs professional geographers: for their educational skills, for their civic involvement, for their analytical spatiotemporal thinking, for their environmental concern, for their ability to link things in place and space, for their awareness that what we do today will affect tomorrow, for their understanding that what happens *here* will affect *there*. None of these skills is unique, others are quite capable of thinking in the same way. But usually they do not have the full range of prior experience, the mental filecard of past examples that lets them transfer their experience of one spatial problem to another, either directly or by analogy. Geographic thinking takes many forms, and shares the subject matters of many other fields, but seems to be a way of looking that an increasingly complex and shrinking world desperately needs.

It has been a tremendous privilege to be a part of all this, to see the way this wonderful eclectic field has developed as a formal discipline, how its students inform the tasks of planning at all urban and regional scales, how their environmental concern translates into concrete ways of bringing a balance back to the natural world, how their social concern illuminates human conditions that barely deserve the name. I suspect the next fifty years will be an even more exciting time of development, involvement, and illuminating scholarship. Who in the 1950s would have foreseen the developments today? And who would dare to predict the state of geography at the next half century? I dare only this: it will be the "spatial century," when many others realize the fundamental importance of the *where* as well as the *when*. I only wish I were going to be around!

Notes

Index

Notes

Introduction

1. David Harvey, "On the History and Present Condition of Geography: An Historical Materialist Manifesto," *Professional Geographer* 36 (1984): 1–11.

2. John Imbrie, "A Theoretical Framework for the Pleistocene Ice Ages," *Journal of the Geological Society of London* 142 (1985): 417–432.

3. Torsten Hägerstrand, "In Search for the Sources of Concepts," in *The Practice of Geography*, ed. Anne Buttimer (London: Longman, 1983), 238–256. The autobiographical diagram appears on page 255.

4. The idea for such a sequence of expanding, generational maps comes from Peter Haggett, "Geographical Aspects of the Emergence of Infectious Diseases," *Geografiska Annaler*, ser. b, 76 (1994): 91–104, who found the idea in D. Bradley, "The Scope of Travel Medicine," in *Travel Medicine*, ed. R. Steffans, H. Lobel, J. Haworth and D. Bradley (Berlin: Springer Verlag, 1988), 1–9.

5. The time and distance estimates of major travels are crude (and who knows what the ordinary daily journeys sum to?), but the estimates for John, Abraham, and Charles are certainly reasonable "ballpark" figures. As for my father Ralph, I can estimate his civilian and wartime travels fairly closely from his own accounts, not the least a moving book about the war years, *One of Many*, that he wrote for his grandchildren shortly before he died. I doubt that it will ever be published, being very much a family book. As for myself, for many years a resident alien in the United States, I had to compile a long and complicated list of all the times I had left the country, from passport stamps going back to 1956, up to the time when I applied for citizenship. So the last two estimates, while still estimates, are fairly close.

6. It was the same with maps. You don't *learn* to read maps. Any child can do it. Which is why I think people like David Stea and Jim Blaut are absolutely correct when they report on the abilities of very young children to make sense out of maps, air photos, and other "spatial models." See, for example, their "Studies of Geographic Learning," *Annals of the Association of American Geographers* 61 (1971): 387–393, and "Toward a Developmental Theory of Spatial Learning," in *Image and Environment: Cognitive Mapping and Spatial Behavior*, ed. Roger Downs and David Stea (Chicago: Aldine, 1973), 51–62.

7. Adding pounds, shillings, and pence is easy for anyone prepared to use the decimal, the duodecimal, and the vigesimal systems simultaneously. You have three columns of numbers, and you start on the right hand one and add up the pence. If they sum to more than twelve, you divide by twelve, put down the remainder, and carry over to the next column of shillings the number of times 12 divides the sum of the first column. You then take the second column of shillings, add those up, together with the number you carried over from the first, and then divide that by 20. Put down the remainder under the shilling column that you are presently working on, and carry the number of times you divided by 20 into the next column of pounds. At this point you're home free, back with civilized people who use the decimal system, so you can just keep counting, letting the pounds add up with no more divisions. Abandoning this simple system for the nasty, complex, and French-influenced decimal system was fiercely resisted by the British, who felt they would be merely slumming with the rest of the world.

8. The one of which Ralph Waldo Emerson wrote:

> By the rude bridge that arched the flood,
> Their flag to April's breeze unfurled,
> Here once the embattled farmers stood
> And fired the shot heard round the world.

9. These experiences, and many others over the next four years, parallel to a quite uncanny degree those of Anthony Baily, who wrote *America, Lost and Found: An English Boy's Wartime Adventure in the New World* (New York: Random House, 1980). I use "uncanny" advisedly: I believe in doppelgängers now.

10. The most beautiful edition I know is Robert Graves, *The Greek Myths* (London: Folio Society, 1996).

11. His marvelous books are still in print as paperbacks, and should form a treasured few inches on the bookshelves of any home with children. That is, if people still have bookshelves?

12. John Wellington, *Southern Africa, a Geographical Study* (Cambridge: Cambridge Univ. Press, 1955).

13. John Gunther, *Inside Africa* (New York: Harper, 1955).

14. Harold Macmillan, a speech to the Parliament of the Union of South Africa, Cape Town, Feb. 3, 1960.

15. Despite its postmod, almost pop, popularity, I still find food for thinking in Gilles Deleuze's and Félix Guattari's notion that concepts appear, and while falling back into history, they are not explained by history. One can offer all sorts of "Yes . . . buts" to this idea, but it still leaves a philosophical itch that needs scratching.

16. Some incidents of the summer field camp are described in "Geography 1957–1997: The Augean Period" pp. 84–100.

17. Peter Gould, *Transportation in Ghana*, Northwestern Studies in Geography, no. 5 (Evanston: Department of Geography, 1960).

18. The full story appears in chapter 16, "The Beltway Bandits," in my book *The Geographer At Work* (London: Routledge, 1985), 180–185.

19. Dylan Thomas, *A Child's Christmas in Wales* (New York: New Directions, 1954).

20. Which, in a somewhat different way, is what the Finnish geographer Reino Ajo had to do, supporting his family as an inspector of automobiles in Turku, while working in the evenings to produce some of the most original scientific geography of its day, geographic research essentially rejected by those in power in the universities of his day. Some of the details are in Gould, "The Trail Blazer: Reino Ajo and Contemporary Geography," *Terra* 84 (1972): 64–66.

21. Alan Janik and Stephen Toulmin, *Wittgenstein's Vienna* (New York: Simon and Schuster, 1973).

22. Clarence Glacken, *Traces on the Rhodian Shore: Nature and Culture in Western Thought from Ancient Times to the End of the Eighteenth Century* (Berkeley: Univ. of California Press, 1967).

African Beginnings

1. Robert Heussler, *Yesterday's Rulers: The Making of the British Colonial Service* (Syracuse: Syracuse Univ. Press, 1963).

Place Preferences in Ghana and Tanzania

1. The whole notion of "modernization" is now considered ethnocentric and Eurocentric, and probably eccentric as well, not to say politically incorrect. Nevertheless, Ghana at Independence in 1957 was very different from the Gold Coast in 1897, sixty years before. Similarly, many European countries, the United States, and Japan were very different in 1897 compared to what they were in 1837. I think we need a general name for this rapid, shattering, and complex process. I call it "modernization"; you may call it what you like.

2. Allan Cardinall, *The Gold Coast 1931*. (Accra: The Government Printer, n.d. [*circa* 1933–34]).

3. Peter Gould, "Tanzania 1920–1963: The Spatial Impress of the Modernization Process," *World Politics* 22, no. 2 (1970): 149–170.

Wheat on Kilimanjaro: The Perception of Choice within Game and Learning Model Frameworks

1. C. West Churchman, *Prediction and Optimal Decision: Philosophical Issues in a Science of Values*. (Englewood Cliffs, N.J.: Prentice-Hall, 1961). The quotation is the title of chapter 4, p. 70.

2. Matthew 9:17.

3. Russell Ackoff, *Progress in Operations Research* (New York: Wiley, 1961), viii.

4. R. Duncan Luce and Howard Raiffa, *Games and Decisions: Introduction and Critical Survey* (New York: Wiley, 1958), 63.

5. Letter from A. F. Palfrey, Chairman, Kilimanjaro West Farmer's Association, to Regional Agricultural Officer, Sept. 16, 1957, contained in File 1/BIR, "Birds *(Grain Eating Quelea)*," vol. 5, Regional Agricultural Office, Arusha, Tanzania.

6. Agricultural Officer, "District Book Part II" (mss.), 5, Regional Agricultural Office, Moshi, Tanzania.

7. Letter from Nduimet Farm to Agricultural Officer, Moshi, Dec. 10, 1951, contained in File 31/2, "Cereals: Wheat, Barley, Oats and Rye," vol. 2, Regional Agricultural Office, Arusha, Tanzania.

8. Senior Plant Breeder, "Report on a visit to the wheat growing areas of Northern Province," File 31/2, "Cereals: Wheat, Barley, Oats and Rye," vol. 2, Regional Agricultural Office, Arusha, Tanzania.

9. Letter from Agricultural Officer, Moshi, Tanzania, to Senior Plant Breeder, Njoro, Kenya, Nov. 16, 1951.

10. Thorpe, H. "Report on a visit to the wheat growing areas of Northern Province, 1952," File 31/2, "Cereals: Wheat, Barley, Oats and Rye," vol. 2, Regional Agricultural Office, Arusha, Tanzania.

11. Thorpe, H. "Report on a visit to the wheat growing areas of Northern Province, 1959," File 31/2, "Cereals: Wheat, Barley, Oats and Rye," vol. 3, Regional Agricultural Office, Arusha, Tanzania.

Mental Maps

1. These results were first reported at a gathering of MICMOG (The Michigan Inter-University Community of Mathematical Geographers), in a back room of a café in Brighton, Michigan, in April 1966. Someone had worked out that this was the PMAT (point of minimum aggregate travel) for Wayne State, Michigan State, and Michigan. These initial findings were written up as a discussion paper (real journals were hard to break into in those days), an essay since reprinted a number of times in books under the original title "On Mental Maps."

2. And what lay behind this particular research possibility is recorded in "Expose Yourself to Geographic Research," 208–19, below.

3. Something I have tried to push along a little out of sheer admiration for his work, and with the help of Antoine Bailly of Geneva, editing some of his essays as the book *Le Pouvoir des Cartes: Brian Harley et la Cartographie* (Paris: Economica, 1995).

4. Brian Harley, "The Map as Biography: Thoughts on Ordnance Survey Map, Six-inch Sheet Devonshire CIX, SE, Newton Abbott," *The Map Collector*, no. 4 (1987): 18–20.

5. To be found as four verses in *The Map Collector*, no. 1 (1988): 50.

Acquiring Spatial Information

1. Herbert Simon, *The Sciences of the Artificial* (Cambridge Mass.: MIT Press, 1969).

2. David Stea, "The Reasons for Our Moving," *Landscape* 17 (1967): 27–28.

3. Bernard Marchand, "Information Theory and Geography," *Geographical Analysis* 4 (1972): 234–257.

4. An average coefficient of multiple correlation, $R_{1.23}=0.85$.

5. Leslie Curry, "A Spatial Analysis of Gravity Flows," *Regional Studies* 6 (1972): 131–147.

6. Purists may insist that there is a distinction between the autocovariance and autocorrelation functions. For pedagogic explication, perhaps they may indulge and tolerate the impure.

Maps as Memory

1. Alfred Tennyson, "Crossing the Bar," his last published poem, much anthologized, written from his house just above the ruins of Salcombe Castle in 1889 when he was eighty-one.

2. After consulting a number of biographies of Captain Joshua Slocum, I do not believe this could be true. Slocum set sail in dirty weather from Vineyard Haven on November 14, 1909, and was last seen by Captain Levi Jackson in his *Priscilla* coming back from fishing cod between Martha's Vineyard and Nantucket. Slocum was never seen again. See Walter Teller, *The Search for Captain Slocum* (New York: Scribner's, 1956). Nevertheless, the decaying hulk in the meadow was about the right size, and many local people took it for granted that it was the *Spray*.

3. John Speed, *The Counties of Britain* (London: Pavilion, 1995), 65–68, a facsimile collection. My interpretation of Chivelstone as Speed's Chilton was challenged by "expert" reviewers, who referred to a standard work, *Place Names of Devon*, and said that Chilton was in the West Budleigh Hundred, 36 miles north-by-east of John Speed's Chilton in the Coleridge Hundred, exactly where modern day Chivelstone lies. Geographers *look* at maps: historians refer to incestuous texts.

4. J. B. Harley and Yolande O'Donoghue, *The Old Series Ordnance Survey Maps of England and Wales*, vol. 2, *Devon, Cornwall and West Somerset* (Lympne Castle, Kent: Harry Margary, 1977), a composite from plates 57, 58, and 65.

5. Brian Friel, *Translations* (London: Faber and Faber, 1981).

6. Elizabeth Jennings, "Map Makers," in *A Way of Looking* (New York: Rinehart, 1955), 28.

Against the Grain

1. As things got better, a number of essays appeared recalling the difficulties, and sometimes the despicable practices, of those early years. See particularly, William Warntz's "Trajectories and Coordinates," and Richard Morrill's "Recollections of the 'Quantitative Revolution's' Early Years: The University of Washington 1955–65," in *Recollections of a Revolution: Geography as Spatial Science*, ed. Mark Billings, Derek Gregory, and Ron Martin (London: Macmillan, 1984), 134–149 and 57–72, respectively. For

an amusing but bittersweet dressing-down of a young Peter Haggett, who was told by Professor Stears of Cambridge that his mathematical analyses presented at the Royal Geographical Society were "bringing the subject of geography into disrepute," see Richard Chorley's "Haggett's Cambridge: 1957–1966" in *Diffusing Geography: Essays for Peter Haggett*, ed. Andrew Cliff, Peter Gould, Anthony Hoare, and Nigel Thrift (Oxford: Blackwell, 1995), 355–374.

2. Compare issues of the *Annals of the Association of American Geographers* 70, no. 1 (March 1980), with vol. 70, no. 2 (June 1980) and vol. 72, no. 2 (June 1982).

3. Association of American Geographers, *Guide to Graduate Departments in the United States and Canada, 1977–78* (Washington, D.C.: Association of American Geographers, 1977).

4. John Hudson, letter to Peter Gould, October 5, 1977.

5. John Hudson, letter to contributors, April 5, 1979.

6. Peter Allen, "Comments on the Paper of P. R. Gould," *European Journal of Operational Research* 30 (1987): 222.

7. Karl Marx, *The Eighteenth Braumaire of Louis Bonaparte* (New York: International Publishers, 1964).

8. Karl Marx, *Revolution and Counter-Revolution of Germany in 1848* (London: Allen and Unwin, 1881).

9. David Harvey, *The Limits to Capital* (Oxford: Blackwell, 1982), and *Consciousness and the Urban Experience: Studies in the History and Theory of Capitalist Urbanization* (Baltimore: Johns Hopkins Univ. Press, 1982).

Geography 1957–1977: The Augean Period

1. The charge was "to write down [your] personal observations, reflections, and second thoughts . . . a series of personal observations . . . leading up to the present . . . an account for future generations." Letter from John C. Hudson, editor, *Annals of the Association of American Geographers*, October 5, 1977.

2. Fredrick Croxton and Dudley Cowden, *Applied General Statistics* (New York: Prentice Hall, 1955); George Snedecor, *Statistical Methods* (Ames: Iowa State Univ. Press, 1956). Both of these books were particularly helpful because the authors had lots of worked examples. They were written in the days when such books were meant to help students, rather than as a foil to show off the author's skill with triple subscripted notation.

3. To whom I refer again in "August Lösch as a Child of His Time," pages 273–82, below.

4. A statement quoted in The Open University, *Mathematics Foundation Course Unit 30, Groups I* (Bletchley: The Open University Press, 1971), 1.

5. Carl Sauer, "The Education of a Geographer," *Annals of the Association of American Geographers* 46 (1956): 287–99.

6. Richard Hartshorne, *The Nature of Geography.* (Lancaster, Pa.: Association of American Geographers, 1939).

7. William Bunge, *Theoretical Geography* (Lund: Gleerup, 1962); and Peter Haggett, *Locational Analysis in Human Geography* (London: Arnold, 1965).

8. And now, twenty years later, it has risen from essentially nowhere to the fourth-ranking graduate program in the country according to the 1995 National Research Council evaluations.

9. Two examples must suffice: Peter Ambrose and Robert Colenutt, *The Property Machine* (Harmondsworth: Penguin, 1974); and John Shepard and Michael Jenkins, "Decentralizing High School Administration in Detroit: An Evaluation of Alternative Strategies," *Economic Geography* 48 (1972): 95–106.

10. A step undertaken brilliantly in David Harvey, *The Limits to Capital* (Oxford: Blackwell, 1982).

11. One of the most marvelous and original scientific books ever written, and a favorite of mine too: D'Arcy Thompson, *On Growth and Form* (Cambridge: Cambridge Univ. Press, 1942).

12. Arthur Robinson and Barbara Petchenik, *The Nature of Maps: Toward Understanding Maps and Mapping* (Chicago: Univ. of Chicago Press, 1976).

13. James Dugundji, *Topology* (Boston: Allyn and Bacon, 1976).

14. And from the date of publication of this book (1999) even longer! Yet many of the same problems are still with us. The academic world is conservative, in both the best and worst sense.

15. Ronald Abler, John Adams, and Peter Gould, *Spatial Organization: The Geographer's View of the World* (Englewood Cliffs, N.J.: Prentice-Hall, 1971).

16. Richard Hartshorne's review of Michael Chisholm's *Human Geography: Evolution or Revolution?* in *The Geographical Review* 58 (1978): 103–105.

A Critique of Dissipative Structures in the Human Realm

1. These statements would not satisfy mathematicians, who would insist that the extension fields be stated. Perhaps, just for once, they could be sympathetic to these attempts to explicate difficult ideas to a nonmathematical audience.

2. Thomas Heath, *A History of Greek Mathematics* (Oxford: Clarendon Press, 1974).

3. Peter Allen and Maurice Sanglier, "A Dynamic Model of Growth in a Central Place System," *Geographical Analysis* 9 (1979): 256–272; "A Dynamic Model of a Central Place System II," *Geographical Analysis* 13 (1981): 149–164; and "Urban Evolution, Self-Organization, and Decision-Making," *Environment and Planning* A 13 (1981): 167–183.

4. Obviously not even experienced geographers can visualize such a multidimensional "state space" through which some dynamic central place or urban system is going to trace a trajectory as its history of development unfolds. So I have squashed all the n points in a regional lattice down to just two, and all the k parameters, whose values control the system, down to just one. All subsequent arguments still hold.

5. To my knowledge, no such statements appear a decade later either. Dissipative structures, in the human realm at any rate, seem to have dissipated, not to say disappeared.

6. The problem refers back to the "Curry Effect" we met in "Acquiring Spatial Information," 59–73.

7. Bernard Marchand, "Quantitative Geography: Revolution or Counter-revolution?" *Geoforum* 17 (1974): 15–23.

8. Joseph Kockelmans, *Heidegger and Science* (Washington, D.C.: Center for Advanced Research in Phenomenology, 1985).

9. Martin Heidegger, "The Age of the World Picture," in *The Question Concerning Technology*, trans. William Lovitt (New York: Harper and Row, 1977), 115–154.

10. Heidegger's concern for etymological meaning is well known, but for a discussion of *mathesis* and *ta mathemata* see Martin Heidegger, *What Is a Thing?* (South Bend: Gateway Editions, 1967); and for specific reference to the mathematical in the rise of "systems thinking" during the past two hundred years, see Martin Heidegger, *Schelling's Treatise on the Essence of Human Freedom* (Athens: Ohio Univ. Press, 1985).

11. Ilya Prigogine and Isabelle Stengers, *Order Out of Chaos* (New York: Bantam, 1984), 117.

12. Prigogine and Stengers, *Order Out of Chaos*, 178.

13. Torsten Hägerstrand, *Innovationsforloppet ur Korologisk Synpunkt* (Lund: Gleerup, 1953).

14. Martin Heidegger, *Being and Time*, trans. John Macquarrie and Edward Robinson (New York: Harper and Row, 1962), 270.

15. Peter Gould, "Things I Do Not Understand Very Well III: Are Algebraic Operations Laws?" *Environment and Planning* A 14 (1983): 1567–1569.

Sharing a Tradition: Geographies from the Enlightenment

1. This advice is in direct opposition to that currently being promulgated by molders of young thinking in many, especially large, research universities today. There is a tremendous drive to produce "team players" (too often a euphemism for "cogs in the machine"), in all disciplines, once again imposing a structure from Big Science, not to say Big Business, blindly and unthinkingly on all scholarship. Too often the result is committee-like, Establishment science, of the sort Thomas Kuhn characterized as "normal science," all perfectly respectable, but perhaps a bit plodding, science. Yet ideas and innovations still originate in an individual's thinking, and seldom from artificially structured committees. So I say to hell with "team players" and their obsequious forelock-pulling to projects that are merely fundable rather than fascinating.

2. All these geographers, touching directly or indirectly on the fragmentation theme, are cited by Ron Johnston in chapter 1, "Introduction: A Fragmented Discipline," in *A Question of Place: Exploring the Practice of Human Geography* (Oxford: Blackwell, 1991), 1–37, in support of his contention that the discipline is in conceptual disarray.

3. An image I used before in "Spaces of Misrepresentation," in *Limits of Representation,* ed. Franco Farinelli, Gunnar Olsson, and Dagma Reichert (Munich: Accedo, 1993), 123–152.

4. Quoted by Edward Said, *Orientalism: Western Conceptions of the Orient* (Harmondsworth: Penguin, 1991, p. 268, commenting on the drive to simplify great complexity, itself indicative of a reluctance to deal with particularistic messiness. Too often such a drive characterizes commentaries by geographers about the century of Enlightenment. Historians know better.

5. A series of joint essays by the geographers Matthew Hannah and Ulf Strohmeyer address precisely such questions of "ground" and moral choice.

6. A distressing clear example appears in David Livingstone, "Never Shall Ye Make the Crab Walk Straight," in *Nature and Science: Essays in the History of Geographical Knowledge,* ed. Felix Driver and Gillian Rose (Cheltenham: Institute of British Geographers, 1992), 37–48.

7. A long-term liaison to which I publicly confessed in *The Geographer at Work.* (London: Routledge, 1985).

8. A nice example is Gustave Flaubert addressing George Sand as *chère Maître,* noted by Julian Barnes in his review "Unlikely Friendship," *New York Review of Books* 40 (1993): 5.

9. Gillian Rose, as one voice (with Steve Pile) in "All or Nothing? Politics and Critique in the Modernism-Postmodernism Debate," *Society and Space* 10 (1992): 123–126.

10. Susan Ford, in "Landscape Revisited: A Feminist Appraisal," in *New Worlds, New Worlds: Reconceptualising Social and Cultural Geography,* ed. Chris Philo (Lampeter: Institute of British Geographers, 1991), 151.

11. Michael Dear and Nigel Thrift, "Unfinished Business: Ten Years of *Society and Space,* 1983–1992," *Society and Space* 10 (1992): 715–719.

12. Timothy Mitchell, *Colonizing Egypt* (Cambridge: Cambridge Univ. Press, 1988).

13. Gunnar Olsson, "Invisible Maps: A Prospectus," *Geografiska Annaler,* ser. b, 73 (1991): 85–91.

14. Derek Gregory, *Ideology, Science and Geography* (New York: St. Martin's, 1978), 63.

15. Tennyson's marvelous poem "Ulysses" is still one of my favorites. There he sits in Ithaca after twenty years of adventure on his way back from "the ringing plains of windy Troy," having taken down the great bow that only he can bend to slaughter the fawning mob of suitors. There he sits . . . absolutely bored out of his skull. His son, Telemachus, is a good lad, and quite capable of carrying out the humdrum duties of kingship. So off Odysseus goes again "To strive, to seek, to find, and not to yield." What splendid stuff, and thoroughly geographic to boot!

16. Ron Johnston, *A Question of Place: Exploring the Practice of Human Geography* (Oxford: Blackwell, 1991), 36.

17. No one has, even seven years later.

18. E. Gruntfest to P. Gould, letter, Nov. 30, 1986.

19. P. Gould to E. Gruntfest, letter, Dec. 10, 1986.
20. D. Lee to P. Gould, letter, Jan. 7, 1987.
21. P. Gould to R. Aangeenbrug, letter, Dec. 1, 1986.
22. S. Brunn to P. Gould, letter, Feb. 4, 1987.
23. R. Aangeenbrug to P. Gould, letter, Mar. 18, 1987.
24. Given the viciousness with which radical feminists attack any woman who does not conform exactly to their point of view, I shall continue to protect the identity of this correspondent.
25. P. Gould to R. Aangeenbrug, letter, May 18, 1987.
26. R. Aangeenbrug to P. Gould, letter, May 27, 1987.
27. G. Demko to P. Gould, letter, May 27, 1987.
28. D. Lee to P. Gould, letter, Jan. 7, 1987.
29. E. Gruntfest to P. Gould, letter, Nov. 30, 1986.
30. Derek Gregory, *Geographical Imaginations* (Oxford: Blackwell, 1994), 351–352.
31. A work that shows us how very difficult it is to struggle with such questions is John Caputo, *Against Ethics: Contributions to a Poetics of Obligation with Constant Reference to Deconstruction* (Bloomington: Indiana Univ. Press, 1993).

Cathartic Geography

1. Aristotle first employs "catharsis" in this sense in the context of music (*Politics*, book 8, 1341b.37–1342b.17), saying "but when hereafter we speak of poetry, we will treat the subject with more precision." In fact, he appears to assume that his reader is already familiar with such usage by the time he discusses it in the context of tragedy (*Poetics*, book 6, 1449b.28). References are to Jonathan Barnes, ed., *The Complete Works of Aristotle*, vol. 2, Bollingen Series 71 (Princeton: Princeton Univ. Press, 1984), 2128–2129, 2320.
2. Alan Sokel, "Transgressing the Boundaries: Towards a Transformative Hermeneutics of Quantum Gravity," *Social Text* 14 (1996): 217–252.
3. David Harvey, "Flexibility: Threat or Opportunity?" *Socialist Review* 21 (1991): 65–77.
4. Jane Wills, "Laboring for Love? A Comment on Academics and Their Hours of Work," *Antipode* 28 (1996): 292–303.
5. Gunnar Olsson, "Invisible Maps: A Prospectus," *Geografiska Annaler*, ser. b, 73 (1991): 85–91.
6. Micheal Watts, "Struggles over Land, Struggles over Meaning: Some Thoughts on Naming, Peasant Resistance and the Politics of Place," in *A Ground for Common Search*, ed. R. Golledge, H. Couclelis, and P. Gould (Santa Barbara: Santa Barbara Geographical Press, 1988), 31–50.
7. Allan Pred, *Lost Words and Lost Worlds: Modernity and the Language of Everyday Life in Late Nineteenth Century Stockholm* (Cambridge: Cambridge Univ. Press, 1990).
8. David Harvey, *Consciousness and the Urban Experience: Studies in the History and Theory of Capitalist Urbanization* (Baltimore: Johns Hopkins Univ. Press, 1985).

9. Anne Godlewska, "Map, Text and Image: The Mentality of Enlightened Conquerors: A New Look at the *Description de l'Egypte*," *Transactions of the Institute of British Geographers*, n.s., 20 (1995): 5–28; and Timothy Mitchell, *Colonizing Egypt* (Cambridge: Cambridge Univ. Press, 1988).

10. Lucy Jarosz, "Agents of Power, Landscapes of Fear: The Vampires and Heart Thieves of Madagascar," *Society and Space* 12 (1994): 421–436.

11. Adam Tickell, "Making a Melodrama Out of a Crisis: Reinterpreting the Collapse of Barings Bank," *Society and Space* 14 (1996): 5–33.

12. "Redlining" refers to a practice of marking certain areas on real estate maps as unacceptable for mortgage or home improvement loans. To the best of my knowledge, the study never appeared in an academic journal.

13. From the last lines of William Butler Yeats's "The Second Coming":

> And what rough beast, its hour come round at last,
> Slouches towards Bethlehem to be born?

William Butler Yeats, *Michael Robartes and the Dancer* (Churchtown: Cuala Press, 1920), 19–20.

14. Neither do Gilles Deleuze and Félix Guattari in their remarkable chapter "Geophilosophy" in *What Is Philosophy*? (New York: Columbia Univ. Press, 1994), 85–113, where they point to the contingency characterizing the emergence of philosophical concepts, concepts that may "fall back into history," but are not explainable in historical terms.

15. Reginald Golledge, "Geography and the Disabled: A Survey with Special Reference to Vision Impaired and Blind Populations," *Transactions of the Institute of British Geographers*, n.s., 18 (1993): 63–85.

16. Rob Imbrie, "Ableist Geographies, Disablist Spaces: Towards a Reconstruction of Golledge's 'Geography and the Disabled,'" and B. Gleeson, "A Geography for Disabled People?" *Transactions of the Institute of British Geographers*, n.s., 21 (1996): 397–403, and 387–396, respectively.

17. Micheal Oliver, *The Politics of Disablement* (New York: St. Martin's, 1990).

18. Imbrie, 398.

19. David Harvey, "Between Space and Time: Reflections on the Geographical Imagination," *Annals of the Association of American Geographers* 80 (1990): 418–434.

20. Gleeson, 391.

21. Gleeson, 394.

22. Reginald Golledge, "A Response to Gleeson and Imbrie," *Transactions of the Institute of British Geographers*, n.s., 21 (1996): 404–411.

23. Imbrie, 402.

24. Golledge, 404.

25. This is a distressingly inadequate shorthand phrase to describe the way Olsson has taken a road that starts in the late sixties and early seventies with a questioning of approaches to the social condition described by algebraic structures, and continues to a never-ending puzzlement with the individual person in a social condition. Many essays are collected in *Birds in Egg/Eggs in Bird* (London: Pion, 1980), *Antipasti* (Göteborg:

Bokförlaget Korpen, 1990), and *Lines of Power/Limits of Language* (Minneapolis: Univ. of Minnesota Press, 1991).

26. Gunnar Olsson, "Chiasm of Thought-and-action," *Society and Space* 11 (1993): 279–294.

27. Matthew Sparke, "Escaping the Herbarium: A Critique of Gunnar Olsson's 'Chiasm of Thought-and-Action,'" *Society and Space* 12 (1994): 207–220.

28. Gunnar Olsson, "Job and the Case of the Herbarium," *Society and Space* 12 (1994): 221–225.

29. Olsson, "Job," 224.

30. Micheal Brown, "Ironies of Distance: An Ongoing Critique of the Geographies of AIDS," *Society and Space* 13 (1995): 159–183.

31. Micheal Brown, review of *The Slow Plague: A Geography of the AIDS Pandemic* (Oxford: Blackwell, 1993), in *Society and Space* 12 (1994): 631–632.

32. Brown review, 632.

33. Such a wealth of insight that I dedicated the book to these two extraordinary "human ecologists" so much in love with *their* city.

34. Brown review, 632.

35. Micheal Brown, "Ironies of Distance: An Ongoing Critique of the Geographies of AIDS," *Society and Space* 13 (1995): 159–183.

36. These are labeled rates per *million*, which would give British Columbia a rate in 1990 of about 0.00026, perhaps the lowest in the world for an area and population this size.

37. Hans George Gadamer, "The Experience of Death," chapter 4 in *The Enigma of Health: The Art of Healing in a Scientific Age* (Stanford: Stanford Univ. Press, 1996), 61–69.

38. Brown, "Ironies of Distance," 162.

39. Brown, "Ironies of Distance," 163.

40. Peter Gould, *The Slow Plague*, 168–177.

41. AIDS rates on a county basis are predictable on the basis of density 0.97 (unlogged) and 0.76 (logged) in California; 0.97 (unlogged) and 0.84 (logged) in New York. Even in sparsely settled and data suspect Wyoming, an expansion model predicts 0.81. All these examples point to the fact that AIDS cases are a direct function of (Population)^2, which simply measures the number of possibilities for pairwise transmission. Why should anyone be surprised?

42. See "As People Go So Do the Viruses," 195–204, below.

43. Peter Gould and Rodrick Wallace, "Spatial Structures and Scientific Paradoxes in the AIDS Pandemic," *Geografiska Annaler*, ser. b, 76 (1994): 105–116.

44. David Bell, "Erotic Topographies: On the Sexuality and Space Network," *Antipode* 26 (1994): 96–100.

45. David Bell, "[screw]ING GEOGRAPHY (censor's version)," *Society and Space* 13 (1995): 127–131.

46. Bell, "Erotic Topographies," 97.

47. Herman Krick, in his now classic, seven-volume work *The Geography of Willow*

Creek, insisted that everything was unique "because nothing is located where anything else is located." His claims are reviewed in my book *The Geographer at Work* (London: Routledge, 1985), 19.

48. David Bell, reviewing Alphonso Lingis's *Foreign Bodies* (London: Routledge, 1994), and P. Rodaway's *Sensuous Geographies: Body: Sense and Place* (London: Routledge, 1994), in *Society and Space* 14 (1996): 621–630.

49. Stephen Gould, "Evolution: The Pleasures of Pluralism," *New York Review of Books* 44, no. 11 (1997): 52.

50. Gillian Rose, "Distance, Surface, Elsewhere: A Feminist Critique of the Space of Phallocentric Self/Knowledge," *Society and Space* 13 (1995): 761–781.

51. Jean Aitchison, in her delightful Reith Lectures, published as *The Web of Language* (Cambridge: Cambridge Univ. Press, 1996), has some pertinent things to say about a hammering repetition of the same word, p. 84.

52. General reviews and debates include on the "scientific" side John Horgan, "Why Freud Isn't Dead," *Scientific American* 275 (1996): 106–111, while on the "humanistic" side Fredrick Crews's two-part article "The Revenge of the Repressed," in the *New York Review of Books*, Nov. 17 and Dec. 1, 1994, produced anguished squeals at the public castration of Daddy. For a thoughtful account pointing to the nonexistence of such posited imaginings as the "unconscious," see Joseph Kockelmans, "Daseinanalysis and Freud's Unconscious," *Review of Existential Psychology and Psychiatry* 16 (1978–79): 21–42.

53. In particular, the Daseinanalytic tradition would appear to be compatible with the respectful and caring attitude voiced by many today. Useful starting points would be Medard Boss's *Existential Foundations of Medicine and Psychology* (New York: Jason Aronson, 1979), and *I Dreamt Last Night* (New York: Gardner, 1977); as well as special issues of two journals "Heidegger and Psychology," ed. Keith Hoeller, *Review of Existential Psychology and Psychiatry* 16 (1978–79), and "Approaches to Dreaming: An Encounter with Medard Boss," ed. Charles Scott, *Soundings: An Interdisciplinary Journal* 60 (1977).

54. See "Think What You Like, But Think for Yourself," 307–15, below.

55. Only briefly treated here to avoid repetition of points made in "Sharing a Tradition: Geographies from the Enlightenment," 118–38, above.

56. Quoted by Peter Hopkirk in his *Quest for Kim: In Search of Kipling's Great Game* (London: John Murray, 1996), 50.

57. Hopkirk, *Quest for Kim*, 141.

58. Charles Trench, *The Frontier Scouts* (Oxford: Oxford Univ. Press, 1986).

59. Alastair Bonnett, "The New Primitives: Identity, Landscape and Cultural Appropriation in the Mythopoetic Men's Movement," *Antipode* 28 (1996): 273–291.

60. Aitchison, *Web of Words*, 72.

61. Geraldine Pratt, "Trashing and Its Alternatives," *Society and Space* 14 (1996): 253–256.

62. Pratt, "Trashing," 254.

63. Sparke, "Escaping the Herbarium," 220.

64. Olsson, "Chiasm of Thought," 222.
65. Pratt, "Trashing," 254.

The Structure of Spaces

1. As *The Structure of Space(s)*, this interest became the title of a symposium held at the Royal Swedish Academy of Sciences, Stockholm, Apr. 24, 1997.

2. Edward Taaffe, "A Map Analysis of United States Airline Competition," *Journal of Air Law and Commerce* 25 (1958): 121–147, 402–427.

3. Details are available in the "double book" *The Structure of Television: Part I Television, The World of Structure: Part II Structure, The World of Television* (London: Pion, 1984), written in collaboration with Jeffrey Johnson and Graham Chapman. Part I was written for broadcasters; Part II assumed a more technical social science background.

Microgeographic and Behavioral Space: The Game of the "Green Revolution"

1. Allan Pred, *Place, Practice and Structure: Social and Spatial Transformation in Southern Sweden, 1750–1850* Cambridge: (Cambridge: Polity Press, 1986).

2. A. Källén, P. Arcuri, and J. Murray, "A Simple Model for the Spatial Spread and Control of Rabies," *Journal of Theoretical Biology* 116 (1985): 377–393.

3. Andrew Cliff, Peter Haggett, Keith Ord, and George Versey, *Spatial Diffusion: An Historical Geography of Epidemics in an Island Community* (Cambridge: Cambridge Univ. Press, 1981).

4. Graham Chapman, *The Green Revolution Game* (Cambridge: Marginal Context, 1982).

5. Graham Chapman, "The Folklore of the Perceived Environment in Bihar," *Environment and Planning* A 15 (1983): 945–968; and Graham Chapman, "The Structure of Two Farms in Bangladesh," in *Understanding Green Revolutions*, ed. Tim Bayliss-Smith and Sudhir Wanmali (Cambridge: Cambridge Univ. Press, 1984).

6. Stuart Corbridge, "The Green Revolution Game," *Journal of Geography in Higher Education* 9 (1985): 171–175.

7. Graham Chapman, "The Epistemology of Complexity," in *The Science and Praxis of Complexity*, ed. S. Aida (Tokyo: United Nations University, 1985), 335–374.

8. Bernard Marchand, "Planning Pedestrian Flows Around a Subway Station: A French Case Study of the Time-Distance Decay Functions," *Geographical Analysis* 9 (1977): 42–50.

9. William Kruskal, *How to Use KYST* (Murray Hill, N.J.: Bell Telephone Laboratories, 1973).

10. Remember, this originally appeared in the French journal *MappeMonde*.

11. Waldo Tobler, "Medieval Distortions: The Projections of Ancient Maps," *Annals of the Association of American Geographers* 56 (1966): 351–360.

12. Waldo Tobler, "Spatial Interaction Patterns," *Journal of Environmental Systems* 6 (1976): 251–301.

Penn State in Postal Space(s)

1. This paragraph was designed to ginger up the then president and provost of the university, both of whom received copies of the study, by pricking their pride and getting them to think about communication properly. The president at the time had said: "the space dimension is global. The traditional barriers of time and space will be removed for students and faculty alike as we attain the capacity to seek and gather information, pursue problems, or communicate and collaborate freely at any time and any place throughout the world" (President Joab Thomas, "Our Strengths and Promises," address to the Board of Trustees, Penn State University). Sometimes it is necessary to pull back administrators from Lalaland to the real world.

2. Like many others, I have long admired Richard Feynmann, from his marvelous physics lectures given to freshmen at Cal Tech, published in three volumes as Richard Feynmann, Robert Leighton, and Matthew Sands, *The Feynmann Lectures on Physics* (Reading, Mass.: Addison-Wesley, 1963), to his autobiographical books such as *What Do You Care What Other People Think?* (New York: Norton, 1988). His ability to cut through claptrap with straightforward thinking and prose is desperately needed as day by day we see language falling apart under the pressures of bureaucratic euphemism and political correctness.

3. I am sure there must be some, but I have been unable to find them; see "As People Go, So Do the Viruses," pages 195–204, below.

4. Two years later, the story had a happy ending as Penn State fired yet another commercial carrier, who bid low, but who took 26 days to deliver an airmail letter to Canberra, Australia. The university now uses IPA, but what was the real price paid over the previous years, and who really paid it?

Skiing with Euler at Beaver Creek

1. From Letter 32 in Peter Huss, *Correspondence Mathématique et Physique de Quelques Célèbres Géometres du XVIIIème Siècle, Tome I* (St. Petersburg: n.p., 1843).

2. Leonhard Euler, "Solutio Problematis and Geometriam Situs Pertinentis," *Commentarii Academiae Scientiarum Imperialis Petropolitanae* 8 (1736): 128–140. It is translated as "The Seven Bridges of Königsberg" in *The World of Mathematics*, vol. 1, ed. James Newman (New York: Simon and Schuster, 1956), 574–580. A nice, and easily accessible, background paper is by Robin Wilson, "An Eulerian Trail Through Königsberg," *Journal of Graph Theory* 3 (1986): 265–275.

3. The number (?) on the right-hand side of $V-E+F=(?)$ is known as the Euler characteristic. It is 1 for a plane, 2 for a sphere, 0 for a torus.

4. Our path is 1-4-5-26-25-19-23-25-22-18-19-15-14-18-17-20-21-24-24-20-16-17-13-12-10-11-8-9-7-3-2-2-1-6-12.

5. Our path is 1-6-16-20-21-24-24-20-17-18-19-25-22-23-25-26-5-4-1-2-2-3-7-9-8-11-10-12-13-14-15-11.

6. For a proof, see Theorem 2.1 in Oystein Ore, *Graphs and Their Uses* (New York: Random House, 1963), 25.

7. Theorem 2.2, in Ore, 26.

8. Leo's consulting fees are quite modest: 50 gold ducats should do it nicely.

9. Our "down to up" path is 1-4-5-8-4-7-9-8-11-10-9-12-10-13-14-15-11-15-19-18-14-B-18-22-25-26-5-A-23-25-A-19-23-22-21-B-17-20-24-24-21-20-16-17-13-12-C-16-6-C-7-3-2-2-1-3-6-1, and we are "home."

10. Our "up to down" path is 1-6-16-20-24-24-21-22-25-A-23-25-26-5-A-19-18-22-23-19-15-11-15-14-B-17-20-21-B-18-14-13-17-16-C-12-13-10-11-8-5-4-8-9-10-12-9-7-C-6-3-2-2-1-3-7-4-1=whew!

11. But it was noticed exactly two centuries later by the great graph theorist Dénes König in his *Theorie der Endlichen und Unendlichen Graphen* (Leipzig: Akademische Verlagsgesellschaft, 1936), p. 58, where he makes a distinction between graphs with such loops ("the broader sense"), and without them ("the narrower sense").

12. Pierre de Fermat posited that $x^n + y^n = z^n$ is impossible if x, y, and z are integers and n is greater than 2, noting, in the narrow margin of his copy of Bachet's edition of Diophantus's *Arithmetica*, that he had a proof "but the margin is too narrow to contain it." August Moebius, of the famous single-edged, one-sided Moebius strip, was the first to pose the four-color problem, that any map in the plane only requires four colors to guarantee that no two adjacent areas will be of the same color. This theorem was proved by bludgeoning it to death with a large computer. Such heavy-handed methods offend both the aesthetic sensibility and the soul of the true mathematician.

Pro Bono Publico

1. Peter Gould, *Fire in the Rain: The Democratic Consequences of Chernobyl* (Cambridge: Polity Press, 1990).

Sources of Error in a Map Series, or Science as a Socially Negotiated Enterprise

1. I believe this is still the case as this book is published nine years later.

2. This is the place to thank once again Joseph Kabel, Ralph Heidl, and William Holliday, all of whom helped in compiling the massive database (when it was readily available all the time under wraps at the Centers for Disease Control), as well as producing the dramatic sequence of maps.

3. "Inexorable March," *Time* 40, no. 9 (1992): 20; "A New Black Death," *Forbes Magazine* 154, no. 6 (1994): 240–250; "At Last Count: The Geography of AIDS," *Atlantic* 271, no. 1 (1993): 90–91; "The Self-Cleaning Oven," *Playboy* 41, no. 2 (1994): 44–45.

4. Peter Gould, David DiBiase, and Joseph Kabel, "Le SIDA: la carte animée comme rhétorique cartographique appliquée," *MappeMonde* 1 (1990): 21–26.

5. New definitions of AIDS, based on impeccable medical criteria, and designed to enhance earlier diagnosing and life-prolonging treatments, were meant to come into

effect on January 1, 1992, a presidential election year. They were delayed until January 1, 1993.

6. The original essay, in French, was dedicated to the memory of Brian Harley, whose many essays on the interpretation and deconstruction of maps has helped us to see with new eyes.

7. AIDS Surveillance Group, *Monthly Report for Washington, D.C.*, June 1991, 2.

8. After a number of requests, the original essay in French (187n), was republished in English, with a few, minor, essentially updating changes, under its present title in *Cartographic Perspectives* 21 (1995): 30–36.

9. Lytt Gardner and John Brundage, "Spatial Diffusion of the Human Immunodeficiency Virus Infection Epidemic in the United States, 1985–87," *Annals of the Association of American Geographers* 79 (1989): 25–43.

10. Andrew Gorlub, Wilpen Gorr, and Peter Gould, "Spatial Diffusion of the HIV/ADIS Epidemic: Modelling Implications and Case Study of AIDS Incidence in Ohio," *Geographical Analysis* 25 (1992): 85–100.

As People Go, So Do the Viruses

1. The classic example, often cited, is that of John Snow, M.D., and his use of maps to investigate the cholera epidemic of 1848 in London. His papers have been brought together as *Snow on Cholera*, introduced by Wade Hamilton Frost (Cambridge, Mass.: Harvard Univ. Press, 1936), reprinted in a facsimile edition (New York: Hefner Publishing Co., 1965).

2. Gerald Pyle, "Diffusion of Cholera in the United States," *Geographical Analysis* 1 (1969): 59–75.

3. Donald Janelle, "Central Place Development in a Time-Space Perspective," *Professional Geographer* 20 (1968): 5–10.

4. See the map sequence, plates 5–9.

5. Peter Gould and Rodrick Wallace, "Spatial Structures and Scientific Paradoxes in the AIDS Pandemic," *Geografiska Annaler*, ser. b, 76 (1994): 105–116.

6. McAndrew was actually extolling the virtues of an engineer if the ship's engines broke down over the long stretch of the Indian and Pacific Oceans between Cape Town, South Africa, and Wellington, New Zealand, with:

> We'll tak' one stretch—three weeks an' odd by any road ye steer—
> Fra' Cape Town east to Wellington—ye need an engineer.

7. John Paul Jones and Emilio Casetti, *Applications of the Expansion Method* (London: Routledge, 1992).

8. Peter Gould, Joseph Kabel, Wilpen Gorr, and Andrew Golub, "AIDS: Predicting the Next Map," *Interfaces* 21 (1991): 80–92.

9. Laurie Garrett, *The Coming Plague: New Emerging Diseases in a World Out of Balance* (New York: Farrar, Straus and Giroux, 1994); Michael Wilson, Richard Levins,

and Andrew Spielman, *Disease in Evolution: Global Changes and Emergence of Infectious Diseases.* (New York: New York Academy of Sciences, 1994).

Trying to Be Honest

1. John Eyles, *Geography in the Making* (Oxford: Blackwell, 1988).
2. National Research Council, *Rediscovering Geography.* (Washington, D.C.: National Academy Press, 1997).
3. Very rare in Pennsylvania, where many of today's well-wooded slopes may have been cut over two or even three times in earlier years. I live on a ridge, and ridge lines, while sometimes forming property boundaries, were often vaguely marked and surveyed. When the iron furnace, now about a mile away, was voraciously demanding trees for charcoal and fuel, the woodcutters stopped about one hundred yards short of the ridge line. One old white oak came down twenty years ago, and we counted its rings back to 1713.

Expose Yourself to Geographic Research

1. The person was John Eyles, who was editing the book *Geography in the Making* (Oxford: Blackwell, 1988), where this essay originally appeared. In general, I am reluctant to republish things that have already appeared in book form, but the opportunities to be as reflective and honest as you can be are rare, and I could not resist making an exception for this one essay—slightly modified and abbreviated for this volume.
2. I recall reading, many years ago, almost the identical words expressing the same opinion of Dudley Stamp. Somehow, I do not think Sir Dudley and I would have found too many professional points and opinions in common, but we certainly would have agreed about the issues of judgment, responsibility, and trust in encouraging disciplined curiosity.
3. For those who have forgotten, Archimedes was lying in his bath thinking of ways to determine the amount of adulterating base metal in Hiero's crown. It suddenly came to him that the density of the metal, what we would call today its specific gravity, was related to the amount of water it would displace. This flash of insight so excited him that he leapt from his bath, and ran naked through the streets of the town crying "Eureka!" ("I have found it!"). Today, of course, he would be arrested immediately for indecent exposure and locked up for hegemonic phallocentricism. We live in such a progressive age.
4. Mark Billinge, Derek Gregory, and Ronald Martin, *Recollections of a Revolution: Geography as Spatial Science* (London: Macmillan, 1984).
5. See "Acquiring Spatial Information," 59–73, above.
6. Brian Berry, "Approaches to Regional Analysis: A Synthesis," in *Spatial Analysis: A Reader in Statistical Geography*, ed. Brian Berry and Duane Marble (Englewood Cliffs, N.J.: Prentice-Hall, 1968), 24–34.

7. Raymond Cattell, "The Data Box: Its Ordering of Total Resources in Terms of Possible Relational Systems," in *Handbook of Multivariate Psychology*, ed. Raymond Cattell (Chicago: Rand McNally, 1966), 67–128.

8. The extraordinary insight that all places are unique, because no place is located where any other place is located, is one that we owe to Professor Herman Krick, a giant of regional geographic research whose contributions and memory I honor in *The Geographer at Work* (London: Routledge, 1985), 19.

9. Peter Gould and Daniel Ola, "The Perception of Residential Desirability in the Western Region of Nigeria," *Environment and Planning* A 2 (1970): 73–87.

10. Once again, see "Acquiring Spatial Information," 59–73, above. The complete study is contained in my monograph *People in Information Space* (Lund: Gleerup, 1975).

11. For empirical examples, see "The Structure of Spaces," 153–84, above.

12. Torsten Hägerstrand, "Aspects of Spatial Structure of Social Communication and the Diffusion of Information," *Papers and Proceedings of the Regional Science Association* 16 (1966): 27–42.

13. Two exemplary studies are Gerald Pyle, "Diffusion of Cholera in the United States," *Geographical Analysis* 1 (1969): 59–75; and Gerald Pyle and Keith Patterson, "Influenza Diffusion in European History," *Ecology of Disease* 2 (1984): 173–84.

14. William Butler Yeats, "The Second Coming," in *The Collected Poems of W. B. Yeats*, ed. Richard Finneran (New York: Collier, 1989), 187.

Thinking like a Geographer

1. This is a theme taken up by Heidegger in a number of works over the course of his life, and is implicit in his essays on the poetry of Hölderlin.

2. Ontario Ministry of Education, *Geography: Intermediate and Senior Divisions* (Toronto: Ministry of Education, 1988).

3. OME, *Geography: Intermediate and Senior Divisions*, A12.

4. I have pointed to this etymological derivation several times, for it seems to me to be important sometimes to stand still and listen to words. Too many people throw words about these days without reflecting upon them, sometimes to the point of giving them a completely different meaning, almost a reversal. Such reflections also open my "Dynamic Structures of Geographic Space," in *Collapsing Space and Time: Geographic Aspects of Communication and Information*, ed. Thomas Leinbach and Stanley Brun (Hammersmith: HarperCollins*Academic*, 1991), 3–30.

5. As the British have done, starting with an Espionage Act in 1915, grounded on foolish panic that wicked German spies were everywhere, and leading to the dreaded D notice that can jail an editor who publishes forbidden material. The trend continues to the present day, resulting in ever-tighter control by the civil (?) and intelligence (?) services, resulting in farces like the recent *Spycatcher* case.

6. As the British, French, and others have done regularly during the past decade or two.

7. Richard Morrill, "Ideal and Reality in Reapportionment," *Annals of the Association of American Geographers* 63 (1973): 463–477.

8. Peter Taylor and Ron Johnston, *Geography of Elections* (Harmondsworth: Penguin, 1979).

9. Michael Jenkins and John Shepard, "Decentralizing High School Administration in Detroit: An Evaluation of Alternative Strategies of Political Control," *Economic Geography* 48 (1972): 95–106.

10. Stan Openshaw, Martin Charlton, Colin Wymer, and Alan Craft, "A Mark 1 Geographical Analysis Machine for the Automated Analysis of Point Data Sets," *International Journal of Geographic Information Systems* 1 (1987): 335–358.

11. Her Majesty's Stationery Office, *Investigation of the Possible Increased Incidence of Cancer in West Cambria* (London: Her Majesty's Stationery Office, 1984).

12. Peter Gould, *Fire in the Rain: The Democratic Consequences of Chernobyl* (Cambridge: Polity Press, 1990).

13. John Pickles, "Geographical Theory and Educating For Democracy" *Antipode* 18 (1986): 136–154.

14. Much of the research appears in nontechnical terms in my book *The Slow Plague: A Geography of the AIDS Pandemic* (Oxford: Blackwell, 1993).

15. The 1992 estimate was close to the mark. As this book is published (1999), the United States will have well over half a million AIDS cases accumulated, with still perhaps a million more people infected.

16. Vincent Van Gogh, *The Complete Letters of Vincent van Gogh* (Boston: New York Graphic Society, 1978).

17. Edward Soja, *Postmodern Geographies: The Reassertion of Space in Critical Social Theory* (London: Verso, 1989).

18. Martin Heidegger, *Being and Time*, trans. John Macquarrie and Edward Robinson (New York: Harper and Row, 1962), 242.

19. Jean Giono, *The Man Who Planted Trees* (Chelsea, Vt.: Chelsea Green, 1985).

20. David Harvey, "On the History and Present Condition of Geography: An Historical Materialist Manifesto," *Professional Geographer* 36 (1985): 1–11.

21. John Imbrie, "A Theoretical Framework for the Pleistocene Ice Ages," *Journal of the Geological Society of London* 142 (1985): 417–432.

22. Derek Gregory, *Ideology, Science and Human Geography* (New York: St. Martin's, 1978).

23. Karl-Otto Apel, "The Problem of a Macroethic of Responsibility to the Future in the Crisis of Technological Civilization: An Attempt to Come to Terms with Hans Jonas's 'Principles of Responsibility,'" *Man and World*, 20 (1987): 3–40.

24. Aristotle, *The Nicomachean Ethics* (Harmondsworth: Penguin, 1955).

25. Peter Sloterdijk, *Critique of Cynical Reason* (Minneapolis: Univ. of Minnesota Press, 1987).

26. An absolute claim made frequently in his writings of the 1980s, for example, in *The Urbanization of Capital,* and its companion volume *Consciousness and the Urban Experience.* (Baltimore: Johns Hopkins Univ. Press, 1985).

27. In this book, let me stand corrected. There is nothing in the Koran, or in the religious teaching of Islam, that requires the physical mutilation of clitoridectomy or vaginal closure, perpetrated solely by women on little girls reaching puberty. This hideous custom is simply that—custom. It has no sanction from the teachings of Islam, or from any other world religion.

28. George Steiner, *Real Presences* (Chicago: Univ. of Chicago Press, 1989).

29. Ernst Bloch, *The Principle of Hope* (Cambridge, Mass.: MIT Press, 1986).

30. So translated by George Steiner in *Real Presences,* 165.

What Is Worth Teaching in Geography?

1. The Open University, *Mathematics Foundation Course, Units 1–36* (Bletchley, Bucks.: The Open University Press, 1970).

2. Some of these examples appear in the chapter on cartography in *The Geographer at Work,* but over the years students have come up with many imaginative examples, and we have all had great fun thinking about configurations of things in spaces other than the conventional geographic. How about toppings in pizza space, or musical notes in Corelli space!

3. This sentence obviously dates the essay to 1977, before the explosion of computer mapping in every undergraduate course in cartography. Yet I submit that we still have a long way to go to open up thinking in genuinely conceptual ways. Too many students sit for days in front of screens making nice maps, but then do not know why they were made—except to hang on the wall as decoration.

4. The Open University, *The Man-Made World* (Milton Keynes: The Open University Press, 1974). Today (1998), 24 years later and after the personal computer revolution, some of these courses are unbelievably imaginative compared to many technical courses in most universities.

5. Jay Forrester, *World Dynamics* (Cambridge, Mass.: Wright-Allen, 1971).

6. These same ideas of flows in a network obviously underpin very complex hydrologic models of river systems, enabling engineers to forecast flood levels downstream, and to give various degrees of warning time to save lives.

7. Much World Bank investment after World War Two went to improving the capacity of existing rail links in the Third World. The "mountains of Kano" in northern Nigeria were gigantic piles of bagged groundnuts awaiting shipment over a railway system severely constrained.

8. Something that Annik Rogier and I tried to get across in a pedagogic mode in our "Famine as a Spatial Crisis: Programming Food to the Sahel," *Geoforum* 17 (1984): 135–154, reprinted in *Famine as a Geographical Phenomenon,* ed. Bruce Currey and Graeme Hugo (Dordrecht: D. Reidel, 1984).

Perspectives and Sensitivities: Teaching as the Creation of Conditions of Possibility for Geographic Thinking

1. David Livingstone, *The Geographical Tradition: Episodes in the History of a Contested Enterprise* (Oxford: Blackwell, 1992).

2. Professional geographers will recognize the words in quotation marks as, respectively, a title of a chapter in Richard Hartshorne, *The Nature of Geography* (Lancaster, Pa.: Association of American Geographers, 1939), or a phrase in a presidential address by Carl Sauer, "Forward to Historical Geography," *Annals of the Association of American Geographers* 31 (1941): 1–24.

3. Peter Jackson, "Berkeley and Beyond: Broadening the Horizons of Cultural Geography," *Annals of the Association of American Geographers* 83 (1993): 519–520.

4. Since I was not totally illiterate before 1980, this may seem a strange statement, but when you spend sixteen weeks reading forty-five pages of Martin Heidegger's "Anaximander Fragment" with a philosopher like Joseph Kockelmans you realize what reading means. Apart from technical and methodological matters, this was one of the few genuine intellectual challenges of my entire academic career.

5. "Complicity" is a favorite word of those presently called to deconstruct and expose the dark and evil ways of others. For a no-hiding-place-down-here-hallelujah effort see Derek Gregory, *Geographical Imaginations* (Oxford: Blackwell, 1994), 62.

6. See "What's Worth Teaching in Geography?" 238–53, above. Charge me with what authorial narcissism you like, but have the courage to admit that we are all like Marcel from time to time, sitting up in bed, leafing with studied casualness through our *Figaro*, pretending we are an ordinary reader who just happens to come across an essay written by . . . our favorite author. The scene appears in Marcel Proust, *A la recherche du temps perdu* (Paris: Gaillimard, 1954).

7. Julie Graham, "Antiessentialism and Overdetermination," *Antipode* 24 (1992): 141–156.

8. Bernard Knox, *The Oldest Dead White European Males* (New York: Norton, 1993).

9. All these examples are contained in Bruce Hewitson and Robert Crane, *Neural Nets: Applications for Geography* (Dordrecht: Kluwer, 1994).

10. I have deliberately borrowed and modified the title of an essay for the resonance of its ambiguity "*Any* Space . . ." and "Any *Space* . . . ." Anthony Gatrell, "Any Space for Spatial Analysis?" in *The Future of Geography*, ed. Ron Johnston (London: Methuen, 1985), 190–208.

11. Edward Lorenz, "The Approaches to Atmospheric Predictability," *Bulletin of the American Meteorological Society* 50 (1969): 345–349.

12. Andrew Cliff and Peter Haggett, *Atlas of Disease Distributions: Analytic Approaches to Epidemiological Data* (Oxford: Blackwell, 1988); and Peter Gould, Joseph Kabel, Wilpen Gorr, and Andrew Gorlub, "Predicting the Next Map," *Interfaces* 21 (1991): 80–92.

13. Arthur Getis and Keith Ord, "The Analysis of Spatial Association by Use of Distance Statistics," *Geographical Analysis* 24 (1992): 189–206.

14. Rodrick Wallace and Deborah Wallace, "Inner City Disease and the Public Health of the Suburbs," *Environment and Planning* A 25 (1993): 1707–1723.

15. Stan Openshaw, "Investigation of Leukemia Clusters by Use of a Geographical Analysis Machine," *Lancet*, 6 Feb. 1988: 272–273.

16. Stan Openshaw and Peter Taylor, "A Million or So Correlation Coefficients: Three Experiments on the Modifiable Unit Area Problem," in *Statistical Applications in the Spatial Sciences*, ed. Neil Wrigley (London: Pion, 1972), 127–144.

17. Graham Chapman, *Human and Environmental Systems: A Geographer's Appraisal* (London: Academic Press, 1977).

18. Waldo Tobler, "Computation of the Correspondence of Geographic Patterns," *Papers and Proceedings of the Regional Science Association* 15 (1965): 131–139.

19. Donna Peuquet and Todd Bacastow, "Organizational Issues in the Development of Geographical Information Systems: A Case Study of US Army Topographic Information Automation," *International Journal of Geographical Information Systems* 5 (1991): 303–319.

A Petition to Judas Gould

1. In classes on spatial analysis I frequently invoke the names of Zeus and Apollo from my own theological framework. One student, in an anonymous evaluation of the course, appreciated this very much because he or she thought I was doing this out of sensitivity for the name of the Christian deity. What condescension!

2. Forgive me, John!

3. In addition to being a university of intellectual excellence, Penn State also plays football occasionally. It happened that the week I received the petition we were playing a small Southern university called Alabama, whose team generally displays promise in an appropriately amateurish sort of way.

Thinking about Learning

1. August Lösch, *The Economics of Location*, trans. Wolfgang Stopler (New Haven: Yale Univ. Press, 1954).

2. Martin Heidegger, "The Anaximander Fragment," in *Early Greek Thinking*, trans. David Krell and Frank Capuzzi (New York: Harper and Row, 1975), 13–58.

August Lösch as a Child of His Time

1. August Lösch, *The Economics of Location* (New Haven: Yale Univ. Press, 1954).

2. Richard Hartshorne, *The Nature of Geography* (Washington, D.C.: Association of American Geographers, 1939).

3. Lösch, *Economics of Location*, 358.

4. The notion of "retrieve" arises at numerous points in the writings of Martin Heidegger. For an account of such efforts, the fine Jesuit interpreter William Richardson is helpful in his *Heidegger: Through Phenomenology to Thought* (The Hague: Martinus Nijhoff, 1974), where the sense of *Weiderholung* is emphasized by hypenhating *re-trieve*.

5. Lösch, *Economics of Location*, 93, 358 n. 84.

6. Lösch, *Economics of Location*, 93.

7. Martin Heidegger, *Schelling's Treatise on the Essence of Human Freedom* (Athens: Ohio Univ. Press, 1985), 35, 39.

8. Lösch, *Economics of Location*, 363.

9. Lösch, *Economics of Location*, 4.

10. Lösch, *Economics of Location*, 218–219.

11. Martin Heidegger, *History of the Concept of Time* (Bloomington: Indiana Univ. Press, 1985), 203.

12. Lösch, *Economics of Location*, 267.

13. Lösch, *Economics of Location*, 184.

14. Lösch, *Economics of Location*, 24.

15. The reasons for this are spelled out in "A Critique of Dissipative Structures in the Human Realm," 101–17, above.

16. Sometimes referred to as a many-valued function. These appear to have come to the fore in the work of Riemann, to be taken up, and investigated intensively, by such Italian algebraic geometers as Levi-Civitas. Unfortunately, a sound and penetrating history of this enlargement of possibility in mathematics does not yet appear to be available.

17. Lösch, *Economics of Location*, 8.

18. Lösch, *Economics of Location*, 29.

19. Lösch, *Economics of Location*, 29, 31.

20. Lösch, *Economics of Location*, 36–37.

21. Lösch, *Economics of Location*, 61.

22. Arthur Eddington, *New Pathways in Science* (Cambridge: The University Press, 1935).

23. Eugene Wigner, "The Unreasonable Effectiveness of Mathematics," *Communications on Pure and Applied Mathematics* 13 (1960): 1–14.

24. Hans-George Gadamer, *Philosophical Apprenticeships*. (Cambridge, Mass.: MIT Press, 1985).

25. Lösch, *Economics of Location*, 29.

26. Lösch, *Economics of Location*, 61, 62.

27. Lösch, *Economics of Location*, 93.

28. Lösch, *Economics of Location*, 349 n. 66.

29. Lösch, *Economics of Location*, 106.

30. Lösch, *Economics of Location*, 194–195.

31. As expressed, for example, in Anthony Giddens, *Central Problems in Social Theory* (London: MacMillan, 1979); Derek Gregory, "Space, Time and Politics in Social

Theory: An Interview with Anthony Giddens," *Society and Space* 2, no. 2 (1984): 123–132; Allan Pred, "Social Predisposition and the Time-Geography of Everyday Life," in *A Search for Common Ground*, ed. Peter Gould and Gunnar Olsson (London: Pion, 1982), 157–186; and Edward Soja, "The Spatiality of Social Life: Towards a Transformative Retheorization," in *Social Relations and Spatial Structures*, ed. Derek Gregory and John Urry (London: Macmillan, 1985), 171–190.

32. Lösch, *Economics of Location*, 326.

33. Lösch, *Economics of Location*, xvii.

34. Lösch, *Economics of Location*, 342–43.

35. Both of whom, from somewhat different perspectives, raise questions of what happens when two systems are brought together. See Alan Wilson, *Entropy in Urban and Regional Modelling* (London: Pion, 1970); and Peter Allen, "Self-organization and the Urban System," in *Self-Organization and Dissipative Structures*, ed. Peter Allen (Austin: Univ. of Texas Press, 1982), 132–158.

36. Jay Forrester, "Counter-intuitive Behavior of Social Systems," in *Towards Global Equilibrium*, ed. Denis Meadows and Deborah Meadows (Cambridge, Mass.: Wright-Allen, 1973), 3–30.

37. Christopher Alexander, *Notes on the Synthesis of Form* (Cambridge, Mass.: Harvard Univ. Press, 1964).

38. These words permeate the works of Martin Heidegger, but perhaps especially in his "Memorial Address for Conradin Kreutzer," contained in the collection of essays *Discourse on Thinking*, trans. John Anderson and Hans Freund (New York: Harper Torchbooks, 1966), 43–57.

Do Foraminifera Assemblages Exist—At Least in the Persian Gulf?

1. Timothy Denham, "Understanding the Kalam, Understanding Ourselves: Reflections on the Re-Presentation of the Kalam, Bismarck Mountain Range, Papua-New Guinea" (master's thesis, Penn State University, 1996).

2. Londa Schiebinger, "The Loves of Plants," *Scientific American* 274, no. 2 (1996): 110–115.

3. Mutation has traditionally meant exchanges between two different viruses; recombination of RNA strands means a single virus can change itself into something different, with different survival abilities in immune systems.

4. Such spaces are invariably defined by orthogonal, independent dimensions, but the "stretch" of each dimension, and therefore the distances between the things to be classified, depend totally on the degree of redundancy of the characteristics chosen by the analyst.

5. Ronald Atkin, *Mathematical Structure in Human Affairs* (London: Heinemann Educational Books, 1974).

6. For a highly, not to say agonizingly, detailed analysis, see my essay "A Structural Language of Relations," in *Future Trends in Geomathematics*, ed. Richard Craig and Mark Labovitz (London: Pion, 1981), 281–312. Pages 300–310 are particularly pertinent.

7. Jacob Houbolt, *Surface Sediments of the Persian Gulf near the Qatar Peninsula* (The Hague: Mouton, 1957).

8. Graeme Bonham-Carter, *Fortran IV Program for Q-mode Cluster Analysis of Non-Quantitative Data Using IBM 7090/7094 Computers* (Lawrence: Kansas State Geological Survey, 1967).

9. Andrew Boyce, "The Value of Some Methods of Numerical Taxonomy with Reference to Hominoid Classification," in *Phenetic and Phylogenetic Classification*, ed. Victor Heywood and James McNeill (London: The Systematics Association, 1964).

Thinks That Machine

1. Martin Heidegger, *The Question Concerning Technology and Other Essays* (New York: Harper Colophon, 1977).

2. Medard Boss, *Existential Foundations of Medicine and Psychology* (New York: Jason Aronson, 1977), 64.

3. Sigmund Freud, *Gesammelte Werke* (London: Imago, 1941), 243, and referenced by Medard Boss in *Existential Foundations*, 62.

4. Joseph Kockelmans, "Daseinanalysis and Freud's Unconscious," *Review of Existential Psychology and Psychiatry* 16 (1978–79): 21–42.

5. Anthony Storr, *The Art of Psychotherapy* (New York: Methuen, 1980).

6. Medard Boss, "Martin Heidegger's Zollikon Seminars," *Review of Existential Psychology and Psychiatry* 16 (1978–79): 7–20.

7. Medard Boss, *Psychoanalysis and Daseinanalysis* (New York: Basic Books, 1963).

8. Jacques Ellul, *Perspectives on Our Age* (Toronto: Canadian Broadcasting Corporation, 1981), 38. A series of public talks given in the impressive annual Massey Lecture Series, itself reflecting the tradition of the Reith Lectures of the BBC.

9. Peter Gould, Jeffrey Johnson, and Graham Chapman, *The Structure of Television: Part I Television, The World of Structure: Part II Structure, The World of Television* (London: Pion, 1984).

10. Joseph Weizenbaum, *Computer Power and Human Reason* (Cambridge, Mass.: MIT Press, 1976).

11. A point expanded upon in "Do Foraminifera Assemblages Exist—At Least in the Persian Gulf?" 289–99, above.

12. As advocated in the whoopee article by Kenneth Colby and Michael McGuire, "Signs and Symptoms," *The Sciences* 21 (1981): 21–23.

Think What You Like, but Think for Yourself

1. Peter Gould and Rodrick Wallace, "Spatial Structures and Scientific Paradoxes in the AIDS Pandemic," *Geografiska Annaler*, ser. b, 76 (1994): 105–116.

2. Peter Gould, "La géographie du SIDA: l'étude des flux de population permet de predire," *La Recherche* 280 (1995): 36–37.

3. Martin Heidegger, *What Is A Thing?* (South Bend: Gateway, 1967).

Epilogue: The Arrival of a Spatial Century

1. The third millennium starts on Jan. 1, 2001.

2. Just to show my intellectual flexibility to all my former students, and to acknowledge again Jean Aitchison's wonderful Reith Lectures, published as *The Language Web: The Power and Problem of Words* (Cambridge: Cambridge Univ. Press, 1997), I have decided to split one infinitive in this book. It really feels quite daring. The standards of the English language are going down everywhere. They will not be raised in our lifetime.

Index